Sustainable Development Goals Series

The **Sustainable Development Goals Series** is Springer Nature's inaugural cross-imprint book series that addresses and supports the United Nations' seventeen Sustainable Development Goals. The series fosters comprehensive research focused on these global targets and endeavours to address some of society's greatest grand challenges. The SDGs are inherently multidisciplinary, and they bring people working across different fields together and working towards a common goal. In this spirit, the Sustainable Development Goals series is the first at Springer Nature to publish books under both the Springer and Palgrave Macmillan imprints, bringing the strengths of our imprints together.

The Sustainable Development Goals Series is organized into eighteen subseries: one subseries based around each of the seventeen respective Sustainable Development Goals, and an eighteenth subseries, "Connecting the Goals", which serves as a home for volumes addressing multiple goals or studying the SDGs as a whole. Each subseries is guided by an expert Subseries Advisor with years or decades of experience studying and addressing core components of their respective Goal.

The SDG Series has a remit as broad as the SDGs themselves, and contributions are welcome from scientists, academics, policymakers, and researchers working in fields related to any of the seventeen goals. If you are interested in contributing a monograph or curated volume to the series, please contact the Publishers: Zachary Romano [Springer; zachary.romano@springer.com] and Rachael Ballard [Palgrave Macmillan; rachael.ballard@palgrave.com].

Teresa J. Kennedy
Editor

Ocean Literacy: The Foundation for the Success of the Ocean Decade, Volume I

Transforming Education, Research, and Engagement

Springer

Editor
Teresa J. Kennedy
University of Texas at Tyler
Tyler, TX, USA

ISSN 2523-3084 ISSN 2523-3092 (electronic)
Sustainable Development Goals Series
ISBN 978-3-031-98144-9 ISBN 978-3-031-98145-6 (eBook)
https://doi.org/10.1007/978-3-031-98145-6

This work was supported by University of Texas at Tyler, Ocean Conservation Trust, Instituto de Educação at the Universidade de Lisboa, Everett Community College Foundation, Kingsborough Community College, Monterey Bay Aquarium Research Institute, and Institutionen för marina vetenskaper.

This Springer imprint is published by the registered company Springer Nature Switzerland AG
The registered company address is: Gewerbestrasse 11, 6330 Cham, Switzerland

Editor's Note

2021 United Nations Decade
2030 of Ocean Science
for Sustainable Development

This book series is an officially Endorsed Activity of the United Nations Decade of Ocean Science for Sustainable Development (2021–2030), submitted by the International Council of Associations for Science Education (ICASE), an NGO in Official Partnership with UNESCO, as the Lead Institution. Volume I embodies this endorsement through its focus on education, research, and engagement—foundations essential for realizing the Ocean Decade's vision of transforming knowledge into action for a sustainable future.

Volume I opens with two forewords by esteemed global leaders whose careers have shaped international Ocean science and policy. Their voices highlight the broad relevance and profound significance of Ocean Literacy within the framework of the United Nations (UN) Decade of Ocean Science for Sustainable Development (2021–2030), referred to as the Ocean Decade (OD).

Foreword 1—Vladimir E. Ryabinin, PhD

Former Executive Secretary of the Intergovernmental Oceanographic Commission (IOC) of UNESCO, Dr. Vladimir Ryabinin offers a visionary reflection on the foundational importance of education, ethics, and societal transformation, framing Volume I as a call to build an Ocean-literate society that begins in early childhood and extends across generations and sectors.

Foreword 2—Margaret S. Leinen, PhD

Director of the Scripps Institution of Oceanography and Vice Chancellor at the University of California San Diego, and former Vice Chair of the UN Ocean Decade Advisory Board, Dr. Margaret Leinen emphasizes that Ocean Literacy is central to achieving the Ocean Decade's societal outcomes, highlighting the essential roles of public engagement, education, and cross-sector collaboration.

In addition to these forewords, Volume I includes reflections from members of the global Ocean and science education community, including perspectives from Early Career Ocean Professionals (ECOPs), whose voices bring fresh insight and passion to the Ocean Literacy movement. Together, their contributions underscore the spirit of global collaboration, shared responsibility, and inclusive knowledge that defines this three-volume series.

International Council of Associations for Science Education

Foreword 1

I see the three volumes of *Ocean Literacy: The Foundation for the Success of the Ocean Decade* as if they were responding to the classic detective questions: *"When?" "Who?" "How?"* Volume I explores the foundations—*When do we begin to build an Ocean-literate society, and how do we empower change-makers across all sectors?*

There may be two answers to the "When?" The first is *"now."* Humanity—*Homo sapiens*, or perhaps more accurately, *Homo consumens*—must urgently transform its ethics and attitudes if our civilization is to endure. Building a more harmonious relationship with the Ocean is not just about the environment; it is about improving who we are as a species.

The second answer is *"as early in life as possible."* To be most effective, this "Ocean vaccination of kindness" should be instilled in human souls from the earliest age—ideally beginning in kindergarten, if not before, through shared experiences such as listening to Ocean stories and songs, engaging with Ocean-themed play and picture books, or exploring water in natural and everyday settings—regardless of whether a child lives near the coast or in a landlocked area. But the reality is, change must happen across all generations and all sectors—education, research, governance, and civil society.

Volume II answers the question *"Who?"*: focusing on communities and the creative drivers of change. Volume III responds to *"How?"*: highlighting communication strategies, technology, and innovation. Although the Ocean Decade is officially about transforming science, it is even more deeply about transforming society.

The 21 chapters in Volume I offer rich insight into the expanding world of Ocean Literacy, why it matters, who it empowers, and how it can lead us toward more inclusive and ethical futures. This volume brings together frameworks for change, inspiring case studies, and examples of real-world impact from across the globe. Some chapters introduce foundational concepts: the theory of change behind Ocean Decade Challenge 10, co-design principles, and systemic strategies. Others present field-based programs and initiatives—from engaging underserved youth to empowering scientists as Ocean advocates. This diverse ecosystem of ideas, ethics, and actions is not just informative; it is transformative. The structure of this volume is like a university course: theory is presented alongside practical experience. The result is both intellectually rigorous and deeply human.

I wholeheartedly recommend this volume to the reader. Begin with Volume I and imagine the ideal Ocean Literacy initiative—whether in a classroom, community, laboratory, or policy space. Compare your vision with the experiences shared in these chapters and set sail! Then continue your journey through Volume II, where communities, cultures, and creative forces come to the forefront, and onward to Volume III, where innovation, technology, and global partnerships steer us toward the Ocean we want.

Vladimir E. Ryabinin, PhD
Former Executive Secretary of the Intergovernmental Oceanographic Commission (IOC) of the United Nations Educational, Scientific and Cultural Organization (UNESCO)
Paris, France

Foreword 2

While I was Vice Chair of the United Nations Decade of Ocean Science for Sustainable Development Advisory Board, I was pleased to see the Ocean Literacy movement take root, with many different initiatives, such as this book series, earning endorsement as official Ocean Decade Activities. Ocean Literacy is not simply a supporting pillar of the Ocean Decade—it is fundamental to its success. Achieving the Decade's societal outcomes requires more than scientific breakthroughs or technological innovations; it depends on public engagement, cross-cultural understanding, and informed action at every level of society.

During the 13th International Forum of NGOs in Official Partnership with UNESCO, held in 2022 in Ottawa, Canada, it was clear that the energy, vision, and commitment of the NGO community would be vital to the success of the Ocean Decade. Ocean Literacy is not only essential to building public understanding; it is key to ensuring that Ocean science delivers equitable and enduring benefits.

Ocean Literacy is not merely an adjunct to the scientific goals of the Ocean Decade—it is foundational to achieving its societal outcomes. Scientific knowledge must be paired with public understanding, cultural resonance, and actionable engagement if we are to succeed in restoring and sustainably managing the Ocean.

The remarkable three-volume series presented here, *Ocean Literacy: The Foundation for the Success of the Ocean Decade*, was developed entirely through voluntary contributions and embodies the commitment of that far-reaching international collaboration. It advances the foundational role of Ocean Literacy in realizing the Ocean Decade's transformative goals. This series reflects a broad vision—bringing together educators, scientists, artists, policymakers, and community leaders to strengthen our shared relationship with the Ocean. Its emphasis on education, community, the arts, innovation, and global collaboration reflects the multifaceted nature of Ocean Literacy and its essential role in driving meaningful progress.

The contributors to this volume have done more than document good practices—they have helped define a collective path forward. By integrating education, engagement, innovation, and global dialogue, they underscore that Ocean Literacy is not a niche pursuit, but a central pillar of sustainable development and Ocean stewardship. I commend the editor, authors, and reviewers for a visionary and timely contribution that will serve as a touchstone for the Ocean Decade and beyond.

Margaret S. Leinen, PhD
Director, Scripps Institution of Oceanography
San Diego, CA, USA
Vice Chancellor, University of California
San Diego, San Diego, CA, USA
Former Chair of the UN Ocean Decade
Advisory Board

Reflections from the Ocean and Science Education Community

The voices featured here offer reflections from both Ocean and science education leaders whose work continues to shape the dialogue, reach, and impact of Ocean Literacy around the world.

"As the Lead Institution for this officially Endorsed Ocean Decade Activity, the International Council of Associations for Science Education (ICASE) is proud to support this transformative contribution to the global Ocean Literacy movement. This series exemplifies the power of international collaboration in advancing science education that is equitable, inclusive, and responsive to the sustainability challenges of our time. It reflects what is possible when educators, scientists, and civil society come together in pursuit of shared goals.

As an organization dedicated to strengthening science education worldwide, ICASE is honored to have played a role in supporting this project as it will inspire educators and learners across generations and geographies. The success of the Ocean Decade depends on more than just data and discovery, it depends on education that connects people to the Ocean in meaningful ways. This three-volume series represents a significant step in that direction and aligns with the mission of ICASE to promote quality science education that supports sustainable futures."

—Declan Kennedy, PhD, President, International Council of Associations for Science Education (ICASE), University College Cork, Cork, Ireland

"This volume provides a comprehensive framework for embedding Ocean Literacy across both formal and informal learning environments, offering actionable strategies that empower educators, researchers, and policymakers to inspire meaningful stewardship of the marine world."

—Niccolò Bassan, Science Officer, Intergovernmental Oceanographic Commission (IOC) of UNESCO, Paris, France

"For us in the Pacific, Ocean Literacy is not an abstract concept; it is a lived experience. The Ocean is not just part of our environment, but part of our identity. We are Ocean people. Our lives are intertwined with its tides, our stories etched into waves. From childhood games on the beach to fishing traditions, chants, and navigation, the Ocean is our classroom, our highway between islands, and a sacred space.

This first volume of *Ocean Literacy: The Foundation for the Success of the Ocean Decade*, focusing on transforming education, research, and engagement, resonates with our worldview. It honors multiple ways of knowing and recognizes that Ocean Literacy is not limited to formal education, but deeply rooted in community, culture, and intergenerational wisdom. To be Ocean literate in the Pacific means to know, feel, taste, and protect the Ocean—with a deep understanding passed down through generations."

—Katy Soapi, Coordinator for Partnerships and Engagement, Pacific Community Centre for Ocean Science, Suva, Fiji Islands

ECOP
Early Career
Ocean Professionals

Contents

1 **Anchoring the Ocean Decade: Building an Ocean-literate World** 1
Teresa J. Kennedy

2 **Actioning the Ocean Decade Challenge 10 White Paper: A Theory of Change** 17
Diz Glithero, Nicola Bridge, Jen McRuer, and Natalie Hart

3 **Delivering Critical Ocean Literacy Initiatives in the Framework of the UN Decade of Ocean Science for Sustainable Development** 25
Francesca Santoro, Paula Barbeito Morandeira, Valentina Lovat, Ana Vitória Tereza de Magalhães, Gaia Maria Sole Intonti, Esteban Gottfried Burguett, Alessia Tombolini, Paola Carluccio, and Jack Coulton

4 **ScienceOnShore (SOS): A Wake-up Call from the Ocean Literacy Movement** 43
Thomas F. Greene

5 **The TransforMAR Project: A Deep Dive into Ocean Literacy** . . . 53
Alexandre Augusto da Silva, Giovanna Scagnolatto, and Karina F. Orsolini

6 **Ocean Literacy in a Land-Locked State** 71
Gregory J. Barord

7 **Engaging Students with Ocean Technology: A Journey from Floats to Ocean Literacy** 83
George I. Matsumoto, Katie M. Lodes, Teresa J. Kennedy, and Jennifer L. Magnusson

8 **Once Is Not Enough: Connecting Students of Color to their Local Waterways** 103
Meghan E. Marrero and Kristen V. Napolitano

9 **Shaping Scientific Literacy: How Underserved Students Conceptualize Marine Scientists and Themselves** 117
Sarah Nuss, Anna Caputo, Lisa Lawrence, Greta Olsen Gardiner, Bethany Smith, and Celia Cackowski

10 Investigating the Longer-term Student Outcomes of a Two-day Ocean-focused Program for Underserved Youth in Cape Town, South Africa . . . 135
B. Troy Frensley, Hannah Eaton, Claire Carretti, Katherine Wallis, Amalia De Abreu, and Peter Marshall

11 Ocean Equity: Unveiling Gender Opportunities and Challenges in Brazil's Ocean Literacy Activities . . . 151
Tatiana Martelli Mazzo, Adriana Lippi, William Remo Pedroso Conti, Andrezza Justino Gozzo Andreotti, Camila Keiko Takahashi, Barbara Lage Ignacio, Marilia Bueno, Ronaldo Adriano Christofoletti, and Leandra R. Gonçalves

12 Cultivating Ocean Guardians: Evaluating a Novel Pedagogical Model for Enhanced Ocean Citizenship in Basic Education . . . 165
Caroline Schio and Pedro Reis

13 The Ocean We Have: A Case Study at the Ocean Research College Academy . . . 187
Josh Searle and Ardi Kveven

14 Whale Tale: Big Challenges Facing the Ocean's Biggest Species . . . 201
Tammy D. Lee, Bonnie Glass, Gail Lemiec, and Fernando Martín Velazco

15 Ocean Literacy in Science Curriculum: A Quantitative Content Analysis in Sweden . . . 227
Géraldine Fauville, Craig Strang, Ewa Skantz, Jens Bjelvenmark, and Anne Solli

16 Development of an Aquaponics STEAM Program for Improving Ocean Literacy . . . 243
Tsuyoshi Sasaki and Kota Sawada

17 Investigating Ocean Literacy Levels of Pre-Service Science and Social Studies Teachers in Türkiye: The Influence of Gender, Ocean Experience, and Environmental Activities . . . 259
Bülent Çavaş, Şermin Açık, Simge Koç, and Mısra Kolaç

18 Insights from Project SEA: Building Ocean Literacy and Enhancing Marine Science Education Through Teacher Professional Development . . . 273
Lauren Madden, Louise Ammentorp, Nathan Magee, Graceanne Taylor, Gabriela Rivera-Peña, and Alexia Schianodicola

19 The Global Marine Curriculum: Academic Pathways for Sustainable Marine Development 289
Ninja Mueller

20 Scientists for Ocean Literacy: Empowering Scientists as Ocean Advocates 305
Dina Eparkhina, Elisa Banchi, Ángel Borja, Donata Canu, Daria Ezgeta-Balić, Elena Giusta, Andrew King, Panayota Koulouri, Marina Locritani, Marta Manca Zeichen, Angela Pomaro, Francesca Petrera, Maria C. Uyarra, and Louise Valestrand

21 Co-Designing Transdisciplinary Science for the Ocean We Want: Actions for Scientists to Enrich Ocean Literacy 331
Lei Li, Minyan Zhao, Min Wei, Shuangfang Hu, Baki Billah, and Ziyi Lin

22 Ocean Literacy at a Global Turning Point: Reflections, Foundations, and Future Pathways 351
Teresa J. Kennedy

Volume I: Contributors (Alphabetical Order) 353

Volume I: Contributors (Chapter Order) 355

Authors' Institutional Affiliations 357

Volume I: Expert Reviewers (44) 359

Volume I: Acronym List 361

1 Anchoring the Ocean Decade: Building an Ocean-literate World

Teresa J. Kennedy

Abstract

The United Nations Decade of Ocean Science for Sustainable Development (2021–2030), the Ocean Decade, presents an unprecedented opportunity to strengthen Ocean Literacy worldwide. This chapter introduces *Ocean Literacy: The Foundation for the Success of the Ocean Decade*, a three volume collection endorsed by the United Nations Educational, Scientific and Cultural Organization (UNESCO) Intergovernmental Oceanographic Commission (IOC) as an official Ocean Decade Activity. The series showcases global efforts to build an Ocean-literate society by bridging education, innovation, engagement, and collaboration.

Volume I focuses on transforming education, research, and engagement to lay the foundation for strengthening Ocean Literacy in formal and informal education settings. Volume II emphasizes the role of communities, the Sustainable Blue Economy, and the Arts in promoting sustainable Ocean engagement. Volume III explores how communication, technology, and global initiatives can accelerate impact. Across all three volumes, the series emphasizes the critical role of Ocean Literacy in addressing the Ocean Decade's societal outcomes and challenge areas, especially Challenge 10: Restore Society's Relationship with the Ocean, and calls for bridging the gap between scientific knowledge, public understanding, and collective action. Drawing on global case studies and research-based strategies, the collection underscores the transformative power of Ocean Literacy in shaping a more sustainable future.

This opening chapter highlights interdisciplinary learning, experiential education, and inclusive approaches, including Indigenous knowledge systems, that promote educational innovation and equity. It frames Ocean Literacy as a key driver for science-informed decision-making, deeper Ocean engagement, and enhanced global collaboration in advancing Ocean sustainability.

Keywords

United Nations Decade of Ocean Science for Sustainable Development · Ocean Decade (OD) · Ocean Decade Challenge 10 · UN

T. J. Kennedy (✉)
University of Texas at Tyler, Tyler, TX, USA

International Council of Associations for Science Education (ICASE), Cork, Ireland
e-mail: tkennedy@uttyler.edu

T. J. Kennedy (ed.), *Ocean Literacy: The Foundation for the Success of the Ocean Decade, Volume I*, Sustainable Development Goals Series, https://doi.org/10.1007/978-3-031-98145-6_1

Sustainable Development Goals (SDGs) · SDG 14 Life Below Water · Ocean Literacy (OL) · Marine science · STEM education · Environmental education · Teacher training · Experiential learning · Sustainability · Human-Ocean connection · Traditional ecological knowledge (TEK) · Citizen and Participatory science · Climate resilience · Policy integration · Scientific communication · Equity · Inclusion · youth empowerment · youth engagement in the Ocean Decade · Non-Governmental Organizations (NGOs) · NGO contributions to the Ocean Decade

1.1 Introduction

Ocean Literacy: The Foundation for the Success of the Ocean Decade is a three-volume collection developed in response to the United Nations Decade of Ocean Science for Sustainable Development (2021–2030), known as the Ocean Decade (UNESCO-IOC 2021a). Each volume advances understanding of and engagement with the Ocean through distinct and transformative approaches.

The first volume focuses on educational innovation, curriculum development, and research, laying the groundwork for strengthening Ocean Literacy within formal and informal education environments. As the initial installment of this series, Volume I equips learners, educators, researchers, and policymakers with foundational knowledge and strategies that support the outreach, governance, and innovation explored in Volumes II and III.

Volume II shifts the focus to community outreach, the Sustainable Blue Economy, and the Arts, highlighting creative and economic pathways to sustainability. Volume III explores technological advances, innovation, and global initiatives, emphasizing cutting-edge solutions and cross-sector collaborations that strengthen Ocean Literacy worldwide. This emphasis on collaboration reflects the Ocean Decade's call for co-designed science that is "solutions-oriented, transformative, and grounded in inclusive and participatory processes" to address society's needs and inform action at all levels (UNESCO-IOC 2021b, p. 5).

Across all three volumes, each chapter concludes with a brief summary organized into key concepts explored in the chapter, followed by recommended resources for further research, ensuring that readers can build upon the insights shared and apply their learning in meaningful ways. Together, these volumes examine the role of Ocean Literacy in addressing the Ocean Decade's societal outcomes and challenge areas, emphasizing the need to bridge the gap between scientific knowledge, public understanding, and collective action.

"The enhancement of Ocean Literacy is critical to the success of the Ocean Decade" (UNESCO-IOC 2021c, p. 7). Ocean Literacy refers to the understanding of the Ocean's influence on us and our influence on the Ocean. It encompasses not only knowledge, but also behaviors, values, and actions that promote sustainable engagement with the Ocean. Initially conceived as a knowledge-sharing tool, Ocean Literacy has evolved into a far-reaching strategy for connecting people, policies, and science through shared understanding and collective responsibility. Today, it represents capacity building and action, promoting a common understanding of global citizens as stakeholders furthering society's relationship with the Ocean. Engaging youth as change agents and future leaders is essential to fostering long-term Ocean sustainability and ensuring intergenerational continuity in Ocean knowledge and stewardship.

This three-volume work demonstrates how Ocean Literacy spans diverse sectors of society, addressing the interconnected challenges outlined by the Ocean Decade. It reinforces a recognition of the Ocean as a living entity and emphasizes the reciprocal relationship between humanity and the Ocean. By shifting away from human-centered narratives, this collection promotes an inclusive, interconnected approach to Ocean sustainability that values diverse knowledge systems and fosters meaningful engagement with the Ocean Decade's goals.

The complexities of Ocean interactions require more than traditional scientific research;

they demand a fundamental change in mindset. "This new level of responsibility requires a paradigm shift in the research and application of Ocean science. Striving for greater Ocean Literacy can promote the vital behavioral change that drives the Ocean Decade. It will ensure that citizens feel compelled and empowered to promote and undertake actions to address the most pressing threats to Ocean health" (UNESCO-IOC 2021c, p. 5).

As noted by the National Academies of Sciences, Engineering, and Medicine (2025), interdisciplinary collaboration and knowledge-sharing are essential in fostering this 'cultural shift' that integrates science, education, and governance (pp. 57–58). The Challenge 10 White Paper, *Restoring Society's Relationship with the Ocean* (Glithero et al. 2024), provides a critical framework for addressing the interconnected human-Ocean relationship. It highlights the need for knowledge-sharing, strategic communication, education, and cultural connections as fundamental drivers for societal transformation. This three-volume collection aligns with these priorities, demonstrating how Ocean Literacy initiatives contribute to restoring humanity's connection to the Ocean and fostering informed decision-making for a sustainable future. Recent policy guidance from the UNESCO-IOC, outlined in its 2025 education brief *Promoting Ocean Literacy*, reinforces this priority, emphasizing the urgent need to embed Ocean Literacy not only across education systems, but also within broader policy frameworks that support sustainability, equity, and informed Ocean governance (UNESCO-IOC 2025).

Ocean Literacy is not merely a branch of science education; it is foundational to scientific and environmental literacy. One cannot fully understand climate change, biodiversity, or even fundamental Earth systems, nor recognize the influence of terrestrial bias without understanding the Ocean's role in regulating global processes (National Science Teaching Association (NSTA) and National Marine Educators Association (NMEA) 2025). As the UNESCO-IOC Framework explains, Ocean Literacy encompasses not only what people know, but also how they feel, think, and act in relation to the Ocean (UNESCO-IOC 2021c, p. 5).

Despite this, Ocean education has historically been limited, particularly for students in landlocked regions and underrepresented communities, contributing to disparities in access to Ocean science careers. Furthermore, the concept of the Ocean watershed underscores that all water is ultimately connected to the Ocean, reinforcing the idea that Ocean Literacy is relevant to everyone, regardless of proximity to the coast. As climate challenges intensify, equipping individuals with a strong understanding of the Ocean's influence on weather, carbon cycles, and biodiversity is essential for informed decision-making and long-term sustainability.

In addition, this collection builds on the guidance presented in the *Ocean Decade Manual for Non-Governmental Organizations (NGOs)* (Kennedy et al. 2024; Kennedy and Mascarenhas 2025) amplifying multidisciplinary research and community-driven actions that raise global awareness of Ocean-related challenges. It also elevates the role of civil society while engaging a broader spectrum of Ocean stakeholders. Both interdisciplinary and transdisciplinary approaches are included, with interdisciplinary referring to collaboration across established academic fields and transdisciplinary emphasizing the co-production of knowledge across academic, policy, and community boundaries to solve real-world Ocean challenges. The books serve a broad audience, including educators, scientists, policymakers, conservationists, Indigenous communities, industry professionals, and NGOs, offering actionable insights for education, policy, and community engagement, and fostering a global network of Ocean Literacy practitioners. Drawing upon diverse perspectives from all stakeholders, these volumes function as both a resource and a call to action, underscoring the urgency of Ocean-related challenges while highlighting scalable, innovative solutions. Ocean Literacy has become more than a capacity-building tool; it serves as a bridge between scientific knowledge and the empowerment of global citizens working toward Ocean sustainability (UNESCO-IOC 2021c, p. 8).

These three volumes are the product of an international collaboration, including over 250 authors, excluding those who contributed to multiple chapters, representing 42 countries and regions based on the affiliation organizations of each author. Contributors are from Argentina, Australia, Bangladesh, Belgium, Bosnia and Herzegovina, Brazil, Canada, Cabo Verde, Chile, China, Croatia, Denmark, Ecuador, England, Finland, France, Germany, Greece, Hong Kong SAR, the Republic of Ireland, Italy, Japan, Kenya, Kuwait, Lebanon, Malta, Mexico, the Netherlands, Norway, Northern Ireland, Peru, Poland, Portugal, Romania, Scotland, South Africa, Spain, Sweden, Taiwan, Türkiye, the United States, and Wales (see Fig. 1.1). These authors bring expertise from a wide range of disciplines and geographic regions, reflecting the global commitment to Ocean Literacy and reinforcing our shared responsibility for fostering a sustainable future. All chapters underwent rigorous double-blind peer review by over 80 international experts, followed by Springer Nature's final peer review process to ensure the highest academic standards.

This Ocean Decade Activity unites experts and practitioners from multiple disciplines, fostering a rich exchange of knowledge and strategies. Leveraging local, regional, and international initiatives, these contributions create a shared knowledge base that strengthens global efforts toward a more Ocean-literate society. Understanding the complexities of Ocean interactions requires more than traditional scientific research, it demands a "cultural shift" toward interdisciplinary collaboration and education (National Academies of Sciences, Engineering, and Medicine 2025, pp. 57–58). All sectors of civil society must work together to integrate multiple knowledge systems, conduct research, and develop predictive models to better understand and forecast Ocean and ecosystem changes. This type of collaborative model "requires including both scientists and nonscientists, especially from historically marginalized communities and under-represented groups in Ocean science research, education, and governance" (National Academy of Sciences, Engineering, and Medicine 2025, p. 62). Marine education networks, media, and the foundational principles of Ocean Literacy (National Oceanic

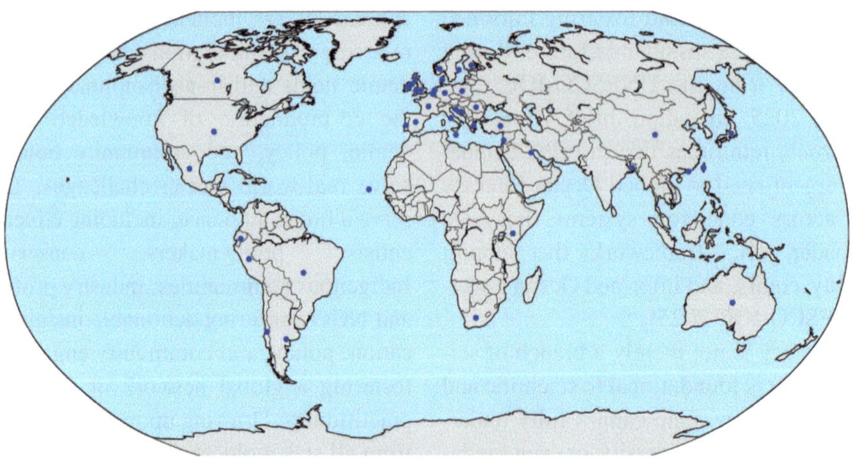

Fig. 1.1 Global Distribution of Author Affiliations Across *Ocean Literacy: The Foundation for the Success of the Ocean Decade*. *Note:* While the map reflects country-level affiliations, additional regional detail is worth highlighting. Contributors from Canada represent four provinces: British Columbia, Nova Scotia, Ontario, and Quebec. Those from the United States span 17 states and Washington, D.C., including Alabama, California, Florida, Georgia, Iowa, Maine, Massachusetts, Missouri, New Jersey, New York, North Carolina, Oregon, Pennsylvania, South Carolina, Texas, Virginia, and Washington. This regional diversity within North America, from the west coast to the east coast, complements the strong international representation, underscoring the truly global and collaborative nature of this Ocean Literacy initiative. Similar patterns of regional representation are evident across many other participating countries, where contributors span a variety of cities, provinces, and institutional contexts

and Atmospheric Administration 2024; UNESCO-IOC 2021c) offer powerful pathways for building transdisciplinary capacities. These mechanisms help engage civil society and other non-academic stakeholders in the co-production of knowledge, advancing the Ocean Decade's vision of inclusive, science-informed decision-making. Contemporary models of Ocean Literacy emphasize not only knowledge acquisition but also active participation in Ocean governance, emotional connection, and the inclusion of diverse ways of knowing, including Indigenous and local perspectives, to drive meaningful behavior change and Ocean sustainability (McKinley et al. 2023, pp. 3–5).

Inclusive and collaborative Ocean knowledge generation must be accompanied by a strong commitment to Ocean Literacy, ensuring that educators and learners at all levels gain the knowledge and skills needed to interpret and apply scientific insights. The Ocean Decade calls for a global effort to enhance education, advance scientific research, and inform decision-making. Building an Ocean-literate society depends on uniting diverse perspectives, equipping community practitioners, policymakers, educators, and the public with the necessary tools to inspire action and innovation across all sectors. Education plays a critical role in this effort, both within formal and informal education environments, empowering individuals to engage meaningfully in Ocean governance, policy, industry, and community initiatives that drive sustainable solutions.

Building on these recommendations, this collection provides evidence-based strategies, educational innovations, and research that empower educators, scientists, policymakers, and civil society, from formal institutions to grassroots organizations, with the knowledge and tools needed to drive lasting, meaningful change in addressing Ocean-related challenges.

1.2 The Ocean and Ocean Literacy

Across all three volumes, the word "Ocean" is intentionally capitalized to emphasize its singular, interconnected nature rather than treating it as a collection of separate basins. Although geographic regions are traditionally identified as the Atlantic Ocean, Pacific Ocean, Indian Ocean, Arctic Ocean, and Southern Ocean, and further distinguished by numerous marginal Seas such as the Caribbean, Mediterranean, and South China (among others), together they form one unified global Ocean that spans over 70% of Earth's surface. For clarity in certain contexts, especially when referencing multiple geographic regions (e.g., "the Pacific and Atlantic Oceans"), the plural form is used. In such cases, "Oceans" refers only to the distinct geographic areas and does not imply that they operate independently; they remain integral parts of a single, interconnected global Ocean system.

Historically, the word *Ocean* was treated as a common noun, reflecting the perception of multiple distinct "Oceans." However, contemporary scientific understanding recognizes the Ocean as a unified entity, functioning cohesively across geographical and ecological boundaries. While capitalization conventions vary by language, most treat *Earth* as a proper noun when referring to the planet, reinforcing its singular identity. Similarly, the Ocean deserves the same recognition, underscoring its central role as a unique, interconnected, and defining feature of planet Earth. There is only one Ocean.

The linguistic shift in capitalizing the word *Ocean*, both when referring to the singular, interconnected system and when naming specific regions such as the Pacific Ocean, aligns with the evolving messaging of the Ocean Decade, reflecting modern scientific consensus and promoting a unified perception of the Ocean's importance. By acknowledging the Ocean as a singular geographic region, we highlight its critical contributions to climate regulation, biodiversity, and human prosperity. This framing is further reinforced by recent high-level scientific and policy initiatives, such as the 2025 One Ocean Science Congress organized ahead of the Third UN Ocean Conference and the joint position paper by NSTA and NMEA describing the importance of teaching and learning about the Ocean, both of which consistently capitalized the word "Ocean" to reflect its singular and essential role.

These efforts not only emphasize the interconnectedness of our one Ocean but also promote a shared understanding among Heads of State, governments, broader society, and Non-Governmental Organizations (NGOs) of the importance of working together to ensure the Ocean's health and future trajectory. They highlight the urgency of science-informed decision-making to conserve and sustainably use Ocean resources, ensuring both Ocean health and human well-being. Recognizing the Ocean as a singular, unified system is not just a matter of linguistic precision; it is a crucial step toward fostering a global mindset that values unity, sustainability, and informed action to ensure the global Ocean's health and its vital role in supporting life on Earth.

Similarly, the term Ocean Literacy has been capitalized throughout these volumes. This decision underscores its significance as a concept that bridges the gap between science and society, while further recognizing Ocean Literacy as vital within education, research, and conservation. Ocean Literacy refers to a specific, internationally recognized framework and growing global movement, meriting capitalization as a proper noun. Many entities, such as UNESCO's Intergovernmental Oceanographic Commission (2025), have embraced this capitalization, reflecting its growing importance in building an Ocean-literate society. Ocean Literacy encapsulates a comprehensive understanding of the Ocean's influence on our lives and the reciprocal impact of human activities on the Ocean. It represents not only knowledge but also a commitment to informed action and shared responsibility, empowering individuals and communities to meaningfully address the challenges and embrace the opportunities presented by the Ocean Decade.

1.3 Building an Ocean-Literate Society for Sustainability

The Ocean Decade emphasizes the need for society to understand, value, and engage with the Ocean to achieve "the Ocean we want"—a healthy, resilient, and sustainably managed Ocean that benefits both people and the planet (Ryabinin et al. 2019, pp. 4–5). It simultaneously focuses on "the science we need for the Ocean we want" (UNESCO-IOC 2020). This vision requires global collaboration across disciplines, sectors, and communities to address seven key societal outcomes: a clean Ocean free of pollution; a healthy and resilient Ocean with well-managed ecosystems; a productive Ocean supporting sustainable food supply and the economy; a predicted Ocean where society can respond to changing conditions; a safe Ocean protecting lives and livelihoods; an accessible Ocean with open and equitable data access; and an inspiring and engaging Ocean valued for its role in human wellbeing and sustainable development (Kennedy 2024; Ocean Decade n.d.).

To guide progress, the Ocean Decade Action Framework outlines three objectives: identifying the knowledge needed for sustainable development; generating comprehensive Ocean understanding; and increasing the use of Ocean knowledge (UNESCO-IOC 2021a, 2021b). These objectives are supported by ten challenges, ranging from tackling marine pollution and restoring ecosystems to fostering an equitable Ocean economy, improving climate resilience, and expanding global Ocean observation and digital modeling.

A foundational Challenge within the Ocean Decade is Challenge 10: Restore Society's Relationship with the Ocean, which underscores the importance of fostering Ocean Literacy as a shared global responsibility. Integrating Ocean Literacy into all aspects of society, it ensures that scientific evidence informs decision-making, governance structures, and collaborative management strategies at local, regional, national, and international levels. Through the co-production and sharing of Ocean knowledge, Challenge 10 bridges the gap between science, policy, and education, empowering individuals and communities to make informed decisions and take responsible actions for a healthy Ocean. Making Ocean knowledge more accessible and engaging is central to this effort, ensuring that complex Ocean-related issues are communicated effectively and that diverse communities, through education, advocacy, and innovative strategies, are equipped to contribute to sustainable solutions.

Achieving "the Ocean we want" depends not only on scientific progress but also on transforming the ways individuals, communities, and institutions engage with the Ocean. Around the world, educators, researchers, policymakers, and advocates are actively contributing to Ocean Decade Challenge 10 by implementing innovative strategies that foster Ocean Literacy and inspire tangible and sustained action. Recent work implementing project-, problem-, and phenomenon-based learning (PBL, PrBL, and PhBL, respectively) in science, technology, engineering, and mathematics (STEM) education, demonstrates how anchoring experiences in real-world phenomena can strengthen understanding of complex concepts and increase engagement during the learning process (Kennedy and Fields, 2023; Kennedy and Odell 2024). The diverse educational approaches highlighted throughout this volume, including curriculum models, instructional strategies, and examples of place-based learning, align with this evidence by offering practical pathways for increasing learner engagement and supporting the transformative educational practices envisioned in Challenge 10.

1.4 The Vision Behind *Ocean Literacy: The Foundation for the Success of the Ocean Decade*

This book brings together an international collection of diverse experiences, providing a platform for sharing impactful initiatives that can be replicated, adapted, and expanded in locally relevant ways. By documenting and amplifying these efforts, we aim to support the collective movement toward a more informed and engaged society, one that values and protects the Ocean for current and future generations.

Each volume explores key themes, including foundational knowledge, innovative practices, diverse communities, teacher training, and curriculum development. The chapters progress from broad global frameworks such as the Ocean Decade, to specific educational applications, spanning introductory concepts to advanced curriculum design. This structure reflects a commitment to geographic diversity and educational equity, with particular attention to reaching underserved populations and advancing inclusive, culturally responsive pedagogies. Achieving equitable science education requires more than accessibility; it demands curriculum relevance, inclusive classroom practices, and the empowerment of learners from all backgrounds (Kennedy and Sundberg 2017). Together, these contributions illustrate the universal relevance of Ocean Literacy in preparing society to address pressing environmental challenges.

Education plays a central role in achieving the United Nations (UN) Sustainable Development Goals (SDGs), providing the knowledge, skills, and global awareness needed to address interconnected environmental and societal challenges (Kennedy and Cherry, 2023). This collection explores educational innovation, curriculum development, and research-driven strategies that strengthen Ocean Literacy, equipping learners with the knowledge and skills needed for informed decision-making and long-term Ocean engagement. Through case studies, success stories, and expert analyses, each volume highlights how these approaches contribute to achieving the SDGs and the broader vision of the Ocean Decade. They also emphasize the importance of cultivating twenty-first-century skills—such as critical thinking, creativity, communication, collaboration, digital literacy, global awareness, ethical responsibility, and lifelong learning—as essential tools for navigating complex environmental issues and preparing learners for a rapidly evolving global workforce (Kennedy and Sundberg 2020; 2025). Through cross-disciplinary collaboration, this volume advances the role of Ocean Literacy as a catalyst for societal transformation, shaping a more informed, engaged, and responsible society.

This evidence-based series brings together voices from around the world to demonstrate how education, research, and community engagement can strengthen society's relationship with the Ocean. It deepens understanding of Ocean Literacy's role in achieving the SDGs and the overarching vision of the Ocean Decade. Beyond informing, it also inspires action by equipping individuals and organizations with the knowledge and tools to implement meaningful Ocean Literacy initiatives. In doing so, it makes a lasting contribution to Ocean Decade Challenge 10 by strengthening efforts to foster a more informed, engaged, and responsible Ocean-literate society.

1.5 Framing Volume I: Ocean Literacy—Transforming Education, Research, and Engagement

Volume I brings together over 90 authors from 19 countries. While Fig. 1.1 presents the global distribution of author affiliations across the full three-volume series, this volume specifically features contributors from across Africa (South Africa); Asia (Bangladesh, China, and Japan); Europe (Belgium, Croatia, England, Germany, Greece, Italy, Norway, Portugal, Spain, Sweden, and Türkiye); North America (Canada—British Columbia and Nova Scotia; and the United States—California, Iowa, Missouri, North Carolina, New Jersey, New York, Texas, Virginia, and Washington); and Latin America (Brazil and Mexico). Collectively, these diverse voices offer a global perspective on Ocean Literacy, showcasing innovative approaches to education, curriculum development, and research that advance the goals of the Ocean Decade. A preview of each chapter follows.

Chapter 2 articulates the foundation for Volume I and, more broadly, for the entire three-volume collection. This opening perspective piece, *Actioning the Ocean Decade Challenge 10 White Paper: A Theory of Change*, sets the stage for the series by exploring the overarching themes of Ocean Literacy and societal transformation. The authors examine Challenge 10: Restore Society's Relationship with the Ocean, highlighting the need to unite diverse knowledge systems, cultures, and professional communities to foster an Ocean-literate world. This chapter introduces a Theory of Change model that outlines key drivers for bridging the gap between scientific knowledge and societal action. It explores how evidence-based strategies and inclusive leadership may contribute to behavioral shifts, potentially strengthening global efforts toward a healthier Ocean and a more sustainable future. As the foundational anchor of this three-volume collection, it provides a theoretical lens for the innovative educational practices and collaborative research explored in the chapters that follow.

Chapter 3, *Delivering Critical Ocean Literacy Initiatives in the Framework of the UN Decade of Ocean Science for Sustainable Development*, demonstrates the transformative efforts of UNESCO's Intergovernmental Oceanographic Commission (IOC) in advancing Ocean Literacy. The authors provide examples of how global initiatives bridge the gap between science and society, fostering engagement through education, outreach, and policy integration. This chapter investigates interdisciplinary approaches, including community-driven programs, school-based initiatives, and international collaborations that seek to enhance public understanding of the Ocean. It considers how innovative strategies and cross-sector partnerships may contribute to shaping a more Ocean-literate world.

Chapter 4, *ScienceOnShore (SOS): A Wake-up Call from the Ocean Literacy Movement*, presents a hands-on, inquiry-based marine science initiative developed by the New York City Department of Education and Kingsborough Community College. Designed for fifth graders (typically 10–11 years old), the program provides students with opportunities to explore local marine environments, conduct lab-based investigations, and develop essential scientific skills. This chapter examines how initiatives like SOS may help address the lack of science laboratories in elementary schools by integrating real-world marine science into early education. It considers how regional EcoLabs and outdoor learning experiences can foster Ocean Literacy, climate awareness, and environmental stewardship among young learners.

Chapter 5, *The TransforMAR Project: A Deep Dive into Ocean Literacy*, investigates how immersive experiences can transform Ocean Literacy by fostering emotional connections and a sense of responsibility for marine conservation. The authors highlight their pioneering initiative in Brazil that engages youth from marginalized coastal communities through experiential learning, including SCUBA diving and field-based education. By integrating marine biology, ecology, and environmental stewardship, the project focuses on participants' Ocean identity to inspire behavioral change toward sustainability. This chapter considers the broader implications of using direct engagement with the Ocean as a tool for fostering environmental justice, social equity, and youth empowerment. As the Ocean Decade emphasizes inclusive and transformative education, this initiative demonstrates how hands-on experiences can shape the next generation of Ocean leaders and advocates.

Chapter 6, *Ocean Literacy in a Landlocked State*, highlights how marine science education can foster connections to the Ocean even in regions far from the coast. The author examines a U.S. high school marine biology program in Des Moines, Iowa, where students engage in hands-on experiences such as maintaining aquariums, conducting research projects, and participating in outreach activities. This chapter investigates how immersive learning opportunities, including field expeditions and community engagement, shape students' understanding of marine ecosystems by making remote connections to the Ocean. It considers how innovative educational models can bridge geographic barriers to Ocean Literacy, inspiring stewardship in inland communities.

Chapter 7, *Engaging Students with Ocean Technology: A Journey from Floats to Ocean Literacy* provides examples of how real-time Ocean data collected at sea and activities occurring on research ships can bring marine science into classrooms, regardless of proximity to the Ocean. The authors highlight the U.S. BioGeoChemical-Argo (BGC-Argo) profiling array, a global network of robotic floats that continuously gather data on Ocean chemistry, carbon cycling, and biological processes. This chapter explores how programs like Adopt-a-Float connect students with ongoing Ocean research, allowing them to analyze real-world data, engage in virtual exploration, and develop a deeper understanding of Ocean systems. It considers how integrating ship-collected data into classroom learning fosters Ocean Literacy and environmental awareness, again demonstrating that students don't need to live near the Ocean to study and contribute to its science.

Chapter 8, *Once Is Not Enough: Connecting Students of Color to Their Local Waterways*, explores how hands-on environmental education programs can influence students' perceptions of the Ocean and nearby aquatic environments, particularly in historically marginalized urban communities in the state of New York. Drawing on a series of experiential learning opportunities, including webinars with marine scientists, estuary cleanups, and immersive field trips, this chapter explores how repeated engagement can build emotional connection, deepen understanding, and begin to shift students' sense of place and responsibility. The authors consider the systemic barriers that have limited access to Ocean Literacy for students of color and highlight how meaningful, community-rooted experiences can help foster a stronger, more inclusive pathway toward Ocean Literacy. This chapter underscores the importance of sustained opportunities for engagement as essential to promoting environmental agency, equity, and broader participation in the marine sciences.

Chapter 9, *Shaping Scientific Literacy: How Underserved Students Conceptualize Marine Scientists and Themselves*, analyzes how students from underrepresented backgrounds in U.S. schools perceive marine scientists and explores the role of informal educational experiences in shaping their science identity and career aspirations. Grounded in situated learning theory, this longitudinal study follows students in a 2-week summer enrichment program designed to expand access to marine science for high-ability, low-income youth. Using a modified version of the Draw-a-Scientist Test tailored to marine science (DAMST), the authors explore students' evolving perceptions of scientists, their understanding of marine science careers, and their self-identification as scientists. This chapter highlights the importance of inclusive learning environments, authentic research experiences, and repeated exposure to STEM pathways in broadening participation and dismantling stereotypes that may limit diversity in the marine sciences.

Chapter 10, *Investigating the Longer-term Student Outcomes of a Two-day Ocean-focused Program for Underserved Youth in Cape Town, South Africa*, examines the potential of short-term immersive education to foster sustained environmental engagement among youth. The Ocean Guardians Workshop (OGW), developed by I AM WATER: Ocean Conservation, provides underserved 11–13-year-old students with a transformative experience that blends snorkeling, mindfulness, and ecosystem exploration within Marine Protected Areas. Through a mixed-methods evaluation, including pre-, post-, and 4-month follow-up surveys, the authors investigate changes in Ocean learning, attitudes, and stewardship behaviors. This chapter offers rare longitudinal insight into the effectiveness of short-duration programs, highlighting how culturally responsive instruction, youth empowerment, and emotional connection to the Ocean can shape environmental identities and support long-term engagement with Ocean sustainability goals.

Chapter 11, *Ocean Equity: Unveiling Gender Opportunities and Challenges in Brazil's Ocean Literacy Activities*, considers the persistent gender disparities in Ocean sciences and the transformative potential of Ocean Literacy to address them. The authors explore historical and structural barriers, such as exclusion from scientific expeditions, limited leadership opportunities, and the "leaky pipeline" effect, that continue to restrict women's full participation in marine fields. Focusing on inclusive education and public engagement strategies developed in Brazil, the chapter highlights how efforts to promote gender equity through Ocean Literacy can support broader goals in education, sustainability, and scientific leadership. Aligned with the Ocean Decade and the UN Sustainable Development Goals, particularly SDGs 4, 5, and 14, this chapter underscores the importance of integrating gender equity into Ocean Literacy initiatives to create more inclusive, resilient, and equitable Ocean futures.

Chapter 12, *Cultivating Ocean Guardians: Evaluating a Novel Pedagogical Model for Enhanced Ocean Citizenship in Basic Education*, introduces an innovative approach developed in Portugal that integrates Ocean Literacy, systems thinking, and student activism into classroom learning. The model engages students and educators in hands-on, citizen science activities, including coastal monitoring and advocacy projects, supported by interdisciplinary collaboration and teacher training. Implemented over two academic years in a range of educational settings, including non-coastal schools, this chapter explores the model's potential for fostering Ocean citizenship and expanding Ocean Literacy across diverse learning environments. It offers insight into how experiential, community-based learning can be developed to support the Ocean Decade's call for inclusive, transformative education.

Chapter 13, *The Ocean We Have: A Case Study at the Ocean Research College Academy* (ORCA), showcases how place-based, inquiry-driven education can reshape students' understanding of the Ocean and their roles as informed stewards. Situated in the Snohomish River estuary in the U.S. Pacific Northwest, ORCA engages high school students enrolled in a dual-credit early college program in hands-on research, fieldwork, and cross-disciplinary learning experiences. The chapter examines how sustained exposure to real-time data, collaborative research, and student-led inquiry fosters deeper ecological awareness and strengthens Ocean Literacy. Grounded in the goals of Ocean Decade Challenge 10, the authors reflect on ORCA's potential to serve as a scalable model for integrating Ocean Literacy into diverse educational contexts while honoring the complexity of human-Ocean relationships.

Chapter 14, *Whale Tale: Big Challenges Facing Ocean's Biggest Species*, describes how systems thinking, experiential learning, and Ocean Literacy converge in an undergraduate Honors College course designed to deepen students' understanding of marine conservation and climate science. Centered at East Carolina University on the U.S. East Coast, the course culminates in a field-based expedition to Baja California Sur, Mexico, where students observe gray whales in their natural habitat and engage directly with local ecotourism experts, marine biologists, and conservationists. The authors highlight how this cross-cultural and cross-coastal learning experience was framed through weekly seminars, advocacy assignments, and guided reflections. The chapter examines how immersive, place-based learning can foster a sense of connectedness, ecological awareness, and agency in marine conservation, and considers the course's broader potential as a replicable model for integrating Ocean Literacy into global climate education.

Chapter 15, *Ocean Literacy in Science Curriculum: A Quantitative Content Analysis in Sweden*, evaluates the extent to which national curricula integrate Ocean Literacy principles and identifies where they are insufficiently addressed. Focusing on Swedish education for students aged 7–19, the authors analyze 182 learning goals to assess alignment with the Ocean Literacy Framework and identify structural factors that may contribute to the Ocean's invisibility in formal education. This chapter introduces a replicable methodology for evaluating Ocean-related content in national curricula and explores the implications of vague learning goals, terrestrial bias, and place-based limitations. Framed within Sweden's role in the Ocean Decade, the chapter raises urgent questions about how curriculum design shapes environmental understanding and equity in Ocean education, offering tools and insights for global application.

Chapter 16, *Development of an Aquaponics STEAM Program for Improving Ocean Literacy*, investigates how an interdisciplinary education program in Japan fosters Ocean Literacy among middle school students and pre-service teachers. Grounded in the Theory of Change outlined in Chap. 2, this chapter describes how a circular aquaponics system connecting aquatic and terrestrial life was used to engage students in scientific inquiry, systems thinking, and nature-connectedness. The program integrates sustainability education with national curriculum goals, emphasizing hands-on learning, relational values, and the co-development of competencies critical to environmental decision-making. The authors investigate how participation in aquaponics activities might support the development of scientific inquiry skills and environmental values, and also explore the program's potential as a scalable model for integrating Ocean Literacy, food sustainability, and pro-environmental learning into formal education systems.

Chapter 17, *Investigating Ocean Literacy Levels of Pre-Service Science and Social Studies Teachers in Türkiye: The Influence of Gender, Ocean Experience, and Environmental Activities*, reflects on how experiential learning may shape Ocean Literacy among future educators. Using a survey tool aligned with the seven essential principles of Ocean Literacy, the authors assess the knowledge, attitudes, and self-perceived competencies of pre-service teachers at a seaside university in Türkiye. This chapter explores key factors influencing Ocean Literacy, including gender differences, prior Ocean experience, and participation in environmental activities. It considers how embedding experiential and place-based learning into teacher education programs may strengthen connections between theoretical knowledge and real-world engagement, preparing educators to integrate Ocean concepts into their classrooms.

Chapter 18, *Insights from Project SEA: Building Ocean Literacy and Enhancing Marine Science Education through Teacher Professional Development*, highlights how targeted professional development may help educators integrate Ocean Literacy into classrooms. The authors detail Project SEA, a multi-year initiative in New Jersey (U.S.), designed to support teachers through hands-on workshops, curriculum revisions aligned with the Next Generation Science Standards (NGSS), and collaborations with formal and informal educators. Emphasizing place-based learning, the program connects participants with local ecosystems to explore the Ocean's role in climate regulation, biodiversity, and human impacts. This chapter provides an example of how professional development can influence instructional practice, strengthen Ocean connections, and foster interdisciplinary learning.

Chapter 19, *The Global Marine Curriculum: Academic Pathways for Sustainable Marine Development*, authored by a researcher based in Germany, presents a global study examining the role of higher education in shaping future Ocean-related decision makers. The author analyzed 3366 Academic Marine Education Programs (AMEPs) across multiple continents to explore how concepts such as sustainability, Traditional Ecological Knowledge (TEK), and speciesism are represented within disciplines including marine biology, fisheries, and aquaculture. The chapter identifies patterns of emphasis and areas for further integration within existing curricula, highlighting opportunities to strengthen interdisciplinary, ethical, and sustainability-oriented approaches to Ocean Literacy. By emphasizing the importance of fostering deeper human-Ocean connections, the author calls for the continued evolution of marine education to incorporate diverse perspectives and better prepare students to address contemporary Ocean challenges.

Chapter 20, *Scientists for Ocean Literacy: Empowering Scientists as Ocean Advocates*, details how scientists can engage with Ocean Literacy to connect their research with the public and policymakers. The authors explore an Ocean Decade project led by the European Global Ocean Observing System (EuroGOOS), which brings together European institutions to integrate Ocean Literacy into research, outreach, and decision-making. Through case studies on public events, citizen science initiatives, and interdisciplinary collaborations, the authors investigate strategies scientists use to communicate their work beyond academia, and consider how training programs and interactive tools may enhance public engagement and influence policy, shaping the role of scientists as advocates for the Ocean.

Chapter 21, *Co-Designing Transdisciplinary Science for the Ocean We Want: Actions for Scientists to Enrich Ocean Literacy*, authored by researchers from China and Bangladesh, proposes how scientists can actively contribute to Ocean Literacy by co-producing research with diverse stakeholders. Anchored by the EPICD framework (Engagement, Partnerships, Interdisciplinary Collaboration, Communication Training, and Digital Platforms), the authors outline a strategic approach for connecting scientific knowledge with public awareness and policy impact. Focusing on marine plastic pollution, especially nano- and microplastics, the chapter presents case studies and international collaborations that illustrate how transdisciplinary efforts can break down barriers between research and societal action. This closing chapter offers a model for integrating scientific inquiry, public engagement, and communication strategies in support of the Ocean Decade's goal of a more inclusive, informed, and sustainable future.

A brief concluding section at the end of the volume entitled *Ocean Literacy at the Global Turning Point: Reflections, Foundations, and Future Pathways*, brings these chapters together, highlighting how they collectively establish the educational, research, and engagement foundations of Ocean Literacy. It reflects on shared themes across the volume, including equitable access to Ocean Literacy, the importance of experiential and place-based education, and the roles of teachers, youth, and scientists in strengthening human-Ocean connections. The concluding section also revisits the volume's central aim of linking knowledge with action and situates these contributions within the broader goals of the Ocean Decade, providing a foundation for the themes explored in the next two volumes.

1.6 Conclusion

As a whole, the 21 chapters in Volume I, *Transforming Education, Research, and Engagement*, present diverse strategies for advancing Ocean Literacy in formal and informal education. The chapters explore the intersections of curriculum development, teacher training, interdisciplinary research, and educational policy integration in fostering a deeper understanding of the Ocean's role in human and planetary systems. They highlight innovative approaches, including hands-on, place-based learning, technology-enhanced instruction, and equity-driven practices. By showcasing global perspectives, this volume underscores the transformative potential of education to equip current and future generations with the knowledge, skills, and sense of responsibility needed to address the challenges identified by the Ocean Decade and shape a more sustainable future.

On behalf of the contributing authors in this Volume, I invite you to explore the ideas, strategies, and experiences presented in these chapters. We hope they offer useful perspectives that inform your work, inspire new collaborations, and strengthen Ocean Literacy efforts in your own communities and beyond. By engaging with these insights, each of us helps to build a more informed, Ocean-literate, and sustainable future.

1.6.1 Summary

Key concepts explored in this chapter include:

- The Ocean Decade (2021–2030) presents a critical opportunity to advance Ocean Literacy globally by fostering collaboration across education, research, policy, and civil society to inform and engage all sectors of the population.
- Recognizing the Ocean as a singular, interconnected system, rather than separate basins, shifts global perspectives and reinforces the Ocean's role in climate regulation, biodiversity support, and sustaining human prosperity.
- Capitalizing the word "Ocean" represents a vital linguistic and conceptual shift from viewing disconnected basins to recognizing a singular, living system that unites Earth's processes. Like "Earth," treating the Ocean as a proper noun affirms its unique, interconnected identity, reshapes how we think about it, and reinforces the urgency of restoring humanity's relationship with the Ocean.
- Ocean Literacy is foundational to the success of the Ocean Decade and serves as a bridge between scientific knowledge, public understanding, and collective action, supporting co-designed, solutions-oriented approaches that benefit both people and the planet.
- Ocean Literacy extends beyond knowledge acquisition, emphasizing emotional connection, shared responsibility, and active participation. It also values diverse ways of knowing, including Indigenous and local perspectives, as essential to building inclusive and effective solutions.
- Building an Ocean-literate society requires interdisciplinary collaboration and knowledge co-production, uniting educators, researchers, policymakers, NGOs, and communities to develop informed, sustainable actions at local, national, and global levels.
- Challenge 10: Restore Society's Relationship with the Ocean stands as a central pillar of the Ocean Decade and highlights the urgent need to rebuild human–Ocean connections through education, cultural engagement, and accessible communication.
- This three-volume series, *Ocean Literacy: The Foundation for the Success of the Ocean Decade,* showcases diverse strategies from around the world that bridge the gap between science and society.
  - Volume I focuses on educational innovation, curriculum development, and research-based approaches that support transformative learning and the development of Ocean literate individuals and institutions.
  - Volumes II and III continue this exploration by examining the roles of community engagement, the Blue Economy, artistic expression, technological innovation, and global partnerships in strengthening Ocean Literacy.
- By providing insights, case studies, and actionable strategies, this collection empowers all sectors of civil society to apply Ocean Literacy in varied contexts and contribute meaningfully to achieving the goals of the Ocean Decade and the UN Sustainable Development Goals (SDGs).

Acknowledgements First and foremost, I extend my deepest gratitude to Claudia Acuña, publishing editor at Springer Nature, whose encouragement, guidance, and unwavering support gave me the confidence to take on this ambitious project. Since early 2023, we worked together to bring this book series to life, and her belief in the importance of this work has been invaluable. I also thank Prasad Gurunadham, production editor at Straive, and the Springer Nature author support team for their invaluable assistance throughout the publication process. Their responsiveness, attention to detail, and commitment to ensuring a smooth editorial experience have been instrumental in bringing this project to completion.

I am also grateful to The University of Texas at Tyler for their support of this project and to the International Council of Associations for Science Education (ICASE) for their ongoing efforts to advance Ocean Literacy and promote inclusive, science-based education within the Ocean Decade. I also wish to express my sincere appreciation to the Intergovernmental Oceanographic Commission of UNESCO for designating this book series as an official Ocean Decade Activity. This recognition underscores the alignment of this work with the goals of the United Nations Decade of Ocean Science for Sustainable Development, and I am honored to contribute to its mission.

My sincere thanks go to my colleague Diana Payne for her engagement in the early phases of this work, particularly for supporting the dissemination of the call for proposals within the Ocean Literacy community. I also extend my appreciation to the many representatives from organizations around the world who helped promote this collaborative effort—including colleagues at AMEA, ICASE, ICoRSA, NAME, NMEA, NOAA, NSTA, and numerous members of the UNESCO NGO community. The collective support of these partners helped make possible the global scope and representation reflected across this three-volume work.

I am also deeply appreciative of Diz Glithero, Centre for Ocean Literacy Collaboration (Canada), for her collegial support and for helping to disseminate the call through her extensive networks. Her unwavering dedication to Ocean Literacy and her remarkable ability to unite people around a shared vision are truly inspiring. I have learned much from her thoughtful approach to international engagement and greatly value the insights she shared throughout this process.

Being part of this global collaboration has been an incredible experience. Authors from 42 countries/regions contributed across disciplines, and their diverse perspectives reflect a shared international commitment to restoring society's relationship with the Ocean and building a more sustainable world. Their insights have profoundly enriched this collection, and I am grateful for all I have learned from them.

Words cannot express my appreciation to the 80+ expert reviewers who dedicated their time and expertise to providing valuable feedback throughout the double-blind peer review process. Their thoughtful critiques played a crucial role in refining the work of the contributing authors and ensuring the academic rigor of this collection. Their efforts reflect the collaborative spirit of the Ocean Literacy community and have greatly strengthened this three-volume series.

In addition to these collective contributions, I also wish to acknowledge the institutions whose support made Open Access publication possible. The author teams of seven chapters in ***Volume I: Transforming Education, Research, and Engagement*** generously contributed to the Open Access publication fee, ensuring that all chapters in Volume I of this series could be freely accessible to readers around the world. This collective commitment to equitable knowledge sharing reflects the spirit of the Ocean Decade and the global Ocean Literacy community. I gratefully acknowledge the support of the following institutions whose authors made this possible:

- Chapter 1: The University of Texas at Tyler (Tyler, Texas, USA)
- Chapter 2: The Ocean Conservation Trust (Plymouth, England, UK)
- Chapter 4: Kingsborough Community College (Brooklyn, New York, USA)
- Chapter 7: Monterey Bay Aquarium Research Institute (Moss Landing, California, USA)
- Chapter 12: Instituto de Educação, Universidade de Lisboa (Lisbon, Portugal)
- Chapter 13: Everett Community College (Everett, Washington, USA)
- Chapter 15: Institutionen för Marina Vetenskaper, Göteborgs Universitet (Gothenburg, Sweden).

And last, but certainly not least, I am especially thankful to my family for their constant support and understanding throughout this journey. Their patience, encouragement, and willingness to share our time while I worked on this project mean more than words can express. Their love continues to fuel my passion for the Ocean, and I am endlessly grateful for their presence in my life and their constant support of my work priorities.

References

Glithero DL, Bridge N, Hart N, Mann-Lang J, McPhie R, Paul K, Peebler A, Wiener C, Yen C, Kelly R, McRuer J, Hodgins D, Curtin F (2024) Ocean Decade Vision 2030 White Papers—Challenge 10: Restoring society's relationship with the Ocean. UNESCO-IOC, Paris. [The Ocean Decade Series, 51.10.]. https://unesdoc.unesco.org/ark:/48223/pf0000390126

Kennedy TJ (2024) Advancing Ocean sustainability for climate-resilient change. In: Hufnagel L (ed) Globalization and sustainability: ecological, social and cultural perspectives (Chapter 2). IntechOpen, pp 7–30. https://www.intechopen.com/online-first/1176877

Kennedy TJ, Cherry A (2023) Sustainable development and science and technology education. In: Akpan B, Çavaş B, Kennedy T (eds) Contemporary issues in science and technology education. (Chapter 10, pp. 131–150). Springer Nature. https://link.springer.com/chapter/10.1007/978-3-031-24259-5_10

Kennedy TJ, Fields DL (2023) Augmenting upper and early learning STEM lessons through phenomenon-based learning. J High Educ Theory Pract 23(18):170–182. https://articlegateway.com/index.php/JHETP/article/view/6630

Kennedy TJ, Mascarenhas T (2025) Empowering NGOs: A Practical Guide to Advancing the Ocean Decade's Vision. OCEANS 2025 Brest, France. https://ieeexplore.ieee.org/abstract/document/11104339

Kennedy TJ, Odell MRL (2024) Integrating STEM: An interdisciplinary approach to preK-12 education. In: Ashraf MA, Tsegay SM (eds) STEM Education: Recent Developments and Emerging Trends. IntechOpen. (ISBN 978-1-83769-256-9). https://www.intechopen.com/online-first/88903

Kennedy TJ, Sundberg CW (2017) International perspectives and recommendations on equity and gender: Development studies in science education. In: Akpan B (ed) Science education: A global perspective (Chapter 15). Springer Nature, pp 295–311. https://link.springer.com/chapter/10.1007/978-3-319-32351-0_15

Kennedy TJ, Sundberg CW (2020) 21st century skills. In: Akpan B, Kennedy TJ (eds) Science education in theory and practice: An introductory guide to learning theory (Chapter 32). Springer Nature, pp 479–496. https://link.springer.com/chapter/10.1007/978-3-030-43620-9_32

Kennedy TJ, Sundberg, CW (2025). 21st century skills. In: Akpan B & Kennedy TJ (eds), Science education in theory and practice: An introductory guide to learning theory (2nd ed.), (Chapter 27, pp. 461–486). Springer Nature. https://link.springer.com/chapter/10.1007/978-3-031-81351-1_27

Kennedy TJ, Dufourneaud O, Castillo MP, Deane S, Durand J, Giaccone T, Jonda S, Mascarenhas T, Mtonga C, Ng K, Palazot S, Paul S, Salazar ER, Santisteban R, Shuwen X, Silva A, Solis Rivera V, van't Land H, Zhou Y (2024) Ocean Decade manual for non-governmental organizations (NGOs). UNESCO-IOC, Paris. [The Ocean Decade Series, 54]. https://unesdoc.unesco.org/ark:/48223/pf0000392198

McKinley E, Burdon D, Shellock RJ (2023) The evolution of Ocean Literacy: A new framework for the United Nations Ocean Decade and beyond. Mar Pollut Bull 186:114467. https://doi.org/10.1016/j.marpolbul.2022.114467

National Academies of Sciences, Engineering, and Medicine (2025) Forecasting the Ocean: The 2025–2035 Decade of Ocean Science. The National Academies Press. https://doi.org/10.17226/27846

National Oceanic and Atmospheric Administration (2024) Ocean Literacy: The essential principles and fundamental concepts of Ocean sciences for learners of all ages (Version 3.2). https://www.marine-ed.org/ocean-literacy/overview

National Science Teaching Association & National Marine Educators Association (2025) The importance of teaching and learning about the Ocean: Joint position statement. https://www.nsta.org/sites/default/files/NSTA-NMEA_Joint-Position-Statement_The-importance-of-teaching-learning-about-the-ocean_FINAL.pdf

Ocean Decade (n.d.) Vision & mission. United Nations Decade of Ocean Science for Sustainable Development. https://oceandecade.org/vision--mission/. Accessed 1 Jan 2025

Ryabinin V, Barberie J, Haugan P, Kullenberg G, Smith N, McLean C, Troisi A, Fischer A, Arico S, Aarup T, Pissierssens P, Visbeck M, Enevoldsen HO, Rigaud J (2019) The UN Decade of Ocean Science for Sustainable Development. Front Mar Sci 6(470):1–10. https://www.researchgate.net/publication/334806579_The_UN_Decade_of_Ocean_Science_for_Sustainable_Development

UNESCO-IOC (2020) The science we need for the Ocean we want: The United Nations Decade of Ocean Science for Sustainable Development (2021–2030). UNESCO-IOC, Paris. https://unesdoc.unesco.org/ark:/48223/pf0000265198

UNESCO-IOC (2021a) The United Nations Decade of Ocean Science for Sustainable Development (2021–2030): Implementation plan, summary. UNESCO-IOC, Paris. [The Ocean Decade Series, 20]. https://unesdoc.unesco.org/ark:/48223/pf0000376780

UNESCO-IOC (2021b) Co-designing the science we need for the Ocean we want: Guidance and recommendations for collaborative approaches to designing and implementing decade actions. UNESCO-IOC, Paris. [The Ocean Decade Series, 29]. https://unesdoc.unesco.org/ark:/48223/pf0000379563

UNESCO-IOC (2021c) Ocean Literacy Framework for the UN Decade of Ocean Science for Sustainable Development 2021–2030. UNESCO-IOC, Paris. [The Ocean Decade Series, No. 22]. https://unesdoc.unesco.org/ark:/48223/pf0000377708

UNESCO-IOC (2025) Promoting Ocean literacy—an educational policy brief. UNESCO-IOC, Paris. [IOC Policy Brief, 1]. https://unesdoc.unesco.org/ark:/48223/pf0000393205.locale=en

Author Contribution and Biographical Profile

[1]*Teresa Kennedy: Conceptualization; Writing—Original Draft, Review, and Editing; Book Project Administration—Supervision, Coordination of Chapter Submission, Oversight of the Double-Blind Peer Review Process, and Editing and Formatting.

*Corresponding Author at tkennedy@uttyler.edu

Teresa Kennedy PhD, is a professor at the University of Texas at Tyler, a three-time Fulbright Scholar, and an active marine oceanographer with extensive experience in international science education. A past president of the International Council of Associations for Science Education (ICASE), she currently serves as the ICASE representative to UNESCO. She has held three elected positions on the NGO-UNESCO Liaison Committee, most recently as vice president and liaison to the Intergovernmental Oceanographic Commission (IOC) and Co-Chair of the IOC Ocean Decade NGO Task Group.

She played a key role in shaping the international science initiatives of the U.S. National Science Teaching Association (NSTA), serving five terms on its International Advisory Board from 2004 to 2024. She also contributed as a member of the writing team for the joint NSTA and NMEA Ocean Literacy position paper (2024–2025). As former Deputy Director of NASA's GLOBE Program, she collaborated with science education leaders in over 100 countries, deepening her commitment to global cooperation and environmental education.

Dr. Kennedy has dedicated her career to advancing global science education, promoting Ocean Literacy, and contributing to international STEM and sustainability initiatives. As the editor and driving force behind this project, she conceived the initiative and guided its development, ensuring a cohesive and impactful collection that unites diverse perspectives to advance Ocean Literacy around the world.

Actioning the Ocean Decade Challenge 10 White Paper: A Theory of Change

2

Diz Glithero, Nicola Bridge, Jen McRuer, and Natalie Hart

Abstract

This contribution serves as a perspective piece that helps to illuminate and situate the chapters across this three-volume publication in relation to one of the Ocean Decade's primary objectives—to connect people and the Ocean. Along with providing a high-level summary of the Ocean Decade Challenge 10 White Paper that was developed through the Ocean Decade Vision 2030 process, this perspective piece explores how the White Paper recommendations can be put into action using a theory of change framework. Challenge 10 focuses on restoring society's relationship with the Ocean by addressing the human behaviors that drive Ocean threats and positioning Ocean Literacy as a key societal outcome. Emphasizing the need for interdisciplinary collaboration and inclusive engagement across sectors, a working theory of change is proposed, structured around four key drivers—Knowledge Systems, Communications, Education, and Cultural Connections—each contributing to a shift in how people value and interact with the Ocean. By outlining a practical foundation for advancing Ocean-positive behaviours, this perspective piece aims to provide a useful lens to apply and help connect the tangible initiatives captured in this important and timely publication. Collectively, these initiatives are helping to create the conditions and opportunities to help inspire and build an Ocean-literate society.

Keywords

Ocean Literacy (OL) · Societal-Ocean connections · Marine social sciences · Human-Ocean relationships · Behaviour change

2.1 Introduction: Vision 2030 Process

The United Nations Decade of Ocean Science for Sustainable Development (hereafter, 'Ocean Decade') provides a collaborative framework and critical opportunity for the global community to rally behind a core mission–transformative Ocean science solutions, connecting people and the Ocean.

Three years into the Ocean Decade, the Vision 2030 process was launched with the aim to achieve a common and tangible measure of success for each of the 10 Ocean Decade Challenges. This ambition-setting process, coordinated by

D. Glithero (✉) · J. McRuer
Centre for Ocean Literacy Collaboration (formerly the Canadian Ocean Literacy Coalition), Dalhousie University, Halifax, Nova Scotia, Canada
e-mail: diz@colcoalition.ca; jen@colcoalition.ca

N. Bridge
Ocean Conservation Trust, Plymouth, UK
e-mail: Nicola.Bridge@oceanconservationtrust.org

N. Hart
Communications Inc., London, UK
e-mail: natalie@communicationsinc.co.uk

T. J. Kennedy (ed.), *Ocean Literacy: The Foundation for the Success of the Ocean Decade, Volume I*, Sustainable Development Goals Series, https://doi.org/10.1007/978-3-031-98145-6_2

IOC-UNESCO and led by 10 Working Groups of diverse experts across geographies, cultures, generations, and genders, began in June 2023. It culminated with the presentation of the White Papers at the Ocean Decade Conference in Barcelona in April 2024, followed by their publication in June (UNESCO 2024).

Collectively, the White Papers helped to identify existing research and gaps; infrastructure, capacity, and resource needs; achievable (rather than aspirational) recommendations and priority activities; and indicators to help measure progress between now and 2030. The White Paper process, although helpful toward these mentioned ends, leaves more to be done on the scales of inclusion, adaptive and distributed leadership, and operational mechanisms and resources to achieve transformative outcomes.

2.2 Situating Challenge 10

The 10 Ocean Decade Challenges collectively focus on addressing the critical issues facing the Ocean. As outlined across the set of White Papers, there exists a clear and causal link between human behaviour and the future health of the Ocean: If we continue to behave as we do on a global scale, there will be catastrophic consequences. Challenge 10 is fundamental to operationalizing the changes and scale of local, regional, and ultimately global action needed across society to avoid this fate. Challenge 10 bridges the science solutions (Challenges 1–6), and the data, technology, infrastructure, and capacity needs (Challenges 7–9) with *society*.

At the onset of the Ocean Decade, Challenge 10 was initially framed as *changing humanity's relationship with the Ocean*. The White Paper (Glithero et al. 2024) shifted this framing to *Restoring society's relationship with the Ocean*. Challenge 10 requires clear recognition that current 'threats' to the Ocean are a direct result of human behaviour. Although existing tools of science, technology, and communication allow us to understand the Ocean in finer detail than ever before and to share that understanding widely, knowledge generation, data sharing, and information exchange alone are not enough. These efforts, despite their important contributions, cannot solely drive the societal transformations required to achieve Ocean sustainability and climate mitigation. The Ocean community, in its broadest and most inclusive form (e.g., scientists, governments, businesses, educators, communicators, fishers, community groups, artists), needs to inspire and generate the motivation, capability, and opportunity for people across all aspects of society to make decisions and behave in ways that ensure a healthy Ocean, sustaining all life–present and future. If threats to the Ocean are a people problem, restoring our interdependent relationship requires people-driven solutions.

Together, across geographies and generations, knowledge systems and cultures, sectors and scales, commitment to tangible actions is needed to help close the knowledge-action gap. Such actions can provide opportunities for society to discover meaningful connections with the Ocean and deepen peoples' understanding of the vital role the Ocean plays in our lives. They can also foster emotional connections that motivate action and create intentional opportunities for capacity exchange, strengthening community-driven engagement. Ultimately this approach can lead to sustained actions that benefit people and planet.

2.3 Ocean Literacy: A Societal Outcome

Ocean Literacy, when positioned as a societal outcome–a society that better understands, values, and cares for the Ocean–is critical to situating Challenge 10 and achieving Ocean Decade goals. Ocean Literacy, as a collective goal, provides a unifying perspective that bridges diverse fields and professional communities. Previously, Ocean research, marine education, and strategic Ocean communications operated in parallel silos. However, as illustrated by the diverse chapter topics and expertise in this three-volume publication, achieving Ocean sustainability and fostering an Ocean-connected society requires a collaborative effort among many essential actors, including but not limited to scientists, social science researchers, Indigenous knowledge holders, educators, communication experts, public and private sector leaders, policymakers, artists, and cultural influencers. Together, these contributors help contextualize human-Ocean relationships and inspire integrated solutions across science, policy, and society (McRuer et al. 2025). The unification of diverse actors to foster collective action does not require everyone to use or identify with the term 'Ocean Literacy;' rather, to recognize the pursuit of a shared outcome.

In striving toward an Ocean-literate society, Working Group 10 drew on the above fields and professional communities to co-identify four integral '*drivers*' that have a major influence on this aim: (1) Knowledge Systems, (2) Communications, (3) Education, and (4) Cultural Connections. The White Paper proposes key recommendations and operational activities to support each driver's strategic aim; and explores the following questions: Who will enable these key recommendations? What resources, tools, or infrastructure will ensure these recommendations and activities (or variations of) are achieved? How will we measure progress? Building on the White Paper's recommendations and proposed activities, we offer a working model of a theory of change below.

2.4 Actioning Challenge 10: A Theory of Change

A theory of change is a means of illustrating how and why a given set of interventions within a particular context (i.e., 4 drivers of Challenge 10 White Paper) can lead to a desired change. For each driver we use the following logic model: If [actors/'users'] do [operational activities] then [driver] results in [strategic aim], contributing to [societal outcome]. Such a framework helps to illuminate and make sure that all drivers are working towards the same desired outcome—a society that makes decisions and behaves in ways that ensures a healthy Ocean. Figure 2.1 below outlines a working theory of change for two driv-

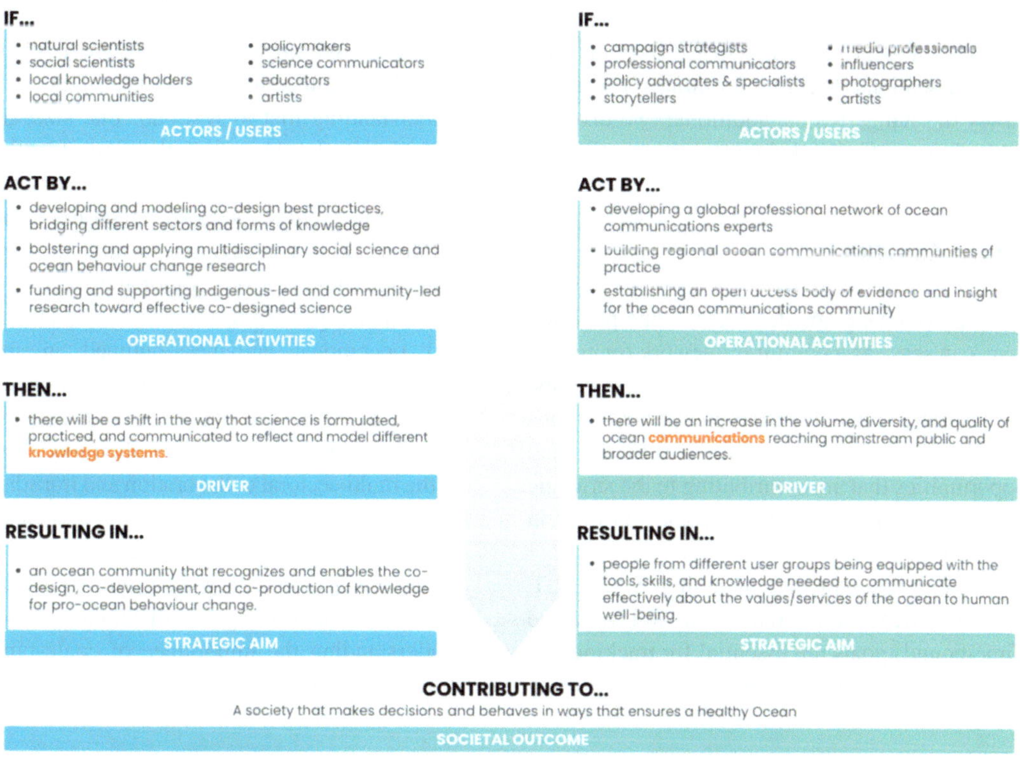

Fig. 2.1 A Theory of Change for two Challenge 10 Drivers–Knowledge Systems and Communications

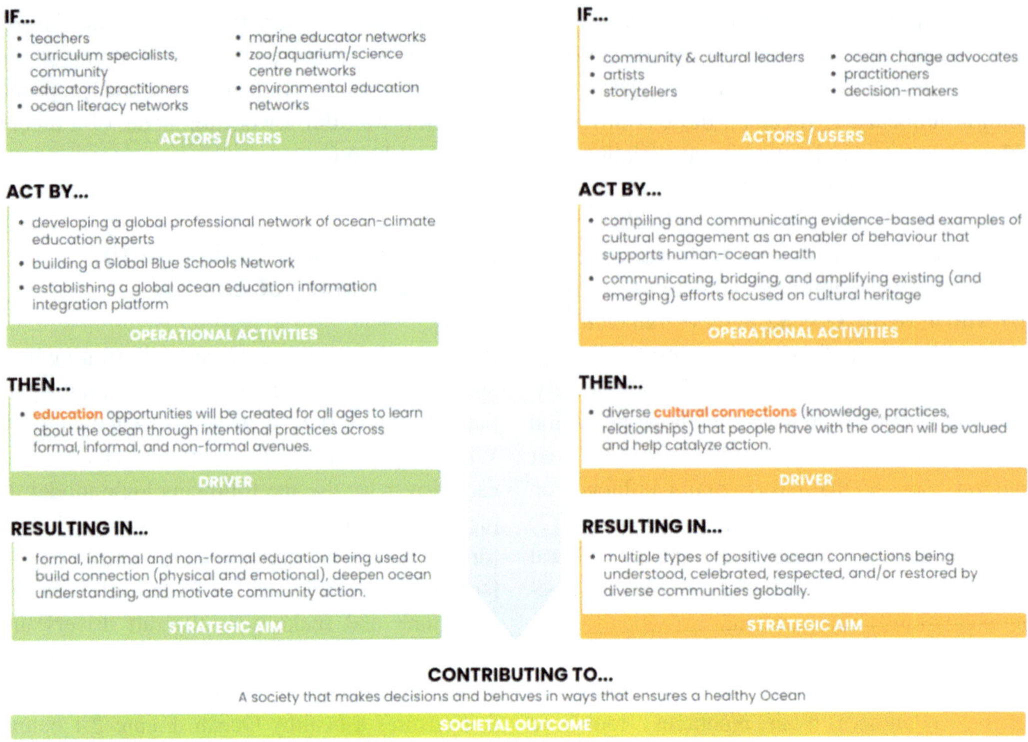

Fig. 2.2 A Theory of Change for two additional Challenge 10 Drivers–Education and Cultural Connections

ers of Challenge 10–Knowledge Systems and Communications. Figure 2.2 outlines a working theory of change for the additional two drivers–Education and Cultural Connections.

2.5 Conclusion

A theory of change model, like the one provided here, can serve as a helpful framework for collective action. Contained within this three-volume publication are tangible examples of how various actors are creating the conditions, activities, and opportunities that are contributing to the strategic aims and overall societal outcome outlined in Figs. 2.1 and 2.2.

Regular assessments, including S.M.A.R.T. (Specific, Measurable, Achievable, Relevant, and Time-bound) goals are essential for tracking and celebrating progress. However, advancing Challenge 10 requires three important realizations by the Ocean community:

1. Understanding that there is no 'one size fits all' solution. Implementing the recommendations from the White Paper necessitates collaboration across sectors, regions, and cultures, each contributing locally relevant solutions while supporting global efforts.
2. Understanding that the key recommendations and operational activities outlined in the Challenge 10 White Paper were intentionally chosen based on (1) evidence-based insights; (2) existing momentum; and (3) their modeling of the multi-sectoral collaboration and transdisciplinary approaches needed to drive societal transformation. And recognizing that there is room for adaptation, elaboration, and expansion as we continue working toward common ends.
3. Understanding the importance of coalescing around the evolved definition of Ocean liter-

acy as a societal outcome, unifying and aligning diverse sectors and professional communities to work collectively toward a healthy Ocean for people and planet.

With these realizations in mind, combined with a willingness and dedication to value diverse and restorative Ocean connections, we need to cultivate together the collective motivation, capability, and opportunity that enables people to take positive and measurable action to ensure a healthy Ocean. To make this vision a reality, robust science-policy frameworks rooted in *society* are essential–not only to guide our efforts and ensure accountability, but also to monitor progress over time, adapting as needed. Theory of change frameworks can help to align and propel transdisciplinary action—starting, for example, with the spearheading efforts outlined in the chapters of this three-volume publication, the cross-cutting ideas outlined in the Challenge 10 White Paper, and the dovetailing endeavours of all other Decade Challenges. Together, we can draw on and mobilize impactful and sustained human-Ocean connections that help to secure a hopeful future—one that celebrates and builds on existing bright spots of how to nurture deep understanding, value, and care for the Ocean.

2.5.1 Summary

Key concepts explored in this chapter include:

- Current 'threats' to the Ocean are a direct result of human behaviour.
- Ocean Literacy, simply put, is a societal outcome—a society that better understands, values, and cares for the Ocean. When positioned as such, Ocean Literacy serves a unifying role, bridging diverse fields and professional communities.
- In striving toward an Ocean-literate society, the Ocean Decade Challenge 10 White Paper identified four integral '*drivers*' that have a major influence on this aim: (1) Knowledge Systems, (2) Communications, (3) Education, and (4) Cultural Connections.
- A theory of change can serve as a helpful framework for collective action, illuminating what actors, activities, and results contribute towards a shared desired outcome.

2.5.2 Recommended Resources

The following resources provide further insights and support on this topic:

- **Ocean Decade Vision 2030 White Papers**: https://unesdoc.unesco.org/search/8d86cc86-b35c-44b3-9b71-4caaa779ce64
- **Challenge 10 White Paper**: https://unesdoc.unesco.org/ark:/48223/pf0000390126
- Glithero, D., Bridge, N., Paul, K., & McRuer, J. (2025). Ocean Decade Challenge 10 Underscores Social Dynamics in Marine Sciences as Critical to Transforming Human-Ocean Relationships. *International Journal of Marine Science.* https://academic.oup.com/icesjms/article/82/3/fsaf030/8075033?utm_source=advanceaccess&utm_campaign=icesjms&utm_medium=email

Acknowledgements Special thanks to Meghan Callon for her contributions to editing and graphic design, including figure development. This perspective piece was derived from the published Challenge 10 White Paper that was part of the Ocean Decade Vision 2030 process. The White Paper was co-authored by a global team of Ocean leaders, including Natalie Hart, Judy Mann-Lang, Romney McPhie, Ken Paul, April Peebler, Carlie Wiener, Chia-Dai (Ray) Yen, and Rachel Kelly. Further thought leadership was provided by additional Challenge 10 Working Group members, including: Paul Anderson, Achare Elvis Ayamba, Easkey Britton, Guadalupe Diaz Costanzo, Louisa Hooper, Olga Mashkina, Sebastian Thomas, and Alexander Turra. Following the publication of the White Paper, work has continued to consider Theory of Change models, led by the co-authors of this paper, that can augment and guide the actioning of the recommendations put forward therein.

References

Glithero DL, Bridge N, Hart N, Mann-Lang J, McPhie R, Paul K, Peebler A, Wiener C, Yen C, Kelly R, McRuer J, Hodgins D, Curtin F (2024) Ocean Decade Vision 2030 White Papers—Challenge 10: Restoring society's relationship with the Ocean. UNESCO-IOC, Paris. [The Ocean Decade Series, 51.10]. https://unesdoc.unesco.org/ark:/48223/pf0000390126

McRuer L, McKinley E, Glithero DL, Christofoletti R, Payne DL (2025) Human-ocean relationships: Exploring alignment and collaboration between Ocean Literacy research and marine conservation. Mar Policy 171:106418. https://doi.org/10.1016/j.marpol.2024.106418

UNESCO (2024) Ocean Decade Vision 2030 White Papers. In: UNESCO Digital Library. UNESCO-IOC, Paris. [The Ocean Decade Series, 51.10]. https://unesdoc.unesco.org/search/9fd6c16f-d470-4de4-983e-e03664dfa781

Author Contributions and Biographical Profiles

[1]*Diz Glithero: Co-conceptualization; Writing—Original Draft, Review, and Editing.

[2]Nicola Bridge: Co-conceptualization; Writing—Review and Editing.

[3]Jen McRuer: Co-conceptualization; Writing—Review and Editing.

[4]Natalie Hart: Co-conceptualization.

*Corresponding Author at diz@colcoalition.ca

Diz Glithero, PhD, is the Executive Director of the Centre for Ocean Literacy Collaboration—a centre for Ocean Literacy collaboration, research, and innovation that is administratively based at the Ocean Frontier Institute at Dalhousie University. Diz is an Adjunct Professor in the Marine Affairs Program at Dalhousie University. She led the co-development of Land, Water, Ocean, Us: A Canadian Ocean Literacy Strategy (March 2021) and co-founded Ocean Week Canada. Internationally, she served as the co-chair and co-author of the Ocean Decade Vision 2030 Challenge 10 Working Group and White Paper—Restoring Society's Relationship with the Ocean. She serves on the Steering Committee of the IOC-UNESCO-led Ocean Decade Coordination Office, Connecting People and the Ocean. Diz's former work as an Adjunct Professor in the Faculty of Education at the University of Ottawa and earlier research focused on environmental sustainability education, civic engagement, and youth agency informs her current Ocean efforts. In 2025, Diz was awarded the King Charles III Coronation Medal for her dedication to Ocean conservation through education and research.

Nicola Bridge is the Head of Ocean Advocacy and Engagement at the Ocean Conservation Trust (OCT) with over 18 years' experience in the field of human-centered conservation, Ocean Literacy, environmental psychology, and environmental education. Responsible for developing, leading and evaluating the Ocean Conservation Trust's formal and informal conservation learning programmes, as well as driving advocacy campaigns to support everyone to think about the Ocean in their daily lives. She is passionate about changing the narrative about what constitutes conservation, recognising that everyone on the planet is responsible for the health of the natural world that sustains us. Nicola is also integrated into the community of Plymouth as a Trustee for Millfields Inspired, a Member of the Connect Multi Academy Trust and a board member for the Plymouth Sound National Marine Park Horizons Programme. Further afield, Nicola is a member of the Defra Ocean Literacy Working Group and a founder of the We Are Ocean network. Nicola is President of EMSEA (European Marine Science Educators Association). She is Co-Chair of an Ocean Decade Working Group, focusing on global sustainable change in human behaviour to support the health of the Ocean.

Jen McRuer, PhD, is the Research Manager of the Centre for Ocean Literacy Collaboration—a centre for Ocean Literacy collaboration, research, and innovation that is administratively based at Dalhousie University (Ocean Frontier Institute). Jen currently leads COLC's public Ocean perceptions research both nationally and internationally—in the co-design of comparative analysis and evaluative tools such as the Ocean & Society Survey (OSS). She also co-leads the Ocean Literacy Research Community (OLRC) and is a co-author of several papers aligning with these efforts and with the broader goal encompassed by the United Nations Ocean Decade Vision 2030 Challenge 10 white paper—Restoring Society's Relationship with the Ocean. Jen's research and facilitation experience stems from participatory action approaches focused on biocultural diversity and sense of place. Through this lens, she has worked collaboratively with global communities to better understand and communicate social and environmental changes and solutions. Jen is from Nova Scotia, and continues to be inspired by working in, across, and with people, culture, and place.

Natalie Hart is Director of the Insight and Strategy Unit at Communications Inc., a communications agency that focuses solely on Ocean issues. She leads the program Advancing Strategic Ocean Communication and advises the OneOcean Flotilla, a network of 100+ Ocean NGOs, on effective communication strategies. Natalie champions a research-led, evidence-based approach to strategic communication. She has extensive experience designing and delivering strategic communications and behavioural change campaigns across the world, including several years spent in the Middle East and North Africa working on conflict and post-conflict issues. Also a writer, she advocates for the power of storytelling in social change. Natalie was a member of the Ocean Decade Vision 2030 Challenge 10 Working Group and White Paper—Restoring Society's Relationship with the Ocean.

Delivering Critical Ocean Literacy Initiatives in the Framework of the UN Decade of Ocean Science for Sustainable Development

3

Francesca Santoro, Paula Barbeito Morandeira, Valentina Lovat, Ana Vitória Tereza de Magalhães, Gaia Maria Sole Intonti, Esteban Gottfried Burguett, Alessia Tombolini, Paola Carluccio, and Jack Coulton

Abstract

The United Nations proclaimed the Decade of Ocean Science for Sustainable Development (Ocean Decade) from 2021 to 2030 with a vision of "the science we need for the Ocean we want" for which the Ocean Decade provides a common framework for diverse stakeholders to generate and use Ocean knowledge to contribute to the achievement of the 2030 Agenda for Sustainable Development. As success of the Ocean Decade depends on public understanding of the Ocean, the Ocean Literacy (OL) programme of the Intergovernmental Oceanographic Commission of UNESCO (UNESCO-IOC) has designed a comprehensive approach that encompasses research, capacity building, project incubation, networking, resource development and programme coordination and has launched several OL initiatives to connect education policies and practices that enhance understanding and transform knowledge into action. This chapter discusses how OL initiatives led by UNESCO-IOC effectively engage different stakeholders through the use of different approaches, methods, and formats. While there is clear evidence of OL's profound potential to drive positive action for the Ocean, accurately quantifying its impact and tracing direct links between Ocean awareness and behaviour change remains a complex task that requires further research.

Keywords

Behaviour change · Capacity building · Community engagement · Education · Ocean Literacy (OL) · Ocean Decade · Stakeholders · Sustainable development · United Nations · UNESCO-IOC

F. Santoro (✉) · P. B. Morandeira · V. Lovat · A. V. T. de Magalhães · G. M. S. Intonti · E. G. Burguett · A. Tombolini · P. Carluccio · J. Coulton
UNESCO Intergovernmental Oceanographic Commission (IOC) Project Office for Ocean Literacy, UNESCO Regional Bureau for Science and Culture in Europe, Venice, Italy
e-mail: f.santoro@unesco.org; p.barbeito-morandeir@unesco.org; v.lovat@unesco.org; av.de-magalhaes@unesco.org; gm.intonti@unesco.org; e.gottfried-burguett@unesco.org; a.tombolini@unesco.org; p.carluccio@unesco.org; j.coulton@unesco.org

T. J. Kennedy (ed.), *Ocean Literacy: The Foundation for the Success of the Ocean Decade, Volume I*, Sustainable Development Goals Series, https://doi.org/10.1007/978-3-031-98145-6_3

3.1 Introduction

UNESCO received a mandate from the UN General Assembly to proclaim the Decade of Ocean Science for Sustainable Development (hereafter the "Ocean Decade") from 2021 to 2030, with a vision of "the science we need for the Ocean we want." The Decade is led by the UNESCO-IOC and aims to generate knowledge for sustainable development while promoting global partnerships. However, conventional science and data alone are insufficient to achieve the kind of societal transformation that will enable humanity to achieve Ocean sustainability and equity (Glithero et al. 2024).

The sustainable management of Ocean resources is a shared responsibility that requires an Ocean-centric perspective and considers the Ocean as a living ecosystem with its own rights. This paradigm shift requires personal, emotional, and cultural connections to the Ocean, making community-engaged science essential (Shellock et al. 2024). A systemic perspective is therefore essential, one that is capable of including and integrating elements related to public educational frameworks, alongside aspects linked to the development of policies and practices in the fields of Ocean economy, science, policy, and management.

This chapter has two main objectives. First, it aims to provide an overview of the various Ocean Literacy (OL) initiatives implemented by UNESCO-IOC, emphasizing the key factors that contribute to their success, identifying the challenges faced, and analyzing the strategies adopted to address them. A central aspect of these initiatives is capacity building, which plays a pivotal role by equipping target groups with both scientific knowledge and interpersonal skills. These competencies are essential for sharing best practices and fostering effective interactions with diverse stakeholders, ultimately enhancing the dissemination of knowledge and supporting the development of more robust conservation policies.

Second, the chapter explores how the insights gained from these initiatives can contribute to improving public understanding of Ocean stewardship and, consequently, to advancing the goals of the Ocean Decade. Beyond the numerous OL initiatives already underway around the world, the Ocean Decade seeks to expand these efforts, creating opportunities for diverse actors to develop and implement OL actions.

To this end, the Ocean Decade Implementation Plan (UNESCO 2021a) was designed to highlight OL as one of the seven outcomes of the Decade and to provide a structured approach to promoting OL globally throughout the Ocean Decade (UNESCO 2021b): an inspiring and engaging Ocean where society understands and values the Ocean in relation to human wellbeing and sustainable development (Outcome 7), as well as specifically addressing two of the ten main challenges: equitable access to Ocean knowledge and technology (Challenge 9) and widespread understanding of the value of the Ocean for human well-being and sustainable development (Challenge 10).

This understanding has driven the evolution of the global OL community, which has expanded from its traditional roots in formal education to reach a broader set of stakeholders through a variety of different methods and approaches. New methods include digital platforms, the promotion of inclusive educational practices, the implementation of transdisciplinary programs, and stakeholder engagement through collaborative efforts, among others (McKinley et al. 2023).

We can now consider OL as a tool capable of increasing our sense of shared responsibility, empowering the global community to protect marine ecosystems and promote sustainable development through a renewed connection to the Ocean. As previously mentioned, we will now delve deeper into OL initiatives and projects with the aim of supporting the OL community in identifying solutions, innovative approaches, and valuable insights.

3.2 Comprehensive Approach to OL Initiatives Led by the UNESCO-IOC

The UNESCO-IOC utilises a wide range of formats to promote OL, including events, artistic and interactive exhibitions, digital campaigns, training sessions, outdoor education, and community-building through networks. This diversity ensures broad reach and effective engagement with diverse audiences. By employing these varied approaches, the UNESCO-IOC has been able to experiment, gather feedback, and evaluate the effectiveness of OL initiatives. This process has facilitated the development of new editions of existing formats and informed the development of strategies to expand the impact of OL on different target audiences, as for the case of "Feel the Change" which will be described in the dedicated section below (Table 3.1).

Table 3.1 Main activities of each of the UNESCO-IOC Ocean Literacy projects

Project	Key Activities
Save the Wave	Restoring marine and coastal blue carbon ecosystems, with a focus on Posidonia Oceanica meadows; education, nature-based solutions, and enhancing Ocean Literacy (OL).
Global Blue Schools Network	Promoting Ocean Literacy (OL) in schools, fostering creative and innovative learning methodologies, supporting Blue Curriculum development, and engaging students, teachers, and communities in 17 countries.
Sea Beyond—Kindergarten of the Lagoon	Ocean education for children aged 4–6 through experiential learning, integrating local culture, and promoting responsibility for Ocean stewardship in Venice's lagoon.
Sea Beyond—Educational Modules	Providing online training for teachers, tailored lesson plans for high schools, and organizing student campaigns on Ocean conservation, engaging schools globally.
Ocean&Climate Village	Interactive, travelling exhibition focused on the connection between the Ocean and climate change, offering multi-sensory experiences, and promoting collaboration and knowledge sharing.
Feel The Change	Sensory journey designed for blind and visually impaired individuals to experience the impact of the climate crisis on ecosystems and marine life through touch and hearing.
Ocean Literacy With All (OLWA)	Designing transformative research-based activities and events for diverse stakeholders to advance Ocean Literacy, including Ocean Literacy Dialogues (OLD) and global participation.
Ocean Literacy Training	Offering free, self-paced training courses for various sectors (government, educators, media, general public, architects, etc.) through the Ocean Teacher Global Academy (OTGA).

3.2.1 Case Studies of Successful Ocean Literacy Initiatives

The method outlined in this chapter involves a careful selection of specific case studies, highlighting examples which had a high impact on target audiences, and notable results. Each case study is presented in a structured format, detailing a brief project description, key activities and methods used, specific outcomes, target audiences and stakeholders, the factors that enabled success, and the challenges encountered along with their solutions. This approach allows for a comprehensive evaluation of the initiatives, showcasing their effectiveness and lessons learned. A comparative analysis has been used to identify commonalities between the factors which enabled success, the challenges and the solutions to those challenges in each of the selected projects. This method is in line with our research objective, as it allows us to have a comparable dataset between the different initiatives, facilitating a comprehensive comparison and allowing us to draw more solid conclusions.

Save the Wave

Brief description of the project

Save the Wave is an initiative dedicated to preserving and restoring marine and coastal blue carbon ecosystems while reconnecting people to the Ocean. The project aims to counter negative human impacts on these ecosystems through education, nature-based solutions, and enhanced OL. It started in Sicily and the Tremiti Islands, Italy, with a focus on restoring *Posidonia Oceanica* meadows, crucial to Mediterranean biodiversity and ecosystem services.

Key activities and methods used

- Site-specific restoration, comprehensive assessments, and community engagement through educational programs
- Knowledge exchange among partners and dissemination among diverse audiences, including the scientific community
- Environmental awareness initiatives on *Posidonia Oceanica* tailored for schools, with classroom lessons and field trips
- Workshops to raise awareness among local communities and tourists
- Capacity building for students and early-career Ocean professionals (ECOPs) through scientific training, hands-on ecosystem restoration activities, and training on essential soft skills for future stakeholder interactions
- Advocacy activities for better conservation policies

Outcomes, target audiences and stakeholders

Outcomes

- Restoration and preservation of marine ecosystems
- Restoration of the connection between people and the Ocean
- Development of technical and soft skills for ECOPs

Target audiences

- Researchers and ECOPs
- Schools, institutions, communities, students, teachers
- Policy makers
- The tourism industry

Stakeholders

- Scientific partners (universities and research centres)
- Donors, whose funding provides long-term financial sustainability
- Local authorities (municipalities, national and regional parks, Marine Protected Areas, the Coast Guard)
- The tourism industry (hotels and restaurants)
- Diving centres: these play a key role in the monitoring phase, and provide essential services during ecosystem restoration and capacity building

Enabling factors for success

- Strong support from local governments and institutions
- Effective scientific assessment, implementation, monitoring, evaluation and reporting system
- Knowledge exchange and capacity building for different target audiences
- Engagement of local communities and tourists

Challenges and Solutions. A synthesis of the challenges and solutions is present in Table 3.2.

> [The lectures during this training] have given me a lot of tools for sharing and communicating my work and overall science and conservation of the Ocean. Seeing other projects has also inspired me to follow their steps and hopefully work in similar conservation projects. [The most important concept I learned] was never giving up on a project if you believe in it, also connecting with people through your experiences and putting all your efforts to share your knowledge with the world. (Testimonial of a participant to one of the Save the Wave editions)

Table 3.2 Synthesis of the challenges and solutions for Save the Wave

Challenges	Solutions
Long-term funding	Mix donations of the private sector with long-term public funding
Policy and authorizations	Work together with the National parks and MPAs
Community and tourist engagement	Local presence throughout the year via local partners, e.g. diving centres

Global Blue Schools Network

Brief description of the project

The Global Blue Schools Network is a community of dynamic hubs taking a proactive approach to OL in their school communities. Blue Schools inspire students to deepen their awareness and understanding of the Ocean through innovative, transversal, and creative learning methodologies across different knowledge fields. There are currently 18 official Blue Schools Networks established, covering 17 countries and the European Union. The network contains 1261 schools, with 8400 students and 2240 teachers involved. The Blue Schools initiative supports the advancement of a Blue Curriculum to support innovation in education, twenty-first-century skills, and interdisciplinary learning.

Key activities and methods used

- Engagement and communication with the global community of Blue Schools
- Mentoring sessions with focal points to facilitate the establishment of committees for each network, as well as contact with UNESCO field offices and government representatives

Outcomes, target audiences and stakeholders

Outcomes

- Equality and opportunity are promoted, as the initiative reduces barriers to Ocean knowledge
- Improved study quality and time spent at school: Blue Schools provide enriched learning experiences and improved pedagogical tools for both teachers and students
- Adoption of eco-friendly school infrastructure, e.g. recycling bins, creating gardens for organic food to be used in the canteen, creation of bicycle parking racks, etc.
- Student empowerment through youth-led activities

Target audiences

- National coordinators
- Schools
- Students

Stakeholders

- Governments
- Local businesses
- Research centres
  - Enabling factors for success

- A highly collaborative ecosystem: fostering good relationships with local communities, parents, school staff, research institutions and governments supports the development of long-term partnerships and engagement
- The establishment of national coordination allows for distributed leadership that is able to consider culturally relevant and appropriate activities for each community

Challenges and Solutions. A synthesis of the challenges and solutions is present in Table 3.3.

> We introduced the Blue Schools project in Luanda, Angola to combat marine pollution. We give lectures in primary schools that are close to the coastal area where we teach the importance of the sea for maintaining life on earth. We work with children aged 9 and 10. (Testimonial of a person from Angola who has established Blue Schools in Luanda, Angola)

Sea Beyond: Kindergarten of the Lagoon

Brief description of the project

Sea Beyond—Kindergarten of the Lagoon is an educational programme that connects children with the Ocean, fostering their healthy growth and promoting a sustainable future. The initiative

Table 3.3 Synthesis of the challenges and solutions for the Global Blue Schools Network

Challenges	Solutions
Many educational systems focus on academic achievements and tests; innovative approaches not seen as priority	National networks need to be familiar with the priorities of their respective educational systems and help Blue Schools to align their activities
Teachers and schools often need funds to develop activities	National grants allocated to schools and networks in order to keep activities sustainable, with renewal of certification to maintain a standard of quality
Busy curriculum	Interdisciplinary approaches and methodologies that can be adapted to fit into diverse curricula

focuses on Ocean education for children aged 4–6, through experiential place-based learning. Inspired by the Blue Schools model, the Kindergarten instils knowledge and responsibility for Ocean stewardship while integrating the context of the local culture. Venice, with its unique lagoon environment and rich cultural landscape, was selected as an ideal location for the pilot project.

The first edition of Sea Beyond—Kindergarten of the Lagoon involved 6 schools, 120 children, and 20 teachers and staff. There were 24 days dedicated to the project, both outdoors, on the island of Torcello, and in the schools themselves.

The second edition involved 2 schools, 72 children, and 15 teachers and staff. Spanning 28 days, the project focused on outdoor experiences and was monitored with a psycho-pedagogical approach. These experiences appeared to improve children's communication and sense of care for the environment, supporting their developmental growth and enhancing their curiosity.

Key activities and methods used

- Outdoor exploration: these free exploration sessions emphasised sensory engagement and interaction with different environments, fostering curiosity and discussion.
- Marine science workshops with interactive activities such as collecting natural samples and observing plankton under a microscope. Simple scientific tools were used to teach basic ecological concepts.
- Creative arts and storytelling sessions: children engaged in art projects focused on the lagoon ecosystem and participated in storytelling with illustrated books. These sessions leveraged their creativity to build emotional connections to the Ocean and reinforce their knowledge of marine ecosystems.

Outcomes, target audiences and stakeholders

Outcomes

- Children gained a comprehensive understanding of the transitional ecosystem, marine citizenship, and improved their scientific and critical thinking skills.
- Creative arts and storytelling sessions fostered an emotional connection to the Ocean.

Target audiences

- Kindergarten children
- Kindergarten teachers
- School coordinators
- Families
- Local people

Stakeholders

- Municipality of Venice
- Artisans
- Researchers
- Schools
- Local NGOs

Enabling factors for success

- Strong support from teachers and coordinators
- Comprehensive and Adaptive Curriculum which includes OL
- Effective Monitoring and Evaluation System

Challenges and solutions. A synthesis of the challenges and solutions is present in Table 3.4.

> I think this project makes my son grow up in a positive way and have a very nice relationship with nature and above all respect nature.

Table 3.4 Synthesis of the challenges and solutions for Sea Beyond—Kindergarten of the lagoon

Challenges	Solutions
Ensuring access to nature for children	Identify practicable areas nearby with the support of the municipality
Engaging teachers in outdoor settings and allowing them to lead independently	Provision of training, resources, collaboration, and support
Securing funds to develop OL activities	Multilateral funding agencies, national agencies, research funds, Environmental, Social and Governance (ESG) and Corporate Social Responsibility (CSR) priorities

Testimonial of one of the parents of a child who has participated in the Kindergarten of the Lagoon project.

> When I have the afternoon shift, I go in and ask my children "What did you do this morning?" And nobody answers. They don't remember. However, when we come back from the Kindergarten of the Lagoon they remember it even later, so maybe the children's interest is more activated there than at school. School is just school. (Testimonial of one of the children's teachers.)

Sea Beyond: Educational Modules

Brief description of the project

Sea Beyond—Educational Modules is an educational program which aims to raise awareness around Ocean conservation. These modules consist of online training sessions for teachers and tailored lesson plans for high schools, as well as a competition for students in participating schools to develop their own awareness-raising campaigns.

The first edition was held online and involved 10 schools. The winning campaign was developed by Agrupamento de Escolas de Vialonga in Vialonga, Portugal.

The second edition, held in 2021–2022, involved 10 schools from Brazil, China, Italy, Mexico, Peru, Portugal, the United Kingdom and South Africa. Each school received visors made from recycled cardboard to be used in immersive virtual reality experiences on Ocean conservation.

The third edition, which is ongoing through 2024, has so far involved more than 200 schools worldwide, representing around 37,000 students.

The program contains lesson plans and a webinar series to guide teachers in integrating OL into their educational frameworks.

Key activities and methods used

- Hackathon-style collaborative problem solving
- Mentoring sessions and a written guide on how to use the provided lesson plans and engage the students during the project
- Teachers are incentivised to empower students through the development of student-led actions

Outcomes, target audiences and stakeholders

Outcomes:

- Teachers empowered to teach OL materials both within the framework of the modules, and beyond
- Schools able to develop their own materials with an interdisciplinary approach and incorporate OL themes across subjects
- Students empowered to work on Ocean themes and propose their own interactive activities for the school community

Target audiences

- Teachers, mainly from high schools
- Students
- School coordinators
- Families

Stakeholders

- SEA BEYONDers: a selection of young people working on high-impact Ocean conservation projects
- Researchers
- Teachers
- Blue Schools
- Scientists

Enabling factors for success

- A space to support teachers and the school community to develop OL activities
- Teachers being incentivised to support the inclusion of OL in the curriculum beyond the project

Challenges and Solutions. A synthesis of the challenges and solutions is present in Table 3.5.

> Six months after finishing the Sea Beyond project, I started my own NGO, called Oceanica, which works for OL. Sea Beyond gave me the skills I needed to start my own journey towards protecting the Ocean as a more educated and self-confident person. All participants in our project came away from the experience with clearer passions and more defined goals. Oceanica became my own lesson to others, of how the Ocean is a key part of our lives, and everything is connected to it. My experience

Table 3.5 Synthesis of the challenges and solutions for Sea Beyond—Educational Modules

Challenges	Solutions
Some teachers face challenges in bureaucratic structures, parent worries and restrictions on outdoor activities	The Sea Beyond educational model can be connected to a whole-school approach and initiatives such as the Blue School networks
Teachers and schools often need grants and funds to develop interactive activities	Grants and funds can be awarded as part of a competition, as happened in Sea Beyond, thus supporting schools to develop tailored activities and continue their OL activities in future
Busy curricula and inflexible structures	The inclusion of OL in the curriculum can be done through interdisciplinary approaches and flexible methodologies through a whole-school approach

> with Sea Beyond allowed me to show students from my own experience that we too have the power to contribute to SDG14. (Testimonial of a Peruvian student who took part in the project Sea Beyond - Educational Modules)

Ocean&Climate Village

Brief description of the project

The Ocean&Climate Village (OCV) is a travelling, interactive educational exhibition originally created in 2021 for Pre-COP26 in Milan, Italy and which has been on tour ever since. Aimed at children, young people, educators, and general society it explores the connection between the Ocean and climate through a multisensory experience that fosters collaboration and knowledge sharing.

The OCV covers topics from Ocean services and climate systems to biodiversity and threats to the Ocean, offering innovative solutions.

Key activities and methods used

- Guided Tours: university students lead interactive tours of the exhibition, simplifying complex concepts and engaging visitors of all ages in meaningful discussions.
- Interactive Workshops: visitors engage in hands-on activities to understand the impact of climate change on Ocean ecosystems and explore sustainable solutions.
- Educational Panels: informative panels provide detailed information on Ocean conservation and climate science
- Experiences: virtual reality stations offer immersive experiences that allow visitors to explore underwater environments and witness the effects of climate change on marine life.
- Panel discussions involving scientists and policy-makers to provide insight into ongoing research and strategies for mitigating the impact of climate change on the Ocean.
- Art Installations: created by local and international artists, these installations highlight the beauty of marine life and the urgent need for climate action, aiming to spark emotional connections.

Outcomes, target audiences and stakeholders

Outcomes:

- Increased awareness and curiosity among different target audiences
- Increased experience for volunteers in science communication and outreach

Target audiences

- Schools (aged 3–18)
- Families
- General public

Stakeholders

- Universities
- Research centres
- Municipalities
- The private sector
- Local artisans
- Policy makers
- Institutes

Enabling factors for success

- An exhibition space designed to make Ocean knowledge accessible for different target groups through a variety of content, from infographics to interactive question boards.

Table 3.6 Synthesis of the challenges and solutions for Ocean&Climate Village

Challenges	Solutions
Securing adequate funding	Involve different sectors and partners that can contribute to funding the exhibition
Building new partnerships and collaborations	Develop targeted outreach and networking strategies to attract more partners for the exhibition.
Developing material for the exhibition and including it all in the space available	Supply additional, digital-only materials that are accessible through QR codes

- The exhibitions received financial support from donors, as well as monetary and in-kind contributions from local municipalities.

Challenges and solutions. A synthesis of the challenges and solutions is present in Table 3.6.

> I participated in a previous edition of this exhibition and I would like to participate again. Raising awareness about the problems of our planet and our Ocean is a big step towards solving them and that's why I think this is a great project to do regardless of age. (Testimonial of a university student who led interactive tours at the Ocean&Climate Village)

Feel The Change: A Case Study

"Feel the Change" is a sensory journey combining the physical and digital worlds, guiding visitors to discover the impact of the climate crisis on different ecosystems and marine organisms. The installation, which combines the physical with the digital, is specially designed for blind and visually impaired, with the aim of engaging an audience of children and youth. For this reason, the main senses considered in the design and development phase of the journey were hearing and touch.

The tangible and tactile nature of the work enabled access to knowledge about acidification processes not only for the intended audience but also offers an enriching experience for children, whose perception of the world is largely shaped by touch. Additionally, the concrete, hands-on elements allow adults to fully understand and compare the effects of acidification, such as the contrast between a healthy tropical coral with vibrant polyps and one damaged by acidification, where the polyps are no longer present.

Ocean Literacy With All (OLWA)

Brief description of the project

OLWA advances OL by designing and implementing transformative locally- and globally-relevant research-based activities and projects by and for diverse stakeholders. Among the most notable initiatives of OLWA is the Ocean Literacy Dialogues (OLD), a series of open, participatory events that take stock of OL across the world through an inclusive and holistic approach, providing opportunities for new action to restore humanity's relationship with the Ocean.

The first edition of OLD was held at the UN Ocean Conference in Lisbon, Portugal, in 2022, engaging more than 700 participants (200 on-site and 500 online), while the second edition, organised by UNIFESP in October 2022 in Santos, Brazil, gathered more than 10,500 participants: Santos is also the first city in the world to sign OL into law. The third edition was held during the Fifth International Marine Protected Areas Congress (IMPAC5) in Vancouver, Canada, in collaboration with the Centre for Ocean Literacy Collaboration, with over 11,900 participants. The fourth edition was held in Dar-Es-Salaam, Tanzania, with a tailored programme for African Ocean professionals, gathering 50 participants. The fifth edition, held in Barcelona, Spain, during the Ocean Decade Conference, supported the consultation process for a subsequent document called the Venice Declaration for OL in Action (UNESCO 2024) through a series of workshops and open dialogues (Fig. 3.1).

Key activities and methods used

- An event format co-designed with members of the OL community, as well as local stakeholders, to ensure an effective programme tailored to the local context.
- Participative workshops and a safe environment for the OL community to propose changes, ideas and identify gaps.

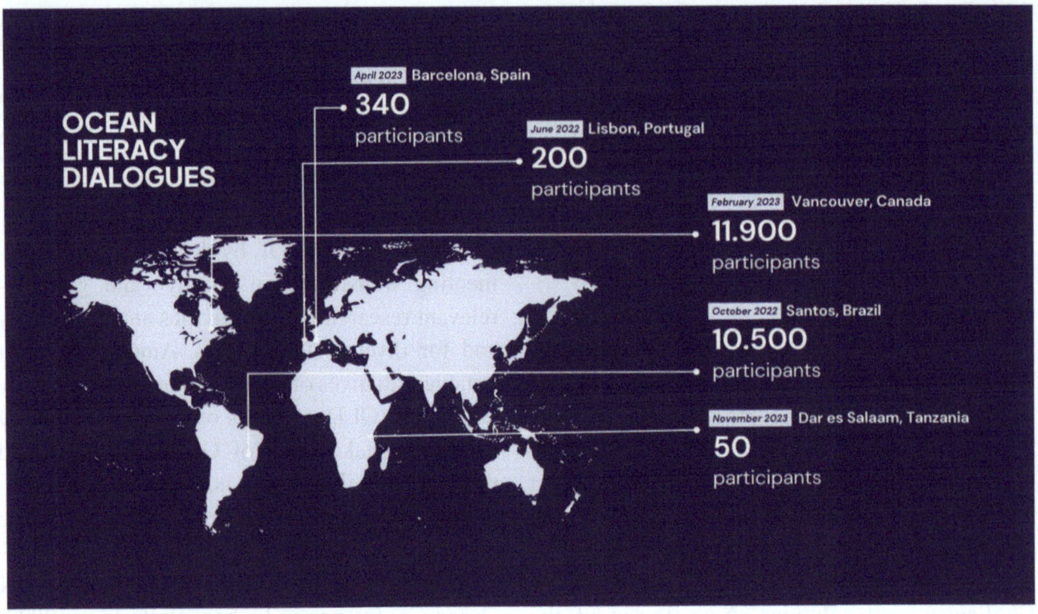

Fig. 3.1 Map with Ocean Literacy Dialogues editions

Outcomes, target audiences and stakeholders

Outcomes

- Expansion of the OL community and inclusion of diverse narratives and sectors of society, contributing to the objectives of the UN Ocean Decade.
- Ocean professionals are able to share their experiences, discuss ways forward on critical Ocean topics, promote their own initiatives and support the implementation of best practices.

Target audiences

- Ocean professionals

Stakeholders

- OLWA partners
- EU4Ocean Coalition
- Early Career Ocean Professionals (ECOPs)

Enabling factors for success

- Community ownership
- Innovation and adaptability where funds are absent
- Creation of initiatives within existing frameworks that can increase interest and drive participation in those activities

Challenges and solutions. A synthesis of the challenges and solutions is present in Table 3.7.

> The event was robust and will help me in the achievement of my Ocean Decade project, which has as its core mandate to increase the level of OL in African youths. […]. The event provided the ground for learning, information and knowledge sharing; it also resulted in the cross-fertilization of ideas and networking with other people across the African continent which is good for the African Youths Sustainable Ocean Campaign. […]. (Testimonial of a participant at the Tanzania edition of OLD)

Ocean Literacy Training

Brief description of the project

Table 3.7 A synthesis of the challenges and solutions for OLWA

Challenges	Solutions
Securing funding	Develop OLD within the context of existing projects, events and conferences
OL is still a marginal topic in high-level discussions	More intensive engagement of the OL community, openness to expand collaborative efforts and act on the feedback collected from the community in order to increase the visibility and relevance of OL

The Ocean Literacy training courses, developed by UNESCO-IOC and made available on the Ocean Teacher Global Academy (OTGA) platform, target diverse sectors. The training courses involved 776 participants from December 2020 to July 2024, including:

- 115 government officials and public administration officers in "Mainstreaming Ocean Literacy in Governance"
- 265 teachers, formal and non-formal educators in "Educating about the Ocean: Waves of change"
- 95 media and press professionals in "iSea Stories - understanding the Ocean for media and press"
- 231 members of the general public in "Why the Ocean should matter to us all"
- 70 architects, urban planners and designers in "Ocean Literacy for Architects, Urban Planners and Designers"

The online courses are free and self-paced; students have the flexibility to finish the course according to their own schedules. During the training, participants spoke to Ocean experts, and understood how their professions connect with the Ocean, exploring future opportunities.

Key activities and methods used

- Learning materials were developed in collaboration with experts from specific fields, offering case-studies and examples of how the Ocean intersects with diverse disciplines
- Participative workshops were developed during the online training sessions, in order to support the engagement of the online community

Outcomes, target audiences and stakeholders

Outcomes

- Improved understanding of the importance of the Ocean, the challenges it faces, and approaches to sustainability for a diverse set of stakeholders and sectors
- Development of an active community of students that understand the Ocean's features, as well how their careers and activities interact with the Ocean

Target audiences

- Ocean professionals
- Education professionals
- Architects
- Urban planners
- Public administrations
- Media and press

Stakeholders

- Ocean Teacher Global Academy
- Contributors to the training sessions (speakers, donors)

Enabling factors for success

- Communication across diverse channels, both Ocean-related as well as regarding relevant sectors (e.g. training for architects promoted in magazines such as ArchDaily)
- Involvement of relevant speakers from diverse sectors

Challenges and solutions. A synthesis of the challenges and solutions is present in Table 3.8.

Following the workshops, my foundation - Amiguitos del Océano - changed its content and started creating a personal methodology for our OL program in Ecuador. We have been involved in the update of the environmental education national

Table 3.8 Synthesis of the challenges and solutions for OL Trainings

Challenges	Solutions
Online training requires a certain level of digital literacy	Once the training sessions are complete the video lessons are uploaded to YouTube, making the content accessible to a broader audience. However, this option does not offer any certification.
Lack of motivation to follow an online course	Webinars, quizzes and forums that can support the students to engage with other peers during the course
Ensuring students have adequate time to dedicate to the course	The application process necessitates the submission of a motivation letter, in which students must confirm their availability for the course.

> strategy, and advocate for the inclusion of OL. We have been working intensely to promote OL and share its benefits through our programs, events, and talks in different sectors. We work with governmental institutions, schools, families, coastal communities, the private sector, academia, other NGOs, fishers, and the tourism and educational sectors. (Testimonial of a participant in the module "Educating about the Ocean: Waves of change")

3.3 Measuring the Impact of Ocean Literacy

The OL initiatives described above focus on education, ecosystem restoration, community engagement and skills development, and are characterised by interactive and participatory approaches. These programmes draw on methodologies inspired by the creative arts, storytelling and direct interaction with the marine environment to foster emotional connections with the Ocean and raise environmental awareness.

Capacity building is a key component of many of the initiatives, as it equips target groups with both the scientific knowledge and the interpersonal skills needed to share best practices. Taken together, these capacities are highly relevant for future interactions with different stakeholders to facilitate the dissemination of knowledge and the promotion of better conservation policies. Key outcomes of these initiatives include the restoration and conservation of marine ecosystems, a stronger connection between people and the Ocean, and enhanced scientific inquiry and critical thinking skills.

The various initiatives described in this chapter target a variety of groups, from kindergarteners to university students and ECOPs, policy makers and the private sector. Stakeholders include academic and research institutions, government agencies, local authorities, artisans, businesses and community organisations.

Among the key factors that have allowed these initiatives to be successful are the support from various stakeholders, including local governments, institutions, school staff, parents and local communities. Initiatives like Save the Wave, the Global Blue Schools Network and OLWA benefit from long-term partnerships and active community engagement. Knowledge sharing and capacity building are also critical, and several initiatives have shown the importance of engaging different audiences through a variety of channels.

Effective monitoring and evaluation systems are essential for continuous improvement and success. Unfortunately, this aspect is not executed to a common standard across initiatives, and some projects lack adequate monitoring and evaluation methods.

Adaptability and innovation are key factors for success, especially in difficult situations. Initiatives that adopt flexible and creative strategies are better equipped to maintain active participation and achieve their goals. These aspects highlight the need for well-supervised, collaborative and adaptive support methods in educational and community projects.

A number of common challenges and solutions have been identified across different projects. One key challenge facing many programs is securing long-term funding and resources. Many initiatives propose mixed-funding strategies combining private sector donations with public funding, as well as collaboration with national parks, local partners and education systems. Political and bureaucratic obstacles, including the need to obtain permits and authorizations, also pose significant barriers. Maintaining a local presence and commitment throughout the year with local partners is essential for sustained success. The involvement of local communities, teachers and students is a widespread problem, further complicated by busy curricula and inflexible educational structures. To address these challenges, many initiatives propose interdisciplinary and adaptive approaches whereby innovative methodologies are integrated within existing structures. Digital solutions, such as online platforms and webinars, are also being used to increase participation and accessibility.

Quantifying the impact of OL initiatives poses a multifaceted challenge. One major difficulty is measuring the impact of any single initiative in isolation from other influencing factors such as existing environmental education initiatives, community participation and media coverage. Behaviour change, a key indicator of OL's success as well as that of the Ocean Decade, often takes place over a long period, well beyond the duration of the initiative. It is therefore difficult to attribute these behavioural changes directly to specific projects. In addition, the lack of standardised evaluation tools further complicates the evaluation process as there is no coherent framework for measuring and comparing results across different OL programmes worldwide.

Nonetheless, opportunities for improvement are plentiful. There is a critical need for interdisciplinary and transdisciplinary research programmes that address the various aspects of OL in a holistic way. Long-term studies are essential to understand the gradual nature of behaviour change and provide a comprehensive picture of programme effectiveness. This is associated with the urgent need to develop innovative assessment methodologies and standardised metrics that accurately measure the impact of OL initiatives in all their nuance. Fostering greater collaboration and community participation may improve the impact of these initiatives by ensuring that they are more inclusive and thus resonate more with diverse target groups. While technology offers innovative ways of delivering content, attracting audiences and tracking progress, thereby improving the reach and accuracy of OL initiatives, the projects led by UNESCO-IOC demonstrate that direct interaction with the marine environment remains the most effective method of forging a greater emotional bond with the Ocean.

3.4 Conclusion

This chapter highlights the progress of UNESCO-IOC OL initiatives by exploring the innovative and inclusive strategies they employ to engage with diverse audiences. These projects emphasise the importance of education with inter- and transdisciplinary approaches, the creative arts, and direct interaction with the natural marine environment to instil scientific knowledge and practical skills. The success of these initiatives depends to a large extent on strong support from stakeholders and strong collaboration among various sectors. These partnerships have proved to be effective for overcoming bureaucratic obstacles and broadening audiences. Ensuring long-term funding is critical to the continuity, scalability and quality of OL initiatives. In addition, the need for standardised evaluation metrics and technology integration highlights opportunities to increase the impact and reach of these programmes.

Support from all stakeholder groups is crucial to ensure the continued advancement of OL initiatives. The private sector can contribute by fostering collaborations with educational institutions to promote awareness and understanding of the Ocean's role in the sustainable blue economy. Meanwhile, the society-science-policy interface must work toward creating a more efficient framework for Ocean governance by bridging the gaps between scientific research, public policy, and societal needs. The blue economy should integrate OL into business practices, ensuring that economic activities understand and adopt the sustainable use of Ocean resources. Lastly, cities need to incorporate OL into their urban planning and decision-making processes to promote sustainable development that respects and preserves marine ecosystems.

By building on these foundations, these initiatives can significantly contribute to preserving marine ecosystems and fostering an informed, engaged society. Ocean Literacy is vital to the success of the UN Ocean Decade, as it promotes a better public understanding of Ocean science and supports informed decisions on sustainable practices that benefit Ocean health.

3.4.1 Summary

Key concepts explored in this chapter include:

- Community-engaged science that integrates personal, emotional and cultural connections to the Ocean is essential for the societal transformation we need to establish a more harmonious, sustainable and equitable relationship with the Ocean.
- Ocean Literacy is vital to the success of the UN Ocean Decade, as it promotes a better public understanding of Ocean science and supports informed decisions on sustainable practices that benefit Ocean health.
- There is a need for educational programs that integrate inter- and transdisciplinary approaches, the creative arts, and direct interaction with the natural marine environment to instil scientific knowledge and practical skills.
- The success of these initiatives depends to a large extent on strong support from stakeholders and strong collaboration among various sectors.
- Long-term funding is critical to the continuity, scalability and quality of OL initiatives.
- There is an urgent need for standardised evaluation metrics to better understand and increase the impact and reach of OL initiatives worldwide.

3.4.2 Recommended Resources

The following resources provide further insights and support on this topic:

- Ocean Literacy for All: a toolkit (2018). https://unesdoc.unesco.org/ark:/48223/pf0000260721.locale=en
- A New Blue Curriculum: A Toolkit for Policy-Makers (2022). https://unesdoc.unesco.org/ark:/48223/pf0000380544.locale=en
- State-of-the-art of Ocean Literacy (2022). https://unesdoc.unesco.org/ark:/48223/pf0000382663.locale=en
- Collaborating with the Ocean: A New Model for Ocean-literate Corporate Action (2023). https://unesdoc.unesco.org/ark:/48223/pf0000389278.locale=en
- Ocean Literacy and the Atlantic Region: A Toolkit for Educators (2024). https://unesdoc.unesco.org/ark:/48223/pf0000391369.locale=en

Acknowledgements As we reflect on the successes of the Ocean Literacy Programme of UNESCO-IOC this year, we extend our heartfelt gratitude to all those who have supported our mission. Your dedication and contributions have been instrumental in making everything we have achieved possible. The innovative and inclusive strategies we employ to engage diverse audiences with Ocean Literacy would not have been successful without your collaboration and commitment. Through inter- and transdisciplinary approaches, the creative arts, and direct interaction with the natural marine environment, our shared goal of instilling scientific knowledge and practical skills is being realized.

We would like to acknowledge the incredible individuals and organizations that have been vital partners in this journey:

Lorenzo Bertelli, Marta Monaco, and the entire team at Prada Group; Sandra Castañer; Olga Mashkina; Andreea Strachinescu and all those at EU4Ocean; Mauro Biraghi and the team at E.ON; Romina Ciferni; all the speakers and guests at the Ocean Literacy Dialogues; the members of Ocean Literacy With All; the Decade Coordination Office for Challenge 10 and the Group of Experts; the members of the Blue Schools Global Network; Professor Mastrototaro at the University of Bari; the Comune di Isole Tremiti; Aquodiving; the Western Norway University of Applied Sciences; the City of Malmö; the Marint Kunskapscenter i Malmö; the Marine Education Center in Malmö; and, most notably, the dedicated members of the ECOPs network who translated the Venice Declaration into 15 languages and have tirelessly supported our global initiatives.

Your unwavering support has been crucial in overcoming bureaucratic obstacles and broadening the reach of Ocean Literacy efforts, ensuring long-term sustainability and quality of our programmes. The strong partnerships we've cultivated demonstrate the power of collaboration across sectors—private, public, and civil society—and help bridge the gaps between scientific research, public policy, and societal needs. The integration of Ocean Literacy into urban planning, business practices, and sustainable blue economy frameworks continues to be a vital step toward fostering informed, engaged communities that support marine conservation.

Thank you for your continued support, which enables us to advance Ocean Literacy and contribute to the UN Ocean Decade's goal of a sustainable and healthy Ocean. Together, we are building a more informed society, fostering solutions for a sustainable future, and ensuring that the Ocean's role in preserving marine ecosystems is understood and respected globally.

References

Glithero LD, Bridge N, Hart N, Mann-Lang J, McPhie R, Paul K, Peebler A, Wiener C, Yen C, Kelly R, McRuer J, Hodgins D, Curtin F (2024) Ocean Decade Vision 2030 White Papers - Challenge 10: Restoring society's relationship with the Ocean. UNESCO-IOC. (The Ocean Decade Series, 51.10), Paris. https://doi.org/10.25607/ekwn-wh61

McKinley E, Burdon D, Shellock RJ (2023) The evolution of Ocean Literacy: A new framework for the United Nations Ocean Decade and beyond. Mar Pollut Bull 186:114467. https://doi.org/10.1016/j.marpolbul.2022.114467

Shellock RJ, Fullbrook L, McKinley E, Cvitanovic C, Kelly R, Martin V (2024) The nature and use of Ocean Literacy in achieving sustainable Ocean futures: A Systematic Map. Ocean Coast Manag 257:107325. https://doi.org/10.1016/j.ocecoaman.2024.107325

UNESCO-IOC (2021a) Ocean Literacy Framework for the UN Decade of Ocean Science for Sustainable Development 2021–2030. UNESCO, Paris. [The IOC Ocean Decade Series, 22]. Intergovernmental Oceanographic Commission of the United National Education, Scientific and Cultural Organization. https://unesdoc.unesco.org/ark:/48223/pf0000377708

UNESCO-IOC (2021b) The United Nations Decade of Ocean Science for Sustainable Development (2021–2030) Implementation Plan. UNESCO, Paris. [The IOC Ocean Decade Series, 20]. https://unesdoc.unesco.org/ark:/48223/pf0000377082

UNESCO-IOC (2024) Venice declaration for Ocean Literacy in action: How to engage society to regenerate the most valuable ecosystem on our planet. In: Ocean Literacy world conference, Venice, 2024. [The IOC Information Document Series, 1540 (6). IOC/INF-1540.]. https://unesdoc.unesco.org/permalink/P--b07bda9a-6070-4553-8357-c477b6907c93

Author Contributions and Biographical Profiles

[1]Francesca Santoro: Conceptualization; Writing—Original Draft, Review, and Editing.

[2]Paula Barbeito Morandeira: Writing—Original Draft.

[3]Valentina Lovat: Writing—Review and Editing; Methodology—Data Collection and Analysis.

[4]Ana Vitória Tereza de Magalhães: Writing—Review and Editing; Methodology—Data Collection and Analysis.

[5]Gaia Maria Sole Intonti: Writing—Review and Editing; Methodology—Data Collection and Analysis.

[6]Esteban Gottfried Burguett: Writing—Review and Editing; Methodology—Data Collection and Analysis.

[7]Alessia Tombolini: Writing—Review and Editing; Methodology—Data Collection and Analysis.

[8]Paola Carluccio: Methodology—Data Collection and Analysis.

[9]Jack Coulton: Writing—Review and Editing.

*Corresponding Author at f.santoro@unesco.org

Francesca Santoro, PhD, is a Senior Programme Officer at the UNESCO Intergovernmental Oceanographic Commission (IOC), where she has been serving since 2011. With a background in climate change research, she previously taught at Ca′ Foscari University of Venice. Francesca has coordinated tsunami alert systems for the Mediterranean and northeastern Atlantic and has worked at prominent international research institutions. She is currently the Ocean Literacy Coordinator at IOC, specializing in the intersection of Ocean governance, education, and scientific communication. Francesca resides in Venice, Italy, and is a leading voice for sustainable blue economy initiatives.

Paula Barbeito Morandeira is an Oceanographer with a master's degree in Sustainability, Society, and the Environment. With over 10 years of experience in artisanal fishing at both European and international levels, she has developed expertise in food systems, social justice, local ecological knowledge, marine ecosystem management, and the blue economy, all through the lens of participatory governance. Her work focuses on integrating these areas to promote sustainable solutions. In recent years, Paula has become increasingly involved in the Ocean Literacy movement and has contributed as an Ocean Literacy consultant with the IOC-UNESCO, advancing efforts in global Ocean awareness and education.

Valentina Lovat is an environmental science professional with a master's in marine sciences, specializing in human geographies, and a master's in International Management for Sustainable Business and Innovation. She is currently working as Associate Programme Specialist for the Ocean Literacy programme at the Intergovernmental Oceanographic Commission (IOC) of UNESCO. In this role, Valentina leads global initiatives to promote Ocean Literacy, with a particular focus on empowering youth and early-career professionals through capacity-building programs. She is passionate about co-designing Ocean-based projects, engaging stakeholders, and developing outreach strategies to further the achievement of the 2030 Agenda for Sustainable Development, especially SDG 14, and the outcomes of the UN Decade of Ocean Science for Sustainable Development (2021–2030).

Ana Vitória Tereza de Magalhães is an Ocean Literacy Associate Programme Specialist at UNESCO-IOC, Ana Vitória Tereza de Magalhães co-leads the coordination of Blue Schools Global Network, the development of educational tools and strategic initiatives to support governments, educators, the private sector, media, and NGOs in raising awareness about the Ocean. Her work contributes directly to advancing the United Nations Sustainable Development Goals (particularly SDG 14: Life Below Water) and the 2030 Agenda. Ana holds a background in Marine Spatial Planning from the International Itinerant Erasmus Mundus Master's. She is deeply passionate about the intersection of Ocean Literacy, culture, and education. With over a decade of international experience, Ana has lived across multiple countries, including Spain, Italy, Belgium, France, and São Miguel Island in the Azores, Portugal. These diverse experiences have shaped her global perspective and strengthened her ability to drive impactful initiatives on a global scale.

Gaia Maria Sole Intonti is an Italian professional with a master's in Marine Biological Resources with a focus on socioecological processes related to environmental awareness. She is currently working at the UNESCO Regional Bureau for Science and Culture in Europe. Within the Ocean Literacy team, Gaia manages and supports the development of educational resources, as well as delivering training and capacity building to different audiences, from educators to young professionals. One of her main focuses is on Outdoor Education, the central pillar of the project Kindergarten of the Lagoon, part of the SEABEYOND educational program.

Esteban Gottfried Burguett is a Marine Sciences professional with a master's degree in the field, currently working at the UNESCO Regional Bureau for Science and Culture in Europe. As an Associate Programme Specialist, Esteban leads initiatives to promote Ocean Literacy and marine conservation, including coordinating the Sea Beyond—Kindergarten of the Lagoon project. Originally from Mexico and living in Italy for over 6 years, he is passionate about creating interactive learning experiences that raise awareness on the importance of the Ocean and its ecosystems. Esteban is committed to advancing marine education globally through innovative educational programs, aligning with the United Nations Sustainable Development Goal 14: Life Below Water.

Alessia Tombolini is an Italian clinical psychologist and group-analytic psychotherapist in training with a master's in European project design. She is currently a consultant for the UNESCO Regional Bureau for Science and Culture in Europe. Over the past 2 years, she has managed and supported the implementation, monitoring, evaluation, and dissemination of the Kindergarten of the Lagoon project in collaboration with Sapienza University and UniBo, as part of the SEABEYOND educational program. She is now also supporting the management of DCO—Connecting People and Ocean and the implementation of the SEABEYOND evaluation project.

Paola Carluccio is a designer with a master's degree in Integrated Product Design, with a focus on strategic design that weaves together sustainability, anthropology of design, and socio-cultural inclusion. Currently, she serves as Project Assistant for the Ocean Literacy programme at the Intergovernmental Oceanographic Commission (IOC) of UNESCO. In this role, Paola develops innovative strategies to engage and raise awareness among diverse audiences, promoting Ocean Literacy through creative and inclusive approaches. She supports the organization of events, exhibitions, and visual materials, contributing to the achievement of the Sustainable Development Goals, particularly SDG 14, Life Below Water, and furthering the objectives of the UN Decade of Ocean Science for Sustainable Development (2021–2030).

Jack Coulton is a writer, editor and communications specialist, originally from Manchester, UK and educated at the University of East Anglia, with over 15 years of experience ranging from scriptwriting for feature films and TV projects in the USA to directing event communications for the Slow Food organization based in Italy. Since 2024 he has been working on all aspects of the communication efforts of the Ocean Literacy programme of UNESCO-IOC, including speechwriting and digital engagement. He lives in Turin, Italy.

ScienceOnShore (SOS): A Wake-up Call from the Ocean Literacy Movement

4

Thomas F. Greene

Abstract

ScienceOnShore (SOS), a successful pilot marine science program created by the New York City Department of Education (DOE) in collaboration with Kingsborough Community College (KCC), engages youth in Ocean education through hands-on Ecolabs. This educational initiative was inspired and motivated by the Ocean Literacy movement which seeks to encourage the inclusion of a marine environmental science curriculum into educational institutions both locally and globally. SOS is a hands-on, inquiry-based marine science program for fifth graders, students 10–11 years old, who, unfortunately, attend one of the vast number of elementary schools nationwide that have no science laboratories. A single science lab on a college campus will not be able to meet the needs of the more than 1000 public elementary schools in NYC without access to science labs. Therefore, SOS hopes to establish Ecolabs located at three different locations along the more than 500 miles of NYC shoreline. The following chapter contains a description of the SOS program, along with a sample lab activity that can be replicated in similar locations. SOS EcoLabs enable students to explore, understand and appreciate their own natural marine environment. The ultimate goal of the SOS program is to promote Ocean Literacy by making climate change marine environmental science an integral part of the science curriculum in NYC's K–5 schools.

Keywords

Ocean Literacy (OL) · Climate change science · ScienceOnShore (SOS) · EcoLab

4.1 Introduction

It's time we listened to the youngsters! In recent years, a number of high school and middle school students have skipped class to protest in the streets about environmental issues. This rising tide of concern from teens and pre-teens, who don't usually protest, should not go unheeded. Young people are fearful about the sorry state of the environment that they will inherit and are pleading for us to do something about it. They are saying—"stop the environmental damage"; "look for new alternative solutions to pollution". "Be resilient".

Regrettably, in the vast majority of elementary schools nationwide, the basic elements of climate change science are not part of the State's educa-

T. F. Greene (✉)
Kingsborough Community College, Brooklyn, NY, USA
e-mail: thomas.greene@kbcc.cuny.edu

T. J. Kennedy (ed.), *Ocean Literacy: The Foundation for the Success of the Ocean Decade, Volume I*, Sustainable Development Goals Series, https://doi.org/10.1007/978-3-031-98145-6_4

tion curriculum in spite of the fact that more than 80% of parents in the U.S. support the teaching of climate change science (Booth et al. 2020). Also, and just as distressing, is the fact that the vast majority of U.S. K–5 schools throughout the country have no science laboratories (Eshack and Fried 2005). This serious lab deficit problem denies students the opportunity to use the tools of science to investigate, understand, and appreciate their own natural marine environment (Kamenetz 2019).

What can be done to bring climate change science into these K–5 schools? In recent years, the environmental education movement known as Ocean Literacy has emerged to raise public awareness about the Ocean's critical role in maintaining global health (State of New Jersey 2022). Ocean Literacy refers to an understanding of how the Ocean influences individuals and communities, as well as how human activities impact the Ocean's health and long-term sustainability. This chapter provides an overview of the ScienceOnShore (SOS) program and research activities currently underway and outlines activities to support its expansion in the future.

4.2 The Pilot ScienceOnShore (SOS) Program in Ocean Literacy

The ScienceOnShore (SOS) Program provides engaging, outdoor, real-world activities that connect directly to pressing environmental issues such as climate change and marine pollution—two essential components of Ocean Literacy and the Next Generation Science Standards (NGSS). Offered free of charge, SOS brings marine environmental science to life for fifth-grade students through experiential learning.

Currently, the program takes place on Friday mornings from 10:00 a.m. to 12:00 p.m. at Kingsborough Community College (KCC), located along the shores of Sheepshead Bay and Jamaica Bay in Brooklyn, NY. This unique coastal setting (see Fig. 4.1) allows students to engage in outdoor exploration of the natural marine environment, followed by indoor, hands-on laboratory activities where they practice using scientific tools and methods. Through these immersive experiences, SOS helps students build foundational science skills while

Fig. 4.1 Kingsborough Community College (KCC) coastal facility

Observing and Recording Temperature

The Fun and Joy of Learning Marine Science

Fig. 4.2 Students participating in group learning

deepening their understanding of the Ocean's critical role in Earth's climate system.

We all learn best by doing, especially children. Group learning in the lab (see Fig. 4.2) teaches students how to communicate and cooperate as a team toward the common goal of investigating and solving problems in marine environmental science.

4.3 ScienceOnShore (SOS) Curriculum

Students learn how to use the Scientific Method by making observations, formulating hypotheses, collecting data, displaying results and drawing conclusions in lab activities that include subject matter in the basic sciences of chemistry, Earth science, physics (engineering), and biology. Students also improve their English language skills in reading and writing, and numeracy skills in math while engaged in inquiry-based science. Our objectives are to ensure that:

- Students will learn how to write up a Lab Report in Scientific Method format. Their reports will be used as the basis of a science fair project or as an oral presentation made to the class at the end of the term.
- Students will develop an understanding and appreciation of living things by observing and caring for marine organisms.
- Students will do beach clean-ups and earn community service credits as evidence of their environmental stewardship.
- Students will do the ScienceOnShore (SOS) labs.

4.3.1 Sample Lab Activities for Fifth Graders

The ScienceOnShore (SOS) program offers a diverse range of hands-on lab activities (Greene 2024) that span multiple scientific and academic disciplines. As illustrated in Table 4.1, these labs engage students in activities that connect Marine Science with subjects like Chemistry, Biology, Physics, Earth Science, Math, English, and Art. For example:

- "How salty is the sea?" incorporates Chemistry, Math, and English to investigate salinity levels.
- "How deep is the Ocean?" integrates Earth Science, Math, and English to explore Ocean depths.
- "How do pollutants affect marine life?" connects Biology, Math, and English to study environmental impacts.
- "Producing electricity from sunlight" introduces Physics, Math, and English while exploring renewable energy.

Table 4.1 SOS Labs and Targeted Disciplines

Labs	Disciplines
How salty is the sea?	Chemistry, Math, English
How deep is the Ocean?	Earth Science, Math, English
How clear is Ocean water	Earth Science, Physics, Math, English
How can we grow an oyster garden?	Biology, Math, English, Art
Does Brooklyn have acid rain?	Chemistry, Math, English
What kinds of creatures inhabit our Ocean?	Biology, English, Art
How do pollutants affect marine life?	Biology, Math, English
What kinds of food do sea creatures eat?	Biology, English, Art
How can we make Ocean water?	Chemistry, Math, English
How fast is a snail's pace?	Biology, Physics, Math, English
	Physics, Math, English
Producing electricity from sunlight.	Physics, Math, English
Producing electricity from a sea breeze.	Physics, Math, English

These interdisciplinary labs not only foster scientific inquiry but also build skills in literacy, numeracy, and creative thinking. By combining outdoor exploration with laboratory-based experiments, the SOS program encourages students to view the Ocean through multiple academic lenses, helping them understand its vital role in Earth's interconnected systems.

4.3.2 Sample Lab Activity and Report

Measuring Ocean depth is essential for understanding both local water conditions and broader environmental changes. Whether it's ensuring safe navigation for boats, protecting swimmers from shallow hazards, or monitoring sea level rise caused by climate change, understanding how deep the water is provides crucial information.

In this lab, **"How Deep is the Sea?"**, students will learn a simple, hands-on technique for measuring water depth. The investigation follows the scientific method, guiding students through the steps of forming a hypothesis, conducting an experiment, and recording data. Using basic materials like a rope, a weight, and a meter stick, students will measure the depth of Sheepshead Bay and connect their findings to real-world applications in marine science.

This lab activity is targeted for fifth graders and not only introduces fundamental measurement skills using the metric system but also encourages critical thinking about environmental phenomena like sea level rise. Through this experience, students gain practical insights into how scientists study the Ocean while reinforcing key concepts from physics, math, and environmental science.

Sample Lab Procedures: How Deep is the Sea

Each lab contains an introduction to set the stage, the formation of a hypothesis, followed by testing and reporting. Materials and procedures to be followed are included, along with a data table for reporting and suggested concluding questions. Labs follow the sample lab layout below.

Introduction: Doesn't it feel good to take a swim at the beach on a hot day! But you need to be careful. The water gets deeper as you walk out further. Boat captains need to know water depth to avoid hitting the bottom. Swimmers need to know how deep the water is before taking a dive. They could risk a serious head injury. Knowing the depth of the water prevents accidents and injuries. Measuring Ocean depth is also needed to see if there is a rise in sea level. A rising sea level is caused by ice melting at the North and South Poles. The ice is melting because of global warming. Scientists use satellites to measure a rise in sea level. Today you will use a much simpler procedure.

Hypothesis: Scientists solve problems like finding Ocean depth by following a procedure called the scientific method. The scientific method is more or less done step by step, like a recipe. The first step in the scientific method is to identify a problem to solve. This is followed by offering a possible solution to the problem, called a hypothesis.

One student suggested the following hypothesis: *If* a rope with a weight attached is dropped into the water, *then* we can measure the depth of the water. A hypothesis is frequently called an "educated guess" because previous knowledge is often used. A hypothesis is often written with the words "if" and "then". When the hypothesis is written in this form, it is easier to create an experiment.

So, let's test this hypothesis. Testing the hypothesis is the third step in the scientific method. You test the hypothesis by doing an experiment by following a procedure. In the following experiment laboratory materials or equipment will be used to solve a problem. In this experiment you will be finding out the depth of water in Sheepshead Bay.

Materials: Meter stick, rope with metal weight, tape, lab apron, gloves.

Procedure: Follow the steps below.

1. Sheepshead Bay contains a floating dock which is next to a wall or bulkhead. See Fig. 4.3. You will measure the depth of the Ocean from the bulkhead by using a rope attached to a weight that is lowered slowly into the water.
2. First, you need to learn how to measure scientifically. Place a meter stick on the table. The meter (m) is the basic unit of length in the metric system. Notice that a meter is subdivided into centimeters (cm) and millimeters (mm). How many meters, centimeters and millimeters is the width of the table? Record your answer in Data Table 1. The metric system is easier to use than the English system because the units are based on multiples of ten.
3. Now you are ready to walk to the bulkhead. To be safe, put on a life vest. Partner up with two or three other classmates. Lower the weighted line slowly until it reaches the bottom. When there is slack in the line you know the weight hit the bottom. Lift the rope slightly to remove slack. Take a good look at the spot where the line meets the top of the bulkhead. Wrap a piece of tape around the line at that spot. Now lift the line out of the water and place it on the ground.
4. Using the meter stick, measure the length of the rope from the piece of tape to the weight. Record the length in meters, centimeters and millimeters on your clipboard.
5. Repeat the above procedure but this time measure from the water surface to the top of the bulkhead. Record the measurements on the clipboard.
6. Return to the lab and record your data in Data Table 2 (see Table 4.2).

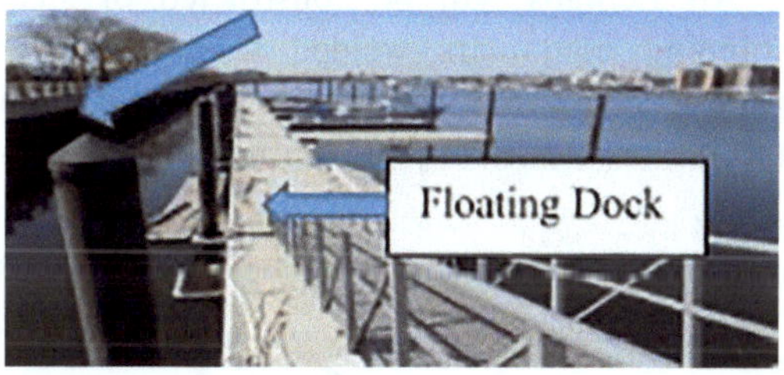

Fig. 4.3 Sheepshead Bay

Concluding Questions: Respond to the questions below.

1. Why is this experiment done many times?
2. Describe briefly how you would measure water depth.
3. Why is the metric system used to measure length instead of the English system?
4. How can you explain different results?

Table 4.2 How Deep is the Sea Lab Activity Data Tables

Data Table 1—Meter Stick Metric Units			
	Meters (m)	Centimeters (cm)	millimeters (mm)
Width of Lab Table			

Data Table 2—Lab Results					
Tape Measurements of Water Depth				Ocean Depth (C) in (cm) and (mm)	
Trials	Ocean Bottom To Bulkhead (A)	Ocean Surface to Bulkhead (B)	Ocean Depth (m) A-B=C	Ocean Depth (cm)	Ocean Depth (mm)
1					
2					
3					
4					
5					
6					
Literacy					
Total					
Mean (Av.)					

4.4 Projected Expansion of ScienceOnShore (SOS) Lab Sites

The SOS program has successfully engaged K–5 students in hands-on marine science activities at the KCC campus, where the coastal setting provides a unique, immersive learning experience. From measuring Ocean depth to exploring marine life and renewable energy, these interactive labs have sparked curiosity and built foundational science skills in young learners. With this success as a foundation, the program envisions expanding to additional locations across New York City, making marine science education accessible to more students throughout the five boroughs. By bringing these engaging, inquiry-based activities closer to schools and communities, SOS aims to inspire the next generation of Ocean stewards while broadening opportunities for all NYC students to connect with the Ocean and understand its critical role in our planet's health.

4.4.1 Regional Ecolabs

The vast majority of the more than 1000 NYC Dept of Education K–5 schools have no science labs. The same is true of public elementary schools nationwide. Hundreds of schools citywide could be serviced by a single regional EcoLab. Plans are currently underway to establish funding to create a regional EcoLab network to additional areas including outside of Denyse Wharf on the Fort Hamilton Army Garrison to serve the vast number of K–5 public schools without science laboratories (Fig. 4.4), on the shore of Coney Island Creek in Calvert Vaux Park Brooklyn that could service the more than 700 NYC K–5 public schools without science laboratories (Fig. 4.5), and expanding existing facilities on the campus of Kingsborough Community College (Fig. 4.6).

All Ecolabs will be constructed of steel shipping containers welded together. Solar panels on the roof of the EcoLab and a wind turbine on a cement platform nearby will fur-

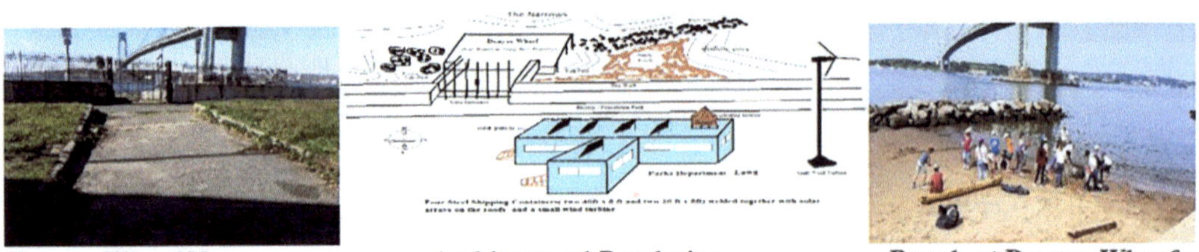

Denyse Wharf Near Verrazzano Bridge — *Architectural Rendering of EcoLab at Denyse Wharf* — *Beach at Denyse Wharf*

Fig. 4.4 Envisioned Denyse Wharf EcoLab

EcoLab from made Shipping Containers — *EcoLab Site in Coney Island Creek* — *EcoLab Floor Plan*

Fig. 4.5 Envisioned Coney Island Creek in Calvert Vaux Park Brooklyn EcoLab

Fig. 4.6 Envisioned EcoLab on the campus of Kingsborough Community College (KCC)

nish the electricity needed to operate the heat, light and cooling systems located inside the EcoLab. Electrical power will also be supplemented from the grid.

4.5 Conclusion

The Ocean Literacy movement comes at a time of great concern about the serious climate change crisis under which we all live. Five Hundred thousand people have died since 2004 in 10 weather disasters (Eshack and Fried 2005). We are currently experiencing numerous destructive hurricanes and their associated tornadoes. The large-scale melting of the polar ice caps, and the subsequent rise in sea level is causing major flooding which leads to destructive global storm surges. And these are only a few of the more dramatic effects of global warming caused by the burning of fossil fuels. Just as insidious is the Ocean acidification and coral bleaching also caused by our Earth warming up.

Educating the general public about these environmental challenges and their causes should begin in the earliest grades. This means introducing children in elementary school to the principles of Ocean Literacy, which involves understanding the critical relationship between humans and the Ocean. Ocean Literacy is defined as recognizing how human activities impact the Ocean's health and, in turn, how the Ocean influences life on Earth. By fostering this awareness early, students can develop a sense of responsibility and curiosity about the Ocean, preparing them to become informed stewards of our planet's future.

The ScienceOnShore (SOS) program described above does this by first getting children out of the classroom and into their own natural environment along the shore. Having students use all of their senses like smelling the Ocean air, feeling the sea breezes, seeing waves crash on the beach, and hearing the seagulls, to name just a few experiences, will open them up to further inquiry. Children are curious. Inquiring minds generate lots of questions. The ScienceOnShore (SOS) lab activities have been fashioned from these questions like, How deep is the sea?, How salty is the Ocean?, How clear is the water? Does Brooklyn have acid rain?, among others.

While Ocean Literacy can be the stimulus that generates interest in doing science, the serious lack of lab facilities in the elementary schools nationwide makes it very difficult to teach students hands-on science activities following the traditional Scientific Method approach employed in doing inquiry-based science investigations.

Citizen science will benefit immensely if government and educational officials seriously consider constructing regional marine environmental science labs or EcoLabs along the shores as described above. This will bring Ocean Literacy to the vast number of elementary schools nationwide that have no science laboratories. In addition, more colleges like Kingsborough Community College that are located along the shore need to open their labs to schools without science labs.

Acquiring a greater understanding of the science of climate change will create a more informed citizenry needed to vote intelligently at the ballot box on environmental issues. Ocean Literacy can lead the way.

4.5.1 Summary

Key concepts explored in this chapter include:

- ScienceOnShore (SOS) is an age-appropriate marine environmental science program for fifth grade students (10–11 years old).
- Ocean Literacy is an educational movement that promotes, recommends and supports the inclusion of marine environmental science into school system curricula locally and globally.
- Ecolabs are regional marine science labs that can provide marine environmental science to schools without science labs.
- Climate Change Science is a body of knowledge about the effect of human activities on weather patterns and on Earth's ecosystem

4.5.2 Recommended Resources

The following resources provide further insights and support on this topic:

- NASA PACE: Plankton, Aerosol, Cloud, Ocean Ecosystem https://science.nasa.gov/mission/pace/
- NOAA: The Ocean Project: Supporting Youth Engagement and Leadership for a Healthy Ocean and a Stable Climate https://theocean-project.org/youth/
- NMEA: A Handbook for Increasing Ocean Literacy https://www.marine-ed.org/ocean-literacy/handbook
- Textbook: Greene, T. F. (2017). *Marine science: Marine biology and oceanography* (3rd ed.). Perfection Learning.

Acknowledgments The initial success of ScienceOnShore (SOS) can be attributed in large part to the dedicated efforts of the National Marine Educators Association (NMEA), which has worked tirelessly over the past two decades to promote Ocean Literacy on a national scale. Special recognition is owed to NMEA for its pioneering work in developing and advancing this critical educational framework, helping communities across the country understand the Ocean's vital role in sustaining life on Earth.

In addition, the New York State Marine Education Association (NYSMEA) deserves acknowledgment for carrying forward NMEA's mission locally, fostering a deeper appreciation for the "world of water" among New York's educators, students, and residents.

Gratitude is also extended to the New York City Department of Education (DOE) and Kingsborough Community College (KCC) of the City University of New York (CUNY) for their innovative partnership. By providing K–5 public school students—many of whom attend schools without dedicated science labs—access to outdoor, hands-on marine science experiences on the KCC campus, this collaboration has created meaningful opportunities to engage with the Ocean environment and promote lifelong Ocean Literacy.

References

Booth AE, Shavlik M, Haden CA (2020) Parents' causal talk: links to children's causal stance and emerging scientific literacy. Dev Psychol 56(11):2055–2064. https://doi.org/10.1037/dev0001

Eshack H, Fried M (2005) Should science be taught in early childhood? J Sci Educ Technol 14(3):315–336

Greene TF (2024) ScienceOnShore (SOS): Marine Science Lab Manual for Elementary Schools. Physical Science Dept., Kingsborough Community College Brooklyn, New York

Kamenetz, A. (2019, October 27). Giving schools and students the tools they need in the fight to save the planet. NPR. https://www.npr.org/2019/10/27/773515557/giving-schools-and-students-the-tools-they-need-in-the-fight-to-save-the-planet

State of New Jersey (2022) New Jersey students enter first school year with K-12 climate change education [Press release]. https://www.nj.gov/governor/news/news/562022/20220906a.shtml

Author Contribution and Biographical Profile

[1]*Thomas F. Green: Conceptualization, Writing—Original Draft, Review, and Editing.

*Corresponding Author at thomas.greene@kbcc.cuny.edu

Thomas F. Greene was born and raised in Brooklyn NY not far from the beaches at Coney Island where he learned how to swim and spent endless hours looking for marine creatures along the shore. He received his BA and MS degrees in biology from NYU and then taught high school science for more than 30 years where he also published a textbook in marine science. Thomas has been an active member of the New York State Marine Education Association (NYSMEA) for more than half a century. He is a former NYSMEA president and a current Board Member. He is a former NYSMEA president, a current Board Member, and an Adjunct Lecturer in the Physical Science Dept. at Kingsborough Community College.

The TransforMAR Project: A Deep Dive into Ocean Literacy

5

Alexandre Augusto da Silva,
Giovanna Scagnolatto, and Karina F. Orsolini

Abstract

Diving beneath the surface, immersive experiences offer a powerful gateway to transforming how individuals connect with the Ocean. By engaging at an emotional level, these experiences cultivate a deep sense of responsibility for marine conservation, shifting attitudes and inspiring collective action. The TransforMAR Project, a pioneering Brazilian initiative, exemplifies this approach by introducing young people—particularly those from marginalized coastal communities—to the wonders of the marine environment through SCUBA diving and hands-on environmental education. Through direct interaction with marine ecosystems, participants develop an "Ocean identity" that fosters long-term stewardship and environmental advocacy. By integrating marine biology, ecology, and conservation with experiential learning, this initiative highlights how Ocean Literacy can drive social equity and environmental resilience, demonstrating that meaningful engagement with the Ocean can be a powerful force for change in the face of climate challenges.

Keywords

Ocean Decade · Ocean Literacy (OL) · Environmental education · Environmental justice · Social equity · SCUBA diving · Youth engagement

5.1 Introduction

The unprecedented global challenges currently affecting the Ocean's balance make it essential to promote an understanding of its role in human life. Ocean Literacy, defined as the understanding of the mutual influence between the Ocean and humanity, is a fundamental concept for educating individuals and communities about the complex interconnections between human activities and marine ecosystems (Kelly et al. 2023; O'Halloran and Silver 2022; Pazoto et al. 2021; Santoro et al. 2018). The urgency of promoting environmental awareness has reached a critical point, given the increasing degradation of the natural world, which may compromise the resilience and recovery capacity of global ecosystems.

The ongoing deterioration of the natural environment is bringing us closer to planetary boundaries, which define the safe limits for humanity within stable planetary systems. Crossing these boundaries risks triggering tipping points—critical thresholds where small disturbances cause irreversible changes in ecosystems or the climate.

A. A. da Silva (✉) · G. Scagnolatto · K. F. Orsolini
Associação TransforMAR, Juiz de Fora, Brazil
e-mail: alexandre@transformar.eco.br; giovanna@transformar.eco.br

T. J. Kennedy (ed.), *Ocean Literacy: The Foundation for the Success of the Ocean Decade, Volume I*, Sustainable Development Goals Series, https://doi.org/10.1007/978-3-031-98145-6_5

Human activities are accelerating this shift, with examples including climate change, Ocean acidification, polar ice melting, and coral reef decline (Rockström et al. 2009). This underscores the need for an informed population capable of addressing political and environmental challenges, protecting marine environments, and promoting Ocean conservation (O'Halloran and Silver 2022).

The TransforMAR Project is a pioneering initiative aimed at fostering an oceanic identity through transformative educational experiences that promote behavioural change. By combining pedagogical activities focused on biology, ecology, and marine conservation with the immersive experience of scuba diving, the project nurtures an emotional connection between youth and the Ocean, instilling a strong sense of belonging through direct contact with vibrant marine life. This approach facilitates the internalization of key concepts needed to cultivate a generation that lives in harmony with nature. By engaging socially disadvantaged youth, TransforMAR enhances environmental awareness and contributes to a fairer, equitable, and sustainable society. This chapter uses the project as a case study to demonstrate strategies for promoting behaviour change, social engagement, and racial equality while building meaningful connections to the Ocean.

5.2 Ocean Literacy and Environmental Education: Foundations and Importance

Throughout history, the Ocean, due to its vastness, has consistently been perceived by humanity as an inexhaustible source of resources, reinforcing continuous exploitation practices without consideration for its ecological limits (Costanza 1999). Despite the recognition of the ecosystem services, it provides, as well as its economic value, human activities have intensified, threatening the integrity of the marine environment. Consequently, we are facing a severe environmental collapse (Bennett et al. 2023).

The environmental crisis has been exacerbated by a lack of awareness of the negative impact of human activities on Ocean sustainability, as well as a misunderstanding of the interdependence between the Ocean and the maintenance of life on Earth (Santoro et al. 2018). Therefore, it is imperative to disseminate knowledge that contributes to this understanding, creating and strengthening the connection between people and the Ocean, while encouraging the behavioural changes needed for its conservation. This is what Ocean Literacy is all about (Beatley 2018; O'Halloran and Silver 2022; Santoro et al. 2018).

The concept of Ocean Literacy emerged in the United States in the 2000s, driven by the concerns and dissatisfaction among scientists and educators about the lack of Ocean-related educational disciplines (Santoro et al. 2018). This gap limited the ability to spread the knowledge about the importance of marine ecosystems to society and to raise public awareness about their conservation and proved to be a significant obstacle to promoting the behavioural changes necessary for Ocean conservation.

Since then, Ocean Literacy has been established as a vital tool for disseminating key knowledge about the functions and importance of the Ocean, influencing both education and advocacy (Pazoto et al. 2021). One of its main goals is to empower and engage society through the production and dissemination of knowledge that enables people to make efficient decisions, aiming at minimizing negative human impact on marine ecosystems (Beatley 2018).

Although Ocean Literacy has been boosted internationally, the movement still faces significant challenges in the Global South due to its late adoption, including in Brazil, where the concept only became widespread in 2009 (Pazoto et al. 2021). In this context, environmental education has become a strong ally in promoting Ocean

Literacy, for which it is essential to encourage learning in innovative ways that go beyond traditional approaches. Hence, education should empower individuals to actively engage with environmental issues and develop a sense of responsible citizenship through critical thinking and collective action.

Environmental education plays a central role in addressing the environmental crisis by fostering significant and lasting changes in values, behaviours, emotions, and collective actions. This commitment involves transforming current society by replacing an unsustainable model that exceeds the planet's limits with one that respects Earth's capacity to regenerate its natural systems. It also emphasizes the importance of celebrating and preserving cultural, ethnic, gender, and biological diversity as essential elements of a sustainable and equitable future (Educação 2007). Only through this shift can we build a society where future generations can live in balance with the planet.

By integrating environmental education and Ocean Literacy through non-formal education approaches, such as workshops and environmental discussions conducted outside traditional school settings, we can foster effective learning. These activities use an engaging and experiential format reminiscent of nature tourism, where learning is facilitated through natural language, enjoyment, and the ease of a guided outdoor experience. Field-based activities like immersion in natural environments and scuba diving further enrich this approach. When combined with school curricula, this strategy empowers a new generation of engaged, aware, and prepared citizens. They become better equipped to face and address environmental, social, economic, and political challenges.

This method ensures that, while the format is informal, the content remains substantive and effectively applied to practical experiences. By equipping young people with scientific literacy and sustainability principles, they can act in an environmentally responsible manner. Additionally, they can disseminate their knowledge and contribute to building a fairer and more sustainable society (Christofoletti et al. 2022). Therefore, it is evident that education plays a central role in promoting behavioural changes and consolidating transformative science, aligned with the objectives of the UN Decade of Ocean Science for Sustainable Development (Ocean Decade).

5.3 The TransforMAR Project: A Journey of Transformation

The TransforMAR Project is an environmental education initiative that utilises SCUBA diving as a pedagogical tool to promote Ocean Literacy among young people from coastal communities. The project is supported by the Sustainable Ocean Alliance and Roots Microgrants and is developed in collaboration with public schools (grades 6–12) in socially vulnerable areas. It combines theoretical and practical activities conducted in natural environments such as beaches, restinga ecosystems (unique coastal forests found primarily in Brazil along the Atlantic coast and northeastern regions), mangroves, and coral reefs. Participants engage in field studies of marine and coastal ecosystems, gaining a deeper understanding of local biodiversity and the associated environmental and social challenges. This immersive approach not only enhances students' scientific knowledge but also actively involves them in the conservation and sustainable management of these ecosystems (Fig. 5.1).

By offering new experiences and knowledge, such actions have the potential to deeply influence responses, reshaping attitudes and creating

Fig. 5.1 TransforMAR student in a scuba experience (Photo by Alexandre Augusto da Silva, 2018)

Fig. 5.2 Field lesson in the mangrove ecosystem (Photo by Alexandre Augusto da Silva, 2022)

the conditions for sustainable change. Lived experiences, in this case, become the foundation for building new actions and decisions that support marine conservation and foster a responsible relationship with the Ocean.

In addition to practical activities, TransforMAR includes workshops aimed at incorporating Ocean Literacy into the formal school curriculum. We encourage teachers, coordinators, principals, and other school professionals to implement structured learning processes and discussions about the Ocean. TransforMAR supports this by engaging educators to integrate Ocean Literacy concepts and promote classroom discussions as part of structured formal education. This approach not only raises awareness among young people about the importance of marine ecosystems but also reinforces the inclusion of this knowledge within formal education (Fig. 5.2).

This endeavour aligns with the challenge highlighted in the UNESCO manual, 'Ocean Literacy for All: A Toolkit,' which states: *'One of the most significant challenges in Ocean education and public engagement is to penetrate the opacity of our global Ocean'* (Santoro et al. 2018). This issue is central to the TransforMAR Project, which uses SCUBA diving as both a literal and metaphorical tool to overcome this opacity, encouraging a deeper and more conscious connection between students and the planet. Additionally, as emphasized in the Ocean Decade framework, a transformative science is achieved when co-designed solutions address societal needs, fostering connections between people and the Ocean (Kennedy et al. 2024). The TransforMAR Project embodies this principle by integrating community engagement with innovative practices to bridge the gap between scientific knowledge and local action.

To motivate students to engage in Ocean conservation, educational programmes must move beyond simply presenting information and content. It is important to explore and value students' emotions, feelings, and sense of belonging, helping them feel part of the environment and understand their connection to the Ocean (Jaksha 2019). This connection is key to developing an Ocean identity that drives meaningful behavioural change, as demonstrated by former participants of the TransforMAR Project.

Emotion has been recognised as a pivotal factor in the internalisation of values linked to environmental education, acting as a catalyst for attitude change (Atchley et al. 2012). Immersive experiences in natural environments can restore mental functions and strengthen emotional connections, both of which are essential for adopting sustainable behaviours. SCUBA diving, in particular, has proven effective in stimulating cognitive, emotional, and behavioural dimensions, promoting a strong environmental identity and encouraging conservation behaviours (Beatley 2018).

During the SCUBA dive, young participants can observe the vastness of the Ocean, explore local marine biodiversity, understand ecological dynamics, and learn about environmental challenges while reflecting on the interaction between the environment and society. This experience, which blends knowledge and emotion in an immersive natural setting, creates meaningful emotional engagement. This combination is essential for reinforcing conservation values among participants and promoting the behavioural changes aimed at by Ocean Literacy.

By consolidating a strong oceanic identity, the TransforMAR Project enhances the impact of education, inspiring young people to become leaders and agents of change within their communities. Throughout this process, participants are encouraged to share the knowledge they have gained, acting as Ocean ambassadors. This leadership extends to the creation of community projects focused on Ocean conservation. Notable examples of these initiatives include the development of a podcast on marine conservation available on major audio streaming platforms, awareness campaigns addressing the issues of shark meat consumption and the impacts of large predator fishing, as well as projects that promote local culture, indigenous traditions, and the way of life of coastal communities.

These results highlight the essential role of Ocean Literacy in driving behavioural change and increasing youth engagement in environmental conservation initiatives within their communities. This is achieved by building emotional connections through practical experiences and community involvement.

5.4 Impact of TransforMAR on Coastal Communities

The TransforMAR Project operates in the coastal regions of the states of São Paulo (SP) and Rio de Janeiro, where 'Caiçara' communities represent a traditional population. These communities originated from the intermingling of indigenous peoples who inhabited Brazilian lands, enslaved Africans brought to Brazil, and Europeans who settled in the region during the colonial period (Gomes De Araujo 2014).

These communities have developed a subsistence culture, with artisanal fishing, agriculture, and extraction as their primary economic activities. Their economic and cultural practices are passed down through generations, maintaining a close relationship with the environment. Their identity is tied to the territory, and nature conservation is vital for the quality of life and sustainability of these communities (Horovitz 2016).

The process of urbanisation in coastal areas, particularly from the nineteenth century onwards, has led to the marginalisation and discrimination of Caiçara communities. The expansion of tourism, real estate speculation, and the construction of large-scale developments have imposed increasing social and economic pressures, resulting in the loss of their traditional means of subsistence and the deterioration of their cultural practices. Despite efforts to preserve their culture, these communities continue to suffer from invisibility and social vulnerability, facing challenges that threaten their identity and economic stability (Branco et al. 2005).

In addition to social inequality and discrimination, the degradation of coastal and marine ecosystems, pollution, climate change, and overfishing jeopardise the livelihoods of these communities (Prado et al. 2015). These environmental factors directly impact communities that rely on natural resources for survival, leading to injustices that heighten their vulnerability (Bennett et al. 2023). The concept of environmental justice, which emerged in the 1980s, recognises that socioeconomically disadvantaged or marginalised groups are the most affected by the environmental crisis and have a reduced capacity to respond to changes, resulting in greater social impacts and widening inequalities (Mohai et al. 2009). Environmental justice also involves the pursuit of equitable access to nature's benefits and fair participation in decision-making processes, aiming to break the cycle in which social inequality determines exposure to environmental risks.

The TransforMAR project, aware of the historical marginalisation and heightened environmental vulnerability faced by these coastal communities, seeks to address this disparity through environmental education and Ocean Literacy. The initiative supports young people in defending their rights, territories, and ways of life while advocating for environmental causes by applying the knowledge they have acquired both individually and collectively. The integration of theoretical and practical elements, along with the confluence and sharing of knowledge with those who have limited access to this type of content, enhances the students' capacity for transformation.

The TransforMAR is a socio-environmental initiative that integrates both environmental and social dimensions, working with public school students in Brazil, a country facing a significant deficit in educational investment, which limits access to knowledge and perpetuates the marginalization of these young people. The project not only addresses the challenges faced by Caiçara communities but also promotes social and environmental transformation by equipping young people with tools to value traditional knowledge. By combining environmental education with Ocean Literacy, the project aims to build a more just and sustainable future, enabling these communities to preserve their identity, resist external pressures, and actively engage in protecting the environment and defending their rights and territories.

5.5 Quantitative Evaluation of the TransforMAR Project

To assess the results of the TransforMAR Project, an analysis was conducted using pre- and post-activity questionnaires designed to evaluate participants' knowledge of topics such as the Ocean, environmental science, climate change, and conservation. These evaluations were carried out in two editions of the project in October 2024, one in the city of Campinas, São Paulo and the other in the city of Ubatuba, São Paulo.

For the Campinas edition, the evaluation comprised 12 equally weighted questions, with final scores calculated as the sum of correct responses, ranging from 0 to 12. This edition involved 11 participants; however, only the responses from nine participants were included, as two individuals diagnosed with autism spectrum disorder were excluded from the analysis. The paired t-test (Gerald 2018) was used for data analysis, with calculations performed using the Jamovi software (Jamovi 2022).

Figure 5.3 shows the statistical analysis conducted using Jamovi, and Fig. 5.4 displays the comparative graph. The initial evaluation showed an average score M = Mean of 6.22, with a standard deviation (SD) of 2.62, which increased to M = 10.56 and SD = 1.50 after the project. The p-value and the Fig. 5.4 comparative graph of pre- and post-intervention evaluations provide strong statistical evidence of a significant increase in participants' knowledge following the TransforMAR Project.

The second edition included 37 youth from Ubatuba, SP, comprising 13 students from Florentina Martins Sanchez State School and 24 members of the Educador Floresta e Mar collective, which includes residents from underprivileged communities and Indigenous villages in

Paired Samples T-Test

Paired Samples T-Test

			statistic	df	p
Pre	Post	Student's t	−5.67	8.00	< .001

Note. H_a $\mu_{Measure\ 1\ -\ Measure\ 2} < 0$

Normality Test (Shapiro-Wilk)

			W	p
Pre	-	Post	0.906	0.286

Descriptives

	N	Mean	Median	SD	SE
Pre	9	6.22	5	2.77	0.925
Post	9	10.56	11	1.59	0.530

Fig. 5.3 Statistical Analysis Campinas exported from Jamovi software

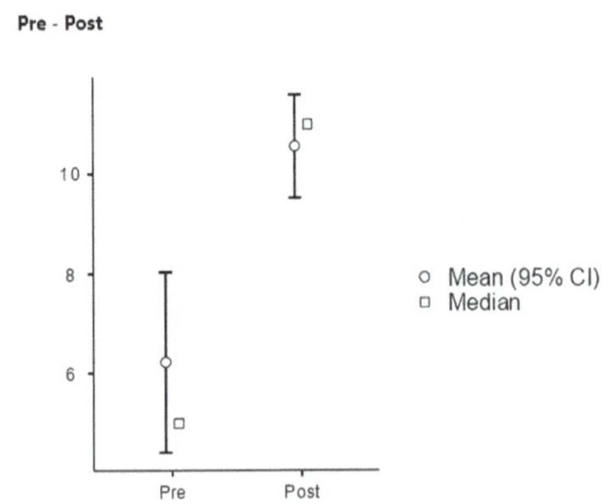

Fig. 5.4 Data Plot from Campinas Statistical Analysis

the region. Participants completed two questionnaires of 10 similar questions, equally weighted questions, with scores calculated as the sum of the number of correct responses, ranging from 0 to 10. The results were analyzed using the paired t-test with Jamovi software, following the previously employed approach (Figs. 5.5 and 5.6).

The statistical evidence obtained in this edition reinforces the effectiveness of the TransforMAR Project's educational intervention. The paired t-test revealed a highly significant dif-

Fig. 5.5 Statistical Analysis Ubatuba exported from Jamovi software

Paired Samples T-Test

Paired Samples T-Test

			statistic	df	p
Pre - Uba	Post Uba	Student's t	−7.30	36.0	< .001

Note. H_a $\mu_{\text{Measure 1 - Measure 2}} < 0$

Normality Test (Shapiro-Wilk)

			W	p
Pre - Uba	-	Post Uba	0.942	0.054

Descriptives

	N	Mean	Median	SD	SE
Pre - Uba	37	5.81	6	1.98	0.326
Post Uba	37	7.49	8	1.63	0.267

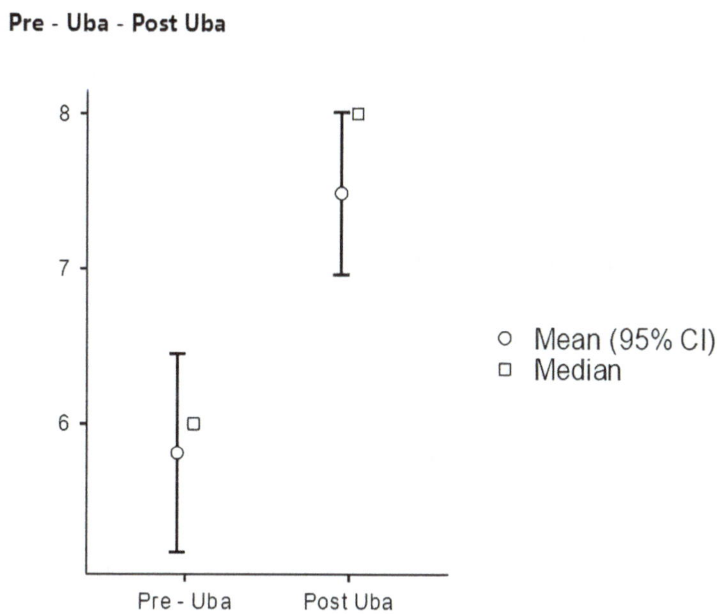

Fig. 5.6 Data Plot from Ubatuba Statistical Analysis

ference between the mean scores of the pre-intervention evaluation M = 5.81, SD = 1.98 and the post-intervention evaluation M = 7.49, SD = 1.63, with $t = -7.30$, $p < 0.001$.

The Shapiro-Wilk test confirmed that the data follow a normal distribution, W = 0.942, p = 0.054, validating the use of the t-test. The 1.68-point increase in the mean scores and the reduction in standard deviation after the intervention indicate greater consistency in the results and the positive impact of the project.

These findings align with the results of the Campinas edition, confirm the initial hypothesis that participants' knowledge would increase, and demonstrate that immersion in a natural environment effectively enhances participants' understanding of environmental.

5.6 Personal Transformation and Empowerment

This section shares stories of personal transformation from interviews conducted by the TransforMAR Project team with teachers from partner schools and students involved in the project's activities. These narratives complement quantitative findings, highlighting long-term impacts and anticipated behavioral changes fostered by Ocean Literacy. They demonstrate the project's positive influence, creating opportunities and contributing to a just and sustainable future. All stories are published with participants' consent.

5.6.1 TransforMAR Project in Alcatrazes Archipelago

The municipality of São Sebastião, located on the coast of São Paulo, holds great environmental significance, encompassing one of the largest remaining areas of the Atlantic Forest in Brazil. Approximately 80% of its territory is protected by various conservation units. Historically marked by disorderly coastal occupation and large infrastructure projects, the city faces significant socio-environmental challenges, including conflicts related to port expansion, which affect biodiversity, traditional communities, and the local economy (Torres et al. 2019).

In this coastal city, marked by conflicts between economic development and biodiversity conservation, many public-school students lack direct contact with the marine environment. Despite living near ecologically significant protected areas and islands important for biodiversity and local lifestyles, most have never been on a boat, explored island ecosystems, or understood the role of conservation units in supporting biodiversity and vulnerable communities.

In 2023, the TransforMAR project, in partnership with the Instituto Chico Mendes de Conservação da Biodiversidade (ICMBio), a Brazilian agency responsible for managing the Alcatrazes Archipelago Wildlife Refuge, conducted an edition of its activities at the Municipal School, Edileusa Brasil Soares de Souza, in São Sebastião. Thirty students participated in the activities under the guidance of public-school teacher Marina da Costa.

Teacher Marina, representing the municipal school, shared that this experience was a transformative journey for both her students and herself. She noted that, despite living near the Ocean, many of her students had never had the chance to explore this environment.

> *They are physically close to the Ocean, but at the same time, so distant in terms of access and experience. Many of these students don't know how to swim, have never been on a boat, and are unaware of iconic places in the region, such as the Alcatrazes Archipelago*, Marina reported.

When the students visited Alcatrazes for the first time, they were astonished. What had previously been just a distant image in books and on screens became a tangible reality. The immediate connection with nature allowed them to grasp, vividly and emotionally, the importance of conservation. This direct experience, according to Marina, profoundly changed the students' perception of the world around them, awakening a sense of responsibility and belonging.

Marina herself underwent a transformation while participating in the project. Although she is an experienced teacher and familiar with Alcatrazes through books and scientific articles, visiting the site added a new dimension to her teaching.

> *The hands-on experience enriched my understanding and allowed me to present the content in a more engaging and authentic way*, Marina shared.

When they returned to school, the students were filled with enthusiasm, eager to share their discoveries with peers and families, becoming ambassadors for the environmental cause.

The TransforMAR Project, by offering young people the unique opportunity to visit the Alcatrazes Archipelago, provided not only an educational experience but also sowed the seeds of environmental awareness and action. This conservation unit is highly relevant to the region. The ecosystem services provided by this refuge, according to the classification of the Intergovernmental Science-Policy Platform on Biodiversity and Ecosystem Services (IPBES) (Díaz et al. 2015), cover a wide range of essential functions for environmental and socioeconomic sustainability. These services include provisioning, such as maintaining fish stocks; regulation, like carbon sequestration; support, including habitat provision, and cultural services, which involve activities such as tourism (Hoff 2015).

The ecological importance of this region goes beyond biodiversity conservation, including the mitigation of human impacts, opportunities for scientific research, identification of sensitive areas, and the provision of ecosystem services essential for community resilience and ecosystem integrity. The socioeconomic well-being of nearby communities depends on the effective preservation of this conservation unit, which relies on local engagement and a sense of ownership over the area.

Fostering an active connection between the community and the protected area is fundamental to helping residents recognize its ecological, economic, and social benefits. Such recognition is essential for sustaining meaningful environmental protection and ensuring the continuity of vital ecosystem services.

This type of initiative, which integrates environmental education into the school curriculum, is essential for nurturing a generation that is both environmentally aware and actively engaged. Marina believes that direct contact with nature significantly helps students understand the urgency of environmental conservation. Through these transformative experiences, TransforMAR not only broadens perspectives but also empowers young people to become proactive agents of change, contributing to a more equitable, sustainable, and Ocean-connected future.

This narrative goes beyond a journey to the Ocean's depths; it opens doors to new opportunities, promotes environmental justice, and strengthens communities. The TransforMAR Project demonstrates how Ocean Literacy can transform lives, inspiring both students and educators to become active advocates for our planet.

5.6.2 TransforMAR Project in Paraty, Rio de Janeiro

Paraty, located on the southern coast of Rio de Janeiro state, is a true sanctuary of the Atlantic Forest, celebrated globally for its biodiversity, cultural diversity, and stunning natural beauty. In 2019, the city was designated a UNESCO World Cultural and Natural Heritage Site, confirming its status as one of the planet's five major biodiversity hotspots. With around 80% of its territory protected by both land and marine conservation areas, Paraty is home to indigenous, quilombola (descendants of escaped African slaves) and Caiçara communities, whose traditions and cultures further enrich the city's unique identity (Conti and Antunes 2012).

Despite its natural beauty, Paraty faces significant environmental challenges, worsened by unplanned urban growth, inadequate wastewater treatment, and inefficient solid waste management. The city also deals with socio-environmental conflicts that affect its traditional communities, which are often marginalised and excluded from decision-making processes (Scagnolatto 2024).

The Ilha das Cobras neighbourhood exemplifies the territorial conflicts that began in the 1950s, driven by infrastructure development in the region. The expansion of highways forced numerous families to relocate without compensation, and the growth of tourism and high-end real estate development worsened these displacements. As a result, residents were pushed into this new neighbourhood, while more central areas were reserved for tourism and urban progress (Corrêa 2021).

Currently, Ilha das Cobras faces high levels of violence, particularly among young people who are often lured into organized crime. The lack of urban planning and basic sanitation, along with real estate speculation and neglect from public authorities, heightens the community's vulnerability, leading to increased crime rates and a growing sense of hopelessness.

In this context, since 2018, the TransforMAR Project has been working to strengthen and engage young people in environmental narratives. Giovani Braga de Oliveira, an entrepreneurship teacher at the State School, CIEP 999, in Paraty, has witnessed the transformative impact of the project on his students' lives. He highlights how TransforMAR provides a unique opportunity to broaden horizons, particularly in scientific knowledge, sustainable development, and social entrepreneurship. For Giovani, the project offers public school students experiences far beyond what the school could provide alone. These activities, in turn, serve as a means of reintegrating young people into society, offering new perspectives and constructive alternatives for the future.

With each new edition of the project, students' interest in science, the environment, and particularly the Ocean grows significantly. As the school year begins, students eagerly ask when they will have the chance to participate in TransforMAR. Giovani emphasizes that:

> [T]he practical experience provided by the project is what creates a strong emotional connection between the young participants and the Ocean.

This direct contact with the environment transforms theoretical learning into something tangible, sparking a keen curiosity about marine life and its interactions.

According to teachers, the project is not only an educational tool but also a means of personal and professional growth. Practical experience, often inaccessible in the school's daily routine, becomes part of the teachers' skill set, enriching their approaches and facilitating knowledge building with students. Moreover, it encourages young people to think creatively and sustainably about the challenges facing their communities.

Giovani also emphasizes the significance of the project in shaping a new generation of social entrepreneurs, dedicated to respecting both the environment and society. Through innovative activities such as podcast production, environmental awareness campaigns, waste management solutions in local communities, and recycling workshops, students are encouraged to enhance their environmental awareness and explore alternatives that promote a more harmonious coexistence with nature.

A particularly memorable experience for the teacher was during the educational diving activity promoted by the project. Despite initial apprehension, Giovani was guided by the TransforMAR team into the waters of Paraty Bay, where he was able to observe marine life in all its richness and complexity for the first time. This experience reinforced the transformative power of the project in opening new perspectives for both students and educators, creating a lasting impact on their lives.

Giovani also highlights the educational activities carried out in the mangroves as equally memorable. The games and dynamics promoted by the project's educators sparked the students' curiosity, prompting them to actively explore the local ecosystem. These practical experiences, which bridge theory with hands-on learning, broaden the students' understanding and deepen their engagement with nature, fostering more meaningful learning.

Similarly, students who engage with the TransforMAR Project undergo a process of transformation, mirroring the experiences of public-school teachers. A significant example of this impact is the story of Alexandre de Carvalho Silva, a 22-year-old from the Ilha das Cobras neighbourhood, one of the most socially vulnerable areas in Paraty. Alexandre recalls spending part of his childhood in the old town until a fire destroyed his home, forcing his family to move to Ilha das Cobras. He lived with his extended family—parents, siblings, and nephews—under difficult conditions on the outskirts of Paraty. His mother, who once worked as a cleaner, is now a massage therapist, while his father works in security. '*I lived in a house with almost ten people and had difficulties studying,*' Alexandre recalls, highlighting the challenges he faced throughout his journey.

Since a young age, Alexandre was involved in traditional Caiçara activities.

> *I am Caiçara; I grew up in Paraty, gathering 'Tarioba' (a type of mollusc found near mangroves), catching crabs, and fishing with my father. My parents are artisans, so I've always had this connection with the environment*, he recalls.

For his family, these practices represented a form of cultural resistance, preserving Caiçara traditions amid external pressures that threatened this heritage. Although this connection with nature was fundamental to his upbringing, Alexandre felt that his understanding of the environment was still limited.

At school, he realized that environmental issues were discussed, but not with the depth he desired, leaving him with few opportunities to expand his knowledge. His participation in the TransforMAR Project marked a turning point, opening new horizons for him. He describes the experience as transformative, expanding his understanding of human impacts on the Ocean and offering a new perspective on the vastness and richness of marine life.

> *The project showed me the vastness of the Ocean and the richness of marine life,* he says, emphasizing the positive impact of the activities provided.

Alexandre also reflects on how environmental issues, though frequently discussed, rarely receive the attention they deserve:

> Even though we hear a lot about environmental impacts, it's often superficial, as very few people explore the topic in a more comprehensive way.

For him, TransforMAR was the first meaningful contact he had with a transformative approach.

The practical experience of SCUBA diving was one of the most remarkable moments for Alexandre.

> *I realized that we are not separate from nature; we are part of it*, he explains, describing how this experience shifted his perception.

The project ignited both an emotional and intellectual awareness of humanity's role in the environment. He also began to understand the impact of large corporations on ecosystems.

> *What big corporations do, how they interfere and will continue to interfere with ecosystems. People realize it's not just their fault; they develop the ability to critically analyses the world around them*, he reflects, highlighting the project's power to provoke changes in participants' critical thinking.

Participating in TransforMAR had a real impact on Alexandre's academic path. After the project, he found his passion for environmental engineering.

> *I don't think I'll be the one to save the world, but I want to go beyond. I want to be involved every day, analysing, researching, and doing what I can, to protect nature*, he says.

He recognizes that the project not only brought him closer to nature but also encouraged him to pursue a career focused on sustainability.

Comparing the project's approach to formal education, Alexandre reflects on the gaps in traditional teaching:

> *I think that's the main point of the project, because in schools, they don't teach it like this. They talk about the environment and how it works, but they don't show its beauty or its real importance. I never felt like I was part of nature.*

According to him, the project filled that gap, connecting him emotionally to nature in a way that school had never been able to.

Proud to be the first member of his family to attend university, Alexandre reflects on his journey:

> *I worked hard, but I was discouraged. When I finally got admitted, it felt liberating*, he recalls.

After overcoming the challenges of life in the urban outskirts, he now dedicates himself to using his knowledge to contribute to environmental conservation and raise awareness among others.

For Alexandre, SCUBA diving was a transformative experience. Although he had always lived near the sea, he had never had such direct contact with marine life. The beauty and diversity of nature left a lasting mark on him.

> I saw all the diversity around me and thought about how many people might never get the chance to see this, because we're losing so much. Even now, when I think about it, it really touches me, and I know I'll remember it forever.

The shared stories show that young people from underprivileged communities often lack access to quality education, reflecting the inequality in Brazil and the lack of investment in education in these communities. This results in their marginalization, both socially and economically. The limited access to quality education also restricts their cultural capital, which is essential for their development, perpetuating the cycle of segregation and inequality.

In this context, the TransforMAR Project offers an opportunity for the future of these young people, focusing on education and knowledge. The project not only changes individual lives but also has the potential to make a positive impact on the communities where it operates.

These young people live with the environmental impacts that affect their livelihoods but often don't realize how much they are affected or know how to respond. The TransforMAR Project helps them understand that they have the power to act. Through environmental education and Ocean Literacy, it shows that the fight for a more just and sustainable world is everyone's responsibility.

5.7 Global Adaptation and Expansion of TransforMAR

While the TransforMAR Project is rooted in the Brazilian context, its core principles can be adapted to various international settings. By focusing on immersive education, emotional connection, and community engagement, the project provides a flexible model to address global challenges in Ocean Literacy.

To replicate TransforMAR internationally, partnerships with local schools, dive centers, organizations, and conservation agencies are important for adapting its methodology to regional cultural and ecological contexts. Incorporating traditional knowledge and addressing specific environmental issues, such as coral reef degradation or mangrove deforestation, ensures relevance and impact.

Additionally, leveraging digital tools like virtual reality can overcome logistical barriers, making Ocean Literacy accessible to landlocked regions. Expanding such initiatives requires consistent investment in financial resources to support NGOs working in Ocean Literacy, as well as funding for training, technology, and educational materials to ensure effective implementation across diverse contexts.

For environmental awareness and knowledge to be effectively developed, education must extend beyond traditional school settings, incorporating the natural environment into learning practices to create meaningful and engaging experiences.

5.8 Concluding Remarks

Ocean Literacy is a fundamental tool for understanding the importance of Ocean conservation and the protection of coastal and marine ecosystems. To maximize its impact, it is essential not only to broaden its dissemination within educational institutions but also to democratize access to high-quality Ocean-related knowledge. Ocean Literacy acts as a transformative instrument for empowering communities and fostering an environmentally engaged and conscious society.

The implementation of projects like TransforMAR shows how Ocean Literacy can shift individual perceptions and create a collective impact on communities. Direct experiences with the Ocean build a strong sense of connection and cultivate a shared responsibility, which are essential for building a society that understands environmental challenges and takes action to address them.

5.8.1 Summary

Key concepts explored in this chapter include:

- The TransforMAR Project demonstrates that Ocean Literacy is essential for addressing global environmental challenges by fostering a deeper understanding of human impacts on marine ecosystems and inspiring behavioral change. Through immersive experiences, the project helps participants develop a strong Ocean identity, reinforcing the need for collective action in shifting toward more sustainable societal models.
- By integrating environmental education with non-formal learning approaches such as workshops and field-based activities, TransforMAR enhances engagement and deepens participants' connection to marine conservation. Its innovative use of SCUBA diving and natural immersion provides students from socially vulnerable communities with meaningful interactions that build emotional connections, leadership skills, and a sense of responsibility toward the Ocean.
- The project supports traditional Caiçara communities by addressing socio-environmental challenges, promoting environmental justice, and fostering cultural preservation. Through hands-on engagement, participants experience personal transformation and become active in their communities, with many launching campaigns, podcasts, and educational initiatives.
- Quantitative assessments show that participants' environmental knowledge increases significantly through the program, and statistical analyses confirm the effectiveness of its educational interventions. These results are reinforced by qualitative insights, with students and educators reporting a shift in their awareness, attitudes, and commitment to Ocean conservation.
- The TransforMAR model offers a scalable approach to Ocean Literacy, with strategies for adaptation in diverse global contexts. Its success highlights the importance of tailoring methodologies to local cultural and ecological conditions while leveraging digital tools and securing financial support to expand educational outreach.
- TransforMAR showcases Ocean Literacy as a powerful tool for conservation and sustainability, demonstrating that direct experiences with the Ocean foster responsibility, drive collective action, and promote an environmentally conscious society.

5.8.2 Recommended Resources

- TransforMAR Project website: An environmental education initiative that utilizes SCUBA diving as a pedagogical tool to promote Ocean Literacy among young people from coastal communities. Available at: http://www.transformar.eco.br
- Sustainable Ocean Alliance (SOA): A global non-profit organization dedicated to empowering youth, developing and implementing innovative solutions, and mobilizing a global

movement to restore Ocean health within our lifetime. Available at: http://www.soalliance.org

- Roots: A program that supports youth-led initiatives focused on addressing climate justice through innovative, intersectional approaches, fostering community-driven solutions for environmental challenges. Available at: http://www.poweredbyroots.org
- Maré de Ciencia: A Brazilian initiative dedicated to promoting science education and Ocean Literacy, fostering engagement to strengthen the interface between science, public policy, and society. Available at: http://www.maredeciencia.eco.br
- Marulho: A social and environmental impact business focused on recycling discarded fishing nets. By transforming them into new products, Marulho prevents these materials from becoming marine waste, contributing to Ocean conservation and promoting sustainable practices. Available at: http://www.fazermarulho.com.br
- Ocean Ride: An innovative project tackling microplastic pollution with electrostatic collection systems for boats and fixed platforms. Committed to education, it engages young people through entrepreneurship programs, inspiring change and raising awareness in public schools. Available at: http://www.instagram.com/oxeanhub
- Euceano: A Brazilian audiovisual project supported by UNESCO and the UN Decade of Ocean Science, dedicated to inspiring Ocean conservation through education, storytelling, and public policy engagement. Available at: http://www.euceano.org.br
- The Blue Planet: A documentary series by the BBC that explores the natural history of the world's Ocean, providing visual insights into marine ecosystems. Available at: http://www.bbc.co.uk/programmes/p04tjbtx
- Breaking Boundaries: The Science of Our Planet: (Netflix Documentary): A powerful exploration of Earth's planetary boundaries, based on the work of Johan Rockström, highlighting the urgent need for sustainable solutions to preserve our ecosystems. Available at: http://www.netflix.com/br/title/81336476

Acknowledgements We extend our deepest gratitude to the Sustainable Ocean Alliance (SOA) and its regional representative Eugénia Barroca for their unwavering belief in our team. Their financial support, training initiatives, and efforts to amplify the reach of the TransforMAR Project have been instrumental in its success since 2020.

Our heartfelt thanks also go to Roots, for believing in our work and supporting our initiatives in 2024, enabling us to carry out transformative sessions and expand the project's impact.

To Professor Carlos Fernando de Andrade, we express our profound appreciation for his intellectual mentorship and steadfast support in shaping the TransforMAR Project.

We are immensely grateful to our dedicated team of volunteers, who have selflessly devoted their time and effort to bring this project to life, to our partner schools, who have always welcomed us with open arms and collaborated enthusiastically in every initiative, and NAUI Brasil for their support in delivering high-quality diving training, equipment, and technical consultancy.

Lastly, we sincerely thank Teresa Sacchetta and Fernando L. Alberto, whose invaluable support during a critical moment ensured the continuation of the TransforMAR Project's activities, allowing us to persist in our mission of fostering Ocean Literacy and environmental awareness.

References

Atchley RA, Strayer DL, Atchley P (2012) Creativity in the wild: improving creative reasoning through immersion in natural settings. PLoS One 7(12):e51474. https://doi.org/10.1371/journal.pone.0051474

Beatley T (2018) Making the marine world visible: fostering emotional connections to the sea. In: Beatley T (ed) Blue biophilic cities: nature and resilience along the urban coast. Springer, pp 57–77. https://doi.org/10.1007/978-3-319-67955-6_4

Bennett NJ, Alava JJ, Ferguson CE, Blythe J, Morgera E, Boyd D, Côté IM (2023) Environmental (in)justice in the Anthropocene Ocean. Mar Policy 147:105383. https://doi.org/10.1016/j.marpol.2022.105383

Branco A, Caseiro F, Branco Z, Flavain E, Natividade H, Kiekebusch A, Sanchez E (2005) Cultura Caiçara resgate de um povo. Editora Etecê

Christofoletti, R. A., Gozzo, A. J., Conti, W. R. P., Takahashi, C. K., Mazzo, T. M., Ignacio, B. L., Martins, F. R., Kitahara, M. V., & Yokoyama, L. Q. (2022). Escola Azul: Educação, ciência, cidadania e a Década da Ciência Oceânica no Brasil

Conti BR, Antunes DDC (2012) Conflitos na gestão do Parque Nacional da Serra da Bocaina: Entraves ao desenvolvimento local na vila de Trindade (Paraty, RJ). Interações (Campo Grande) 13(2):213–223. https://doi.org/10.1590/S1518-70122012000200008

Corrêa APS (2021) Juventude e violência: Cenários turísticos e políticos de Paraty. https://rima.ufrrj.br/jspui/handle/20.500.14407/12582

Costanza R (1999) The ecological, economic, and social importance of the Oceans. Ecol Econ 31(2):199–213. https://doi.org/10.1016/S0921-8009(99)00079-8

Díaz S, Demissew S, Carabias J, Joly C, Lonsdale M, Ash N, Larigauderie A, Adhikari JR, Arico S, Báldi A, Bartuska A, Baste IA, Bilgin A, Brondizio E, Chan KM, Figueroa VE, Duraiappah A, Fischer M, Hill R et al (2015) The IPBES Conceptual Framework—connecting nature and people. Curr Opin Environ Sustain 14:1–16. https://doi.org/10.1016/j.cosust.2014.11.002

Educação MD (2007) Vamos Cuidar Do Brasil: Conceitos E Práticas Em Educação Ambiental Na Escola. Sec. de Educ. Continuada, Alfabetiz., Diversid. e Inclusão

Gerald B (2018) A brief review of independent, dependent and one sample t-test. Int J Appl Math Theor Phys 4:50. http://www.wjph.org/article/10.11648/j.ijamtp.20180402.13

Gomes De Araujo L (2014) A pesca costeira artesanal de Paraty, RJ: Uma análise multiescalar sob o enfoque da cogestão de recursos comuns. Doutora em Ambiente e Sociedade, Universidade Estadual de Campinas. https://doi.org/10.47749/T/UNICAMP.2014.939814

Hoff NT (2015) Integridade biótica dos ecossistemas na região do Arquipélago dos Alcatrazes, São Sebastião—SP. Mestrado em Oceanografia Biológica, Universidade de São Paulo. https://doi.org/10.11606/D.21.2015.tde-22092015-135056

Horovitz G (2016) A Pesca Artesanal em Paraty: Conflitos e Desafios Para a Gestão Ambiental

Jaksha AP (2019) Leveraging Ocean identity in education to impact students' conservation practices. In: Fauville G, Payne DL, Marrero ME, Lantz-Andersson A, Crouch F (eds) Exemplary practices in marine science education: A resource for practitioners and researchers. Springer, pp 93–107. https://doi.org/10.1007/978-3-319-90778-9_6

Jamovi (2022) The jamovi project (Versão 2.3) [Software]. https://www.jamovi.org

Kelly MR, Kasinak J-M, McKinley E, McLaughlin C, Vaudrey JMP, Mattei JH (2023) Conceptualizing the construct of Ocean identity. NPJ Ocean Sustain 2(1):1–3. https://doi.org/10.1038/s44183-023-00025-7

Kennedy TJ, Dufourneaud O, Castillo MP, Deane S, Durand J, Giaccone T, Jonda S, Mascarenhas T, Mtonga C, Ng K, Palazot S, Paul S, Salazar ER, Santisteban R, Shuwen X, Silva A, Solis Rivera V, van't Land H, Zhou Y (2024) Ocean Decade manual for non-governmental organizations (NGOs). UNESCO-IOC, Paris. [The Ocean Decade Series, 54]. https://unesdoc.unesco.org/ark:/48223/pf0000392198

Mohai P, Pellow D, Roberts JT (2009) Environmental justice. Annu Rev Environ Resour 34(1):405–430. https://doi.org/10.1146/annurev-environ-082508-094348

O'Halloran C, Silver M (2022) Awareness of Ocean Literacy principles and Ocean conservation engagement among American adults. Front Mar Sci 9. https://doi.org/10.3389/fmars.2022.976006

Pazoto CE, Silva EP, Andrade LAB, Favero JMD, Alô CFS, Duarte MR (2021) Ocean Literacy, formal education, and governance: A diagnosis of Brazilian school curricula as a strategy to guide actions during the Ocean Decade and beyond. Ocean Coast Res 69(suppl 1):e21041. https://doi.org/10.1590/2675-2824069.21008cep

Prado DS, Seixas CS, Berkes F (2015) Looking back and looking forward: Exploring livelihood change and resilience building in a Brazilian coastal community. Ocean Coast Manag 113:29–37. https://doi.org/10.1016/j.ocecoaman.2015.05.018

Rockström J, Steffen W, Noone K, Persson Å, Chapin FSI, Lambin E, Lenton TM, Scheffer M, Folke C, Schellnhuber HJ, Nykvist B, De Wit CA, Hughes T, Van Der Leeuw S, Rodhe H, Sörlin S, Snyder PK, Costanza R, Svedin U et al (2009) Planetary boundaries: Exploring the safe operating space for humanity. Ecol Soc 14(2):art32. https://doi.org/10.5751/ES-03180-140232

Santoro, F., Santin, S., Scowcroft, G., Fauville, G., & Tuddenham, P. (2018). Ocean Literacy for all: A toolkit (O. V. and R. B. for S. and C. in E. UNESCO & I. O. Commission, Orgs.). https://unesdoc.unesco.org/ark:/48223/pf0000373449

Scagnolatto G (2024) Os Impactos Socioambientais do lixo no mar para os pescadores artesanais de Paraty RJ (Biblioteca UNIFESP). Tese de Mestrado, Universidade Federal de São Paulo

Torres PHC, Ramos RF, Gonçalves LR (2019) Conflitos Ambientais Na Macrometrópole Paulista: Paranapiacaba e São Sebastião. Ambient Soc 22:e0101

Author Contributions and Biographical Profiles

[1*]Alexandre Augusto da Silva: Co-conceptualization; Writing—Original Draft, Review, and Editing; Methodology—Data Collection and Analysis; Graphical Design; Photography.

[2]Giovanna Scagnolatto: Co-conceptualization; Writing—Original Draft, Review, and Editing; Funding Acquisition.

[3]Karina F. Orsolini: Writing—Review and Editing.

*Corresponding Author at alexandre@transformar.eco.br

Alexandre Augusto da Silva is an Engineer, entrepreneur, scuba diving instructor, and founder of the TransforMAR Association. Originally from São Paulo, Brazil, he is passionate about the Ocean and committed to environmental education. He holds a specialization in strategic management and business administration alongside a master's degree in Ocean Science and Technology. With significant project management experience, Alexandre founded the TransforMAR project in 2018, an initiative endorsed by UNESCO as part of the United Nations Decade of Ocean Science for Sustainable Development, which delivers programs that inspire and educate youth from underrepresented and traditional communities along Brazil's southeast coast. Through TransforMAR, he works with young people and coastal communities to foster Ocean stewardship and drive positive environmental change.

Giovanna Scagnolatto is an environmental engineer and holds a master's degree in marine science and technology. Giovanna's research focused on the impact of marine debris on traditional coastal communities in Brazil, combining scientific analysis with close collaboration with local leaders. She currently serves as the executive director of the TransforMAR Association, in addition to working as an environmental educator and sustainability content reviewer. As a young leader at the Sustainable Ocean Alliance, she is responsible for the SOA-São Paulo hub in Brazil, leading youth initiatives that connect Ocean literacy, community engagement and public policy. She is also a passionate Ocean advocate and a certified scuba diving instructor, as well as a speaker on topics related to sustainability, the environment, and Ocean literacy.

Karina F. Orsolini holds a degree in Biology from the Universidade Federal de São Carlos, an MBA in Project Management from the Universidade de São Paulo and a master's degree in Marine Biological Resources from the European IMBRSea Programme. Passionate about nature, she focused on marine conservation during her master's degree, seeking to raise environmental awareness through education. She has experience in communication and education, marine monitoring, scientific diving and environmental consulting. Currently based in São Paulo, Brazil, she is studying remote sensing for environmental studies and looking for socio-environmental projects to get involved with.

Ocean Literacy in a Land-Locked State

6

Gregory J. Barord

Abstract

Communities that are landlocked may not have as much of a direct connection to the Ocean as coastal areas, but it is increasingly evident that all communities are facing the effects of a changing climate and a changing Ocean. If anything, it may be even more critical to ensure that Ocean Literacy is valued and supported the further one gets away from the coast. At Central Campus Regional Academy in Des Moines, Iowa, United States, a Marine Biology program is providing high school students aged 15–18 with connections to the marine ecosystem through aquarium husbandry, research projects, as well as opportunities to first learn about marine topics and then share what they learned with the public. Students within this program are not only becoming leaders within their own community but are also taking on additional responsibilities and roles at state, national, and even international levels as members of the Iowa Junior Academy of Sciences, engaging in the National Oceanic and Atmospheric Administration Ocean Guardian Youth Ambassador Program, and participating in field conservation expeditions to Fiji. This chapter highlights how the program fosters Ocean Literacy in a landlocked state, illustrating the transformative impact of hands-on marine education and community engagement in preparing students to address global Ocean and climate challenges.

Keywords

Ocean Literacy (OL) · Education · Aquarium · Student research · Ocean leadership · Aquaculture · Research · Work-based learning · Project-based learning (PBL) · Career-based learning

6.1 Introduction

The implementation of effective Ocean Literacy programs depends on a suite of variables that must be specific to the community, the learner, and the intended goals and outcomes. Together, these actions provide students with the knowledge and experience to support marine conservation (United Nations Decade of Ocean Science for Sustainable Development n.d.-b). Educational outreach for coastal communities in Fiji would be vastly different than those methods employed along the California coast of the United States. Likewise, outreach programs for communities

G. J. Barord (✉)
Central Campus Regional Academy, Des Moines, IA, USA
e-mail: gregory.barord@dmschools.org

T. J. Kennedy (ed.), *Ocean Literacy: The Foundation for the Success of the Ocean Decade, Volume I*, Sustainable Development Goals Series, https://doi.org/10.1007/978-3-031-98145-6_6

that are inland from the coast would, and should, look quite different.

The Central Campus Regional Academy in Des Moines, Iowa USA serves students throughout central Iowa by providing them with high-quality courses tailored to their career goals and aspirations ranging from welding and robotics, to nursing, graphic design, and fashion. At Central Campus, a Marine Biology program has been an option for high school students aged 15–18 to participate in for over 30 years. Over the last 10 years, the Marine Biology program has been reimagined (Seafriendlyreef 2024) to engage the new generation in not only learning about the newest information gleaned by scientists, but to become active leaders in marine conservation efforts through leadership, outreach, and basic research. Here, students are asked to leverage their personal curiosities within marine biology to address controversial topics and explore paradigm shifts to learn about marine conservation (United Nations Decade of Ocean Science for Sustainable Development n.d.-a) to improve their own Ocean Literacy and that of their community.

6.2 The Department of Marine Sciences

6.2.1 The Facility

The Marine Biology program at Central Campus (Science Center of Iowa 2015) is part of the Career and Technical Education (CTE) options that students can participate in within the Des Moines, Iowa area. Alongside an Aquarium Science program that makes up the Department of Marine Science, the Marine Biology program is not only designed for students to learn about the Ocean, but to also gain employability and lifelong skills to support their future career pathways, whether in the marine biology field or another discipline (Peetz 2024).

The approximate 9000 square foot (836 square meters) facility includes between 130–140 saltwater aquariums ranging from 10 to 2400 gallons (9100 L), totaling approximately 15,000 total gallons (57,000 L) of saltwater (Fig. 6.1). The aquariums hold 130 different marine species from clownfish, corals, and clams to lobsters, cuttlefish, and sting rays. Themes of the different aquariums include ghost-fishing, aquaculture, fish schooling, and kelp forests, to name a few.

The space also includes two designated classrooms, two classroom lab spaces, one fish rearing room, one shrimp culture room, two water quality labs, and a diet kitchen. Additionally, a total of 10 “dry exhibit” displays are present alongside the hallways of the school to highlight different topics of marine biology to share these with other students as they go through their day.

6.2.2 The Marine Biology Program

Each year, over 150 students enroll in one of the two programs (Aquarium Science or Marine Biology) as a first-year student or for students in their second or third years in the programs, as a Laboratory Assistant through an Internship program. The Marine Biology course is a dual-enrollment course in the state which means that in addition to receiving high school credit for graduation, students also receive eight college credits in the field of biology through a partnership with a local community college. Students in

Fig. 6.1 Several larger marine aquariums that are within the Central Campus Marine Biology program

the Internship program receive additional college credits. The students come from 10–15 different high schools (ages 15–18) in the area.

Many of the students are interested in pursuing careers within the marine sciences to work in aquaculture, at public aquariums, or as research scientists. Students who are planning careers in other fields outside of marine biology enroll in the course to learn about the Ocean, care for the organisms, and get to know different people. Together, these students work independently as well as in small and large groups to ensure the organisms are cared for well, each day of the year. While the facility is advanced and provides many opportunities for learning, connecting students with real-time research efforts and work of current scientists around the world is heavily leveraged to show students examples of how conservation works (Reyna-Rodriquez 2024).

6.2.3 First-Year Students

Students may register for the Marine Biology program as a sophomore (15–16 years of age), junior (16–17 years of age), or senior (17–18 years old) based on their standing in high school. With the varied ages of the students comes a diverse background of not only educational knowledge, but also cultures, socioeconomic status, race, and learning styles. For example, one student may enter the program having cared for aquariums for several years prior to this whereas another student in their cohort may have no experience with aquariums. These first-year students are in class for 90 min (about 1.5 h) each day and will spend about one-third of their time caring for the aquariums (part one), one-third in lecture settings to learn about different topics (part two), and one-third of their time working on research projects and different labs (part three).

For the students to be prepared to care for the aquariums, it takes about 1 month to provide enough information and time to learn and practice the many distinct aspects of marine aquarium husbandry. Every day, students must ensure lab spaces are clean, feed the diverse organisms in each aquarium, and ensure the aquariums are clean and in normal working order. For some students, this day involves feeding two sting rays while other students may be doing a water change, and yet other students are counting shrimp hatchlings. After the first month, students engage in more independent/group work to ensure the lab functions well and is continually progressing.

After students have the knowledge and experience to care for the aquariums, they begin the second part of the program, which incorporates advanced, college-level coursework to learn about the different topics within marine biology and how the Ocean functions. These topics range from the history of marine biology, marine geography, and marine geology to the comparative anatomy of marine invertebrates and vertebrates and impact of a changing planet on marine ecosystems. What makes this program so unique is that as students learn topic to topic, they are also immersed in caring for many of the same or similar organisms in the aquariums (Fig. 6.2) which inevitably helps students to appreciate things impacting the Ocean.

By combining aquarium care with academic work, the goal is for students to then be able to develop academic questions from their aquarium observations that they make each day. After students have demonstrated proficiency in the first two aspects of the program, aquarium husbandry and the coursework, they progress to completing different laboratory activities and develop their own research projects.

Fig. 6.2 Example of a smaller aquarium in the Central Campus Marine Biology program whose theme is to highlight the unique mangrove ecosystem in the tropics

For these research projects, the only requirement for students is that they must collect some form of data related to their project, Hence, these projects are not intended to only be papers submitted with information learned about and cited from scientific sources. In addition to this, students need to plan out their experimental design, manage data collection, and analyze these data to test their hypotheses. Some examples of past projects were to contrast the swimming speed of different fish species, analyzing the probability of using soy-based food for aquarium fishes, and determining the impact of educational activities to inform people of the impact of different marine conservation policies.

6.2.4 Second and Third-Year Students

Although there is not currently a second-year course in Marine Biology, students who wish to remain in the program, for several reasons, may register as a Laboratory Assistant during their second year and/or third year. Many students who choose this option are certain that they intend to pursue a career in marine biology and want to gain more experience in the laboratory and aquarium facility at Central Campus.

As Laboratory Assistants, these students are expected to maintain the lab during the first month of school, while the first-year students are being trained, mentor the first-year students throughout the year, as well as develop their own advanced projects that facilitate their continued learning process as individuals and help to advance the overall program. These students are functioning on their own each day with little direct oversight and must ensure that their various tasks and projects are completed and progressed.

6.3 Inside the Classroom

6.3.1 Non-Traditional Dissections

For centuries, the best way to truly learn about the anatomy of an organism was to have it in your possession and be able to dissect it while illustrating the external and internal characteristics of the specimen. This method continues to still be a valuable tool for learners of all ages to engage in the process of comparative anatomy while also being able to make connections to how that anatomy has allowed for the adaptations to the habitat the species lives in.

However, dissections with actual specimens require procuring specimens from the wild or aquacultured sources, handling preservatives like formaldehyde, maintaining dissection supplies, and planning for the disposal of the specimens after dissections, all of which have a budgetary cost to account for each year. As technology has advanced, though, it has provided additional opportunities to learn about marine organisms through virtual dissections that utilize online platforms to learn about different marine organisms, thereby reducing pressure on wild/aquacultured populations and eliminating the need to dispose of specimens and their fixatives.

Of course, these online platforms also depend on students having their own technology, or school-issued technology, to access the software and even the most advanced virtual reality dissections are still not perfect representations of actual specimens. Should dissections continue as they always have by using preserved specimens? Should dissections be solely virtual through online platforms?

At Central Campus, the answer is somewhere in the middle. The new model being implemented combines the positive aspects of both options, utilizing online dissection platforms to support traditional dissections with preserved organisms. During the course's first semester, students use the online dissection platforms (Fig. 6.3) in partners or in groups of three on one computer laptop or another technological device. Students utilize "smelly books" to record their illustrations and label the different anatomical parts, with their function, just as they would using preserved specimens.

During the second semester, students move on to more advanced, in-depth dissections with different specimens given the foundational information and knowledge they obtained from the first semester. For example, rather than all students

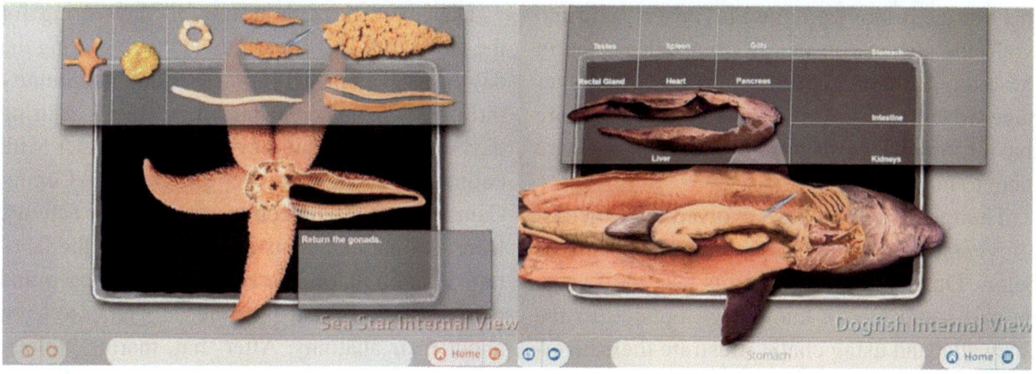

Fig. 6.3 Screen shot examples of the online dissection software being used for the hybrid dissection model at Central Campus with a starfish specimen (left) and dogfish specimen (right)

dissecting every different specimen under the traditional model, students now select three to four specimens to conduct dissections based upon the learning they did in the online platform.

This hybrid model of dissection ensures that all students have opportunities to learn about a diverse array of invertebrates and vertebrates through virtual dissections and then gives students the opportunity to plan their own learning based on their interest level in which specific specimens they would like to learn more about in semester two.

6.3.2 Night at the Aquarium

Many marine organisms display different behaviors and coloration patterns between day and night. During the school day, students can consistently observe behaviors of species in different aquariums, beginning to learn their distinct set of behaviors that they exhibit. However, there are many behaviors, and even species, whose natural behaviors are missed when the lights are on. Thus, the Night at the Aquarium was born on one evening in the first semester of the school year.

On this evening, students arrive at school in the evening to work in the dark to observe the behaviors of the more nocturnal species in the aquariums. Using different light wavelengths and filtered lenses, students observe and record which species are active and if there are any that exhibit fluorescence at night. The evening not only involves planning by the students to know which aquariums and species they need to cue in on, but also an idea of how they plan to share this evening with others that may not be able to participate, due to other commitments such as family obligations, extracurricular activities, or work schedules for distinct reasons.

Some students share photos and videos through their own social media and the program's social media. Other students are being photographed by a district photographer to post to the school's website. And one group has decided that they plan to prepare artifacts and activities from this event to share at the end of year Central Campus Ocean Week. After the nightly observations, students come together in the classroom to share what behaviors they observed at night and, most critically, how those behaviors differed to those observed during the school day.

6.3.3 Central Campus Ocean Week

Each year, Marine Biology students plan out a week of Ocean-related activities and events for the entire school to engage in, called Central Campus Ocean Week. The inception behind this event was because World Ocean Day occurs in June, after the school year in Iowa ends, so students wanted a way to share their favorite things about the Ocean and engage their peers in different activities. One of the overarching themes of this Marine Biology program is conservation and while the students in the program all engage in these topics throughout the school year, another huge aspect of Ocean Literacy is to then share that information with others to continue to inspire them to make changes.

Though the events have evolved over the last 10 years, the general schedule of events is below. On the first day, students take their artistic skills and Ocean knowledge outside to the sidewalks and using chalk, illustrate their favorite marine organisms and the marine conservation issues threatening them and their habitat. By the end of this first day, about 200 feet (243 meters) of blank sidewalk (Fig. 6.4), is then transformed into an Ocean ecosystem for all students to walk by as they come into the building and for the community to view (Des Moines Public Schools 2020; Des Moines Public Schools–TV 2021).

On day two, students set up their different research projects and outreach activities to engage their school community and visitors in learning about the Ocean. These include scientific research posters, sing-along options with Ocean-themed songs, and puppet shows for younger-aged students. These different projects combine the students' interest in the Ocean with other aspects of their personality and truly give them the opportunity to be excited about whatever type of project they want to do.

Day three incorporates a full day caring for the laboratory and aquarium and then sharing the day-to-day work with our virtual community through various social media platforms. During day four, marine biology students partner with culinary students (another program at Central Campus) for a "Dissections for Dinner" activity. Here, culinary students are invited into the laboratory and aquarium area to tour the space and then dissect different specimens together to learn about their anatomy. After that, marine biology students go into the culinary space and learn about different methods of seafood preparation and sustainability efforts as the culinary students prepare a meal, which all students share in and spend time together.

On the last day of Central Campus Ocean Week, students participate in the Ocean Olympics where all students in the program compete in various Ocean-themed events. In teams of four to five, the students must work together to complete the various activities throughout the day. Some of the events include a scientific name match game, salinity determination calculations, and a water siphon relay race, to name a few.

The Ocean Olympics ends with a medal ceremony (Fig. 6.5) for the first, second, and third place teams. This week is a celebration for students of the determined work they put into the program throughout the year. It is a way to highlight different, less well-known topics in marine

Fig. 6.4 Examples of several sidewalk science chalk drawings outside of Central Campus along the road (left) and one specific drawing of a nautilus (right)

Fig. 6.5 Medals awarded to the top three teams after the Central Campus Ocean Olympics

conservation and a way to give other students a pathway to learning about the Ocean for themselves, by either taking the course on their own the following year or just by learning about something new.

6.4 Outside the Classroom

6.4.1 Field Studies Expeditions

During the spring semester (January to May) of each school year, a group of up to 40 students participates in a Field Studies Expedition to a coastal destination within the United States, and most recently, to Fiji to collaborate on nautilus population surveys. The Fiji Expedition was also endorsed by the United Nations Decade of Ocean Science for Sustainable Development as a Decade Activity, furthermore, connecting landlocked students to global Ocean initiatives. The goal of these expeditions is to provide students with authentic opportunities to practice marine science field methods, network with professionals in the field, and learn about the history and culture of the area.

Students partner with local conservation organizations to learn about marine protected areas and fisheries while also assisting with field surveys to aid data collection. One unique aspect of these expeditions is that the students also prepare a presentation for different business partners and colleagues that they meet during the expedition. The presentation is about the Central Campus Marine Biology program and a way for students to learn how to organize their thoughts and orally present them to industry professionals, most of whom are working within the field of marine conservation. During the school year, students are consistently sharing their information with younger people, such as kindergarteners, and these opportunities to network challenge the students to consider different forms of communication to improve their Ocean Literacy.

6.4.2 No School Days

On a normal school day, there are over 100 students between the three classes that are present and able to care for the aquariums. On weekends, holidays, and breaks, the same students volunteer their time to come up and care for the lab on their own, albeit in much smaller numbers. On any given no school day, rather than having 100 students, there are between three to four students who work together to care for the lab on their own. They need to coordinate the timing of when to meet and to call the school's security team to be allowed in the school.

Once in the school, the two-page checklist of things to do to care for all aquariums must be completed. These days provide students with more opportunities to learn about marine organisms and their care and additional ways to build their leadership, teamwork, and other employability skills. These no school days are preparation for students to take the laboratory and aquarium over during the longer summer (June to August) break when school is not in session. In this setting, students are tasked with managing each other's schedules for weeks at a time, again with little direct oversight, to ensure that all organisms and aquariums are cared for properly. On the extremely rare occasions when students are not able to volunteer, the instructor serves as the "plan B" to come up and care for the lab.

6.5 Industry-Based Learning

6.5.1 Shrimp Cultures

As part of the practice to support sustainability, students developed methods to culture various species of shrimp within the laboratory. In particular, the grass shrimp, *Palaemon pugio*, mysid shrimp, and *Americamysis bahia*, are common species found throughout marshes and estuaries of the gulf coast of the United States. They are also common food sources for many species within this area and an ideal food source for many organisms in aquariums, such as cuttlefish and octopus. To reduce the reliance on outside sources to provide live shrimp for aquariums, students at Central Campus developed a protocol to culture these species in-house over multiple generations. Additionally, the results of one of these aquaculture projects were published (Des Moines Public Schools 2017) in an aquarium journal and presented at local and national professional conferences.

6.5.2 Jellyfish Training Videos

Many of the aquariums at Central Campus require advanced husbandry skills and experience to ensure the organisms are thriving. One of these species is the moon jellyfish, *Aurelia aurita*, that requires a specific type of aquarium with delicate flow and consistent food availability to grow normally and reproduce. Culturing jellyfish species in aquariums is a widespread practice to reduce the need for wild collection. One group of students were particularly interested in not only learning these practices but developing a demo video for future students to be able to utilize for their own learning. The student team took it upon themselves to learn husbandry skills, prepare a script for the demo, and record themselves completing the various tasks to care for the jellyfish. The result was a video that is currently shown to students, and the entire school, in one of the external display areas.

6.5.3 Sting Ray Conditioning

Elasmobranchs held in aquariums can benefit from conditioning and enrichment programs for several reasons. At Central Campus, two yellow sting rays, *Urobatis jamaicensis,* were received from a nearby institution that were bred in an aquarium. Prior to accepting the sting rays, one student in the program was tasked with developing a protocol to set up a large, 1000-gallon (3785 L) aquarium for the sting rays that mimicked their natural habitat and provided enough room for the sting rays to display natural behaviors. Coupled with this, the student also needed to plan out a conditioning program for the sting rays to be able to feed each sting ray from a different area of the aquarium, using a target/sign. This type of conditioning allows for routine veterinary health checks to be done and ensures that each sting ray is getting their proper amount of food during each feeding. To accomplish this, the student worked with other students, aquarium professionals, and veterinarians to ensure that the methods were correct, safe, and effective. These practices were further advanced by a future student, who continued working with the veterinarian to develop methods to weigh, draw blood, and test fertility throughout the year to ensure they are healthy. The project resulted in streamlined protocols for future students to follow and these data are also currently being prepared for publication in an aquarium journal.

6.6 Educational Outreach

Students remaining in the program for over 1 year can advance their own learning by combining their varied interests with the Ocean. One such project was to incorporate student interest in social media with the Central Campus Marine Biology program by creating the program's first online account, which was just the first step. Their goal was to share the day-to-day student work and animal behaviors in the lab widely to impact Ocean Literacy within the virtual community.

While doing this, the students were utilizing the scientific method and collecting data to determine which types of posts had the highest engagement, and potential improvement on Ocean Literacy, to refine future posts. The social media accounts originally created by these students are currently in use, being transferred to different students each year to maintain and continue to monitor. Each year, there are usually one to two Laboratory Assistants who not only enjoy learning about the Ocean but also want to develop outreach activities for younger students in the community.

One such student was interested in visiting local elementary schools (6–12 years old) to engage them in different Ocean topics and concerns. To accomplish this, the student contacted local schools and teachers, organized different presentations, and determined what type of "show and tell" and "listen, interact, and feel" artifacts may be useful to bring to each classroom presentation. To learn how students engage in the different presentations, the student developed a questionnaire for students to take after the presentation to guide future visits. As of this writing, the student is coordinating their visits to different classrooms and their goal is to leave them with a greater love of the natural world so that they too will want to further explore their own interests in marine life or the environment.

6.7 Conclusion

"Ideally, the inclusion of Ocean Literacy in the statutory curriculum considers regional differences in curriculum content. This topic is particularly important for landlocked, rural and remote regions with little or no contact with the sea, but still influenced by the Ocean (e.g. transport, climate regulation, temperature, economy, etc.)" (IOC-UNESCO 2022, p.100), as it provides a "remote connection to the Ocean" (Kennedy et al. 2024, p. 25). The Marine Biology program at Central Campus is rooted in conservation and a commitment to being engaged in what is going on in the Ocean, through aquarium husbandry, scientific inquiry, and educational outreach. We do not simply learn about the impact of ghost fishing on ecosystems or read about the changes occurring on distant islands resulting from human activity. Our students create aquarium exhibits to show how ghost fishing impacts ecosystems and engage in critical thought on analogizing habitat loss on islands with changes in our classroom space. These models are then shared with the whole school community and the outside world.

The main thing that we have not only learned, but embraced, over the last 10 years at Central Campus is to be creative in our approach to discussing marine conservation both inside and outside of the classroom. Perhaps the most important aspect of this program is the connection, or the drive, that students must share their passion and what they have learned with others. One passionate, informed teacher can make a difference and spark some excitement and curiosity in a classroom by sharing stories of their own scientific discovery (WHO13 2023). But that impact can become exponential when those students find their own way to connect with the Ocean so much so that they are compelled to share what they know with others.

One student may love the complexity of how a mesophotic reef habitat functions, another student just learned about the mortality of hermit crabs for the pet trade, and another student simply wants to talk more about their favorite cowfish in the lab. At the same time, a fourth student is working with a group to draft a letter to local governmental representatives requesting that they act to protect the environment. People need opportunities to succeed, and even more opportunities to fail along their way to success. Students at Central Campus have opportunities to follow their own path in the field of marine biology and to see themselves not just as spectators, but as stakeholders in what happens to the Ocean.

6.7.1 Summary

Key concepts explored in this chapter include:

- A 100% student-run marine science laboratory and aquarium in a land-locked state fosters Ocean Literacy through authentic experiences that connect students to the Ocean's global significance, even being so far from the shoreline.
- Career exploration and guidance in marine biology for up to 3 years enhance Ocean Literacy by immersing students in marine science pathways,

fostering a lifelong connection to the Ocean and the importance of marine conservation.
- Industry-based opportunities for high school students (ages 15–18) related to the marine sciences ensure that once students graduate from the program, they can take their skills and experience in Ocean Literacy to the next steps in their professional development.
- Hybrid dissection models bridge the traditional techniques and modern innovations to support varied learning styles of students to ensure they all have opportunities to learn about different organisms that call the Ocean their home.
- At Central Campus, students become marine science leaders inside and outside the classroom, not only learning about how important the Ocean is to their own way of living but sharing this with others in their community.

6.7.2 Recommended Resources

The following resources provide further insights and support on this topic:

- Central Campus Regional Academy: https://centralcampus.dmschools.org/welcome/
- Central Campus Marine Biology Program: https://centralcampus.dmschools.org/agriculture-environmental-science/marine-biology/
- Central Campus Marine Biology Program Social Media Accounts:
  - https://www.facebook.com/CCMarineScience
  - https://www.instagram.com/ccmarinescience/
  - https://x.com/ccmarinescience
- Publication, Authored by Student in the program
  - Walker, H., & Barord, G. J. (2017, January). Culture methods of grass shrimp. *DRUM and CROAKER, 48*, 21–27. http://drumandcroaker.org/pdf/2017.pdf
- Online Forum for Cephalopod Husbandry Resources: https://tonmo.com/
- Aquaculture Reference Book:
  - Moe, M. A. (1989). *The marine aquarium reference: Systems and invertebrates.* Green Turtle Publications.
- NOAA Ocean Guardian Youth Ambassador Program: https://sanctuaries.noaa.gov/education/Ocean_guardian/youth-ambassador.html

Acknowledgments I would first like to thank my wife, Karen, for supporting me and helping to troubleshoot the various ideas and changes I had for this program over the 11 years that we have been here. Without her own perspective and creativity, I would not have been able to be as confident in making changes and updates.

The students who have participated in this program are the true secrets to its continued success because they were all, and still are, willing to step outside their comfort zone to care for dozens of aquariums, team up with students from all different high schools, and be open to learning in new, and sometimes cumbersome, ways.

I would also like to thank my administration (Dr. Gary McClanahan, Julie Rosin, Aiddy Phomvisay, Gretchen Watznauer, Tascha Brown, Dr. Lisa Hill, Kelsie Mouchka, and Jessie Masteller) over the years for always supporting my unorthodox instructional practices, like slowly taking chairs and desks out of the classroom as an analogy to habitat loss. My colleagues at Central Campus and outside of it have always been willing to help our students in any way possible, whether it was a remote presentation or interdisciplinary research project.

In particular, Kevin Anderson and Sam Oppel have challenged me with new ideas and thoughts to bring to the classroom and to improve my own teaching pedagogy.

Finally, managing such an advanced, complex program would simply not be possible without the Aquarium Science Instructor, Kirk Embree, who has always provided critical feedback and assistance over the years.

References

Des Moines Public Schools (2017) Central student publishes marine bio research. https://www.dmschools.org/2017/01/central-student-publishes-marine-bio-research-2/

Des Moines Public Schools (2020) Aquatic lessons written on stone by marine biologist. https://www.dmschools.org/2020/07/aquatic-lessons-written-on-stone-by-marine-biologist/

Des Moines Public Schools—TV (2021) Central Campus Sidewalk is 'Chalk'—full of Marine Invertebrates [Video]. YouTube. https://youtu.be/kuD6EjBvEpE

IOC-UNESCO (2022) A new blue curriculum: A toolkit for policy-makers. UNESCO. https://unesdoc.unesco.org/ark:/48223/pf0000380544

Kennedy TJ, Dufourneaud O, Castillo MP, Deane S, Durand J, Giaccone T, Jonda S, Mascarenhas T, Mtonga C, Ng K, Palazot S, Paul S, Salazar ER,

Santisteban R, Shuwen X, Silva A, Solis Rivera V, van't Land H, Zhou Y (2024) Ocean Decade manual for Non-Governmental Organizations (NGOs). UNESCO-IOC, Paris. [The Ocean Decade Series, 54]. https://unesdoc.unesco.org/ark:/48223/pf0000392198
Peetz C (2024) A marine science program in a surprising place shows students new career options. Education Week. https://www.edweek.org/teaching-learning/a--marine-science-program-in-a-surprising-place-shows-students-new-career-options/2024/09
Reyna-Rodriquez V (2024) Landlocked Iowa has a marine biology expert. Why a local teacher was picked for UN research. Des Moines Register. https://www.desmoinesregister.com/story/news/education/2024/02/25/greg-barord-des-moines-central-campus-united-nations-marine-environment-Ocean-research/72711384007/
Science Center of Iowa (2015) STEM in DSM: Marine biology teacher empowers students to save the nautilus. https://www.sciowa.org/about-sci/sci-blog/stem-in-dsm/?page=5
Seafriendly (2024) This is a school for aquarium enthusiasts?!—*best school in the world* [Video]. YouTube. https://youtu.be/GxMKbBhG71s
United Nations Decade of Ocean Science for Sustainable Development (n.d.-a) Ocean Decade challenges. https://Oceandecade.org/challenges/
United Nations Decade of Ocean Science for Sustainable Development (n.d.-b) Vision 2030. https://Oceandecade.org/vision-2030/
WHO13 (2023) Des Moines Public Schools marine biology teacher brings new discovery to classroom [Video]. YouTube. https://youtu.be/RIWLgD74Q5I

Author Contribution and Biographical Profile

[1]*Gregory J. Barord: Conceptualization; Writing—Original Draft, Review, and Editing.
*Corresponding Author at gregory.barord@dmschools.org

Gregory J. Barord, PhD, received his B.S. in Marine Biology at Texas A&M University at Galveston (2005) and completed his M. Phil. and PhD at City University of New York Graduate Center (2015). He is the Marine Biology Instructor at Central Campus (Des Moines, Iowa USA) and a Conservation Biologist for the Save the Nautilus Organization. Greg serves in different leadership roles within Central Campus and at the national and international level for conservation initiatives. His research focus is nautilus biology, behavior, and conservation of nautiluses and including his students in this work.

7 Engaging Students with Ocean Technology: A Journey from Floats to Ocean Literacy

George I. Matsumoto, Katie M. Lodes, Teresa J. Kennedy, and Jennifer L. Magnusson

Abstract

The U.S. BGC-Argo (BioGeoChemical) profiling array is a transformative initiative designed to deepen our understanding of the Ocean's role in climate regulation and biogeochemical cycles. With over 500 robotic floats equipped with chemical and biological sensors, this array provides a sustained, open-access data stream, offering unprecedented insights into Ocean chemistry, carbon cycling, and biological processes. This chapter explores the deployment and impact of BGC-Argo floats in synergy with the Core-Argo and Deep-Argo systems, while also highlighting the array's educational potential through programs like the Global Ocean BioGeoChemical Array (GO-BGC) Adopt-a-Float. These initiatives advance Ocean Literacy by connecting classrooms and communities with real-time scientific data, fostering hands-on, inquiry-driven learning experiences. As a critical component of global Ocean monitoring, the BGC-Argo profiling array contributes to long-term scientific research and empowers the next generation of Ocean stewards through its educational outreach. The program's goals closely align with the United Nations Decade of Ocean Science for Sustainable Development (Ocean Decade) Challenge 10, which seeks to foster an Ocean-literate society capable of making informed decisions to ensure sustainable Ocean stewardship. This chapter introduces the BGC-Argo profiling array, explains how educators can join the Adopt-A-Float program, and provides lesson plans for integrating float data into curricula.

Keywords

Adopt-a-Float · BGC-Argo profiling array · Climate change · Ocean Literacy (OL) · One-Argo system · Real-time data in classrooms · STEM education · Marine curriculum · Marine science education

7.1 Introduction

The Ocean plays a crucial role in regulating Earth's climate and supporting complex biogeochemical processes, yet many aspects of its chemical and biological cycles remain poorly understood. The U.S. Biogeochemical Argo

G. I. Matsumoto (✉)
Monterey Bay Aquarium Research Institute (MBARI), Moss Landing, CA, USA
e-mail: mage@mbari.org

K. M. Lodes
St. Joseph's Academy, St. Louis, MO, USA
e-mail: klodes@stjosephacademy.org

T. J. Kennedy
The University of Texas at Tyler, Tyler, TX, USA
e-mail: tkennedy@uttyler.edu

J. L. Magnusson
MBARI GO-BGC, Victoria, BC, Canada

T. J. Kennedy (ed.), *Ocean Literacy: The Foundation for the Success of the Ocean Decade, Volume I*, Sustainable Development Goals Series, https://doi.org/10.1007/978-3-031-98145-6_7

(BGC-Argo) program, an extension of the international Argo array of profiling floats, is poised to transform our scientific understanding of oceanic biogeochemical cycles. By equipping floats with sensors that measure variables such as oxygen, pH, nitrate, chlorophyll, and suspended particles, BGC-Argo provides a sustained source of high-quality data on the Ocean's chemistry and biology, enhancing our ability to monitor and predict changes in marine ecosystems (Biogeochemical-Argo n.d.).

The array, consisting of over 500 robotic profiling floats equipped with cutting-edge chemical and biological sensors, is part of an international effort to deploy 1000 BGC floats worldwide. These floats are distributed in open Ocean waters deeper than 2000 meters, where they continuously monitor key variables such as carbon, nitrate, and oxygen from 2000 meters to the surface. This ambitious program marks a pivotal advancement in observing the Ocean's biogeochemical dynamics on a global scale, providing unprecedented insights into its complex systems.

The BGC-Argo profiling array represents a transformative tool, not only enhancing scientific research into the Ocean's role in climate change and biogeochemistry but also fostering greater public understanding of these critical processes. By making these data open source and accessible to researchers, educators, and the public, the array provides an invaluable tool for advancing Ocean Literacy. As the world faces the challenges of climate change, this trusted, sustained source of data offers a critical baseline for studying the Ocean's carbon cycle and its evolving chemical and biological processes.

In the classroom, data generated by BGC floats present a unique opportunity to engage students in real-world science. Educators can incorporate float data into their lessons, enabling students to track changes in the Ocean's chemistry and biology over time. By connecting students directly to this research, Ocean Literacy transforms from an abstract concept into a hands-on, inquiry-driven learning experience. Programs like Adopt-a-Float allow students to follow the journey of individual floats, analyze data, and explore the interconnected relationships between the Ocean and global climate systems. These programs and their data are accessible to formal and informal educators, as well as community programs, after-school clubs, and other educational venues. While formal educators currently make up the majority of users, over 10,000 students participate annually in the program, primarily from the United States. Similar initiatives are also being implemented in other nations, such as Canada (Adopt-a-Float Canada n.d.) and France (Ifremer n.d.).

7.1.1 The One-Argo System: A Unified Global Effort

The U.S. BGC-Argo array is a major component of the Argo network of approximately 4000 robotic floats (Riser et al. 2016). The Core-Argo array, sustained by the United States through long-term support from the National Oceanic and Atmospheric Administration (NOAA), has fundamentally transformed our understanding of Ocean heat content and sea level rise by measuring the temperature and salinity of the upper 2000 meters of the Ocean. Argo data have been used in over 6000 PhD theses or papers referenced in refereed journals since the program's inception and are incorporated in all major operational analyses of the Ocean's physical state (Morris et al. 2024).

The mission of the U.S. BioGeoChemical (BGC-Argo) profiling array is to maintain this robotic network of BGC floats at a global scale, providing unprecedented data streams that will drive transformative studies of biogeochemical cycling. These data not only deepen scientific understanding but also enhance public awareness, empowering the next generation of Ocean stewards to engage in meaningful action toward sustaining the health of the Ocean.

The U.S. GO-BGC array operates synergistically with a parallel, international array of BGC profiling floats, as outlined in the BGC-Argo Science and Implementation Plan (Biogeochemical-Argo Planning Group 2016). Together, these Core-Argo and BGC-Argo floats, as well as Deep-Argo floats and the BGC floats built and launched under the Southern Ocean Climate Carbon and Climate Observations and Modeling (SOCCOM) program, function as a unified system termed *One-Argo* (Roemmich et al. 2019).

This highly integrated system enables techniques such as machine learning to extend the relatively sparse BGC-Argo observations to much greater spatial resolution by linking them to the higher-resolution physical measurements of the Core-Argo array (Mignot et al. 2023; Sharp et al. 2023; Xue et al. 2024).

7.2 What Is a Float?

Floats are autonomous data-collection instruments that revolutionized how scientists gather information about the Ocean. Before their development, data on Ocean properties such as temperature, salinity, and nutrients were difficult to obtain because these measurements relied heavily on scientific research vessels or volunteers aboard commercial ships, limiting data collection to specific locations and timeframes.

Autonomous profiling floats changed this paradigm. These long, cylindrical instruments are designed to operate independently in the Ocean, gathering data on temperature, salinity, pressure, and in some cases, biogeochemical properties like dissolved oxygen, chlorophyll, and nutrients. Floats are equipped with sensors and advanced technology to collect data at various depths, typically descending to depths of 2000 meters or more before returning to the surface.

One of the float's most remarkable features is its ability to control buoyancy, allowing it to adjust its vertical position in the water column. While floats can navigate vertically, they drift horizontally with the Ocean's currents, making them ideal for studying Ocean circulation and large-scale processes. Once at the surface, floats transmit their data via satellite to researchers around the globe, providing near-real-time insights into the Ocean's physical and chemical conditions. See Fig. 7.1 for a diagram of a SOCCOM float with parts labeled.

Today, networks like the Argo Program and the Global Ocean Biogeochemical Array deploy thousands of floats worldwide, creating a comprehensive and continuous picture of Ocean dynamics. These floats are invaluable for understanding Ocean–atmosphere interactions, tracking climate change impacts, and supporting sustainable Ocean stewardship.

7.2.1 What Types of Floats Are Being Used by Researchers?

The Argo Float program, initiated in 2000, currently brings together scientists from 30 countries with the primary goal of collecting data about the world's Ocean to improve understanding of climate and weather (Scripps Institution of Oceanography n.d.). The program utilizes nearly 4000 autonomous floats worldwide, with some capable of descending to depths of 6000 meters. As older floats cease to operate, new ones are continuously deployed to maintain the network.

Most Argo floats collect temperature and salinity, while some also gather additional biogeochemical (BGC) data, providing critical insights into the Ocean's dynamics and its influences on the global climate system. Detailed information about BGC floats and the overall program, including classroom applications for educators, is available on the BGC-Argo website listed in the resources section of this chapter.

7.2.2 Southern Ocean Carbon and Climate Observations and Modeling (SOCCOM)

The Southern Ocean Carbon and Climate Observations and Modeling (SOCCOM) project is also augmenting conventional Argo floats with biogeochemical sensors to measure carbon (pH), nutrients (nitrate), and oxygen (Princeton University 2024). SOCCOM began in 2014, with the goal of improving our understanding of the Southern Ocean's role in global climate, particularly in terms of carbon cycling, Ocean circulation, and biogeochemistry. SOCCOM uses autonomous profiling floats equipped with biogeochemical sensors to gather data in the Southern Ocean and was the world's first large-scale biogeochemical Argo deployment (see Fig. 7.2). The project has increased the number of

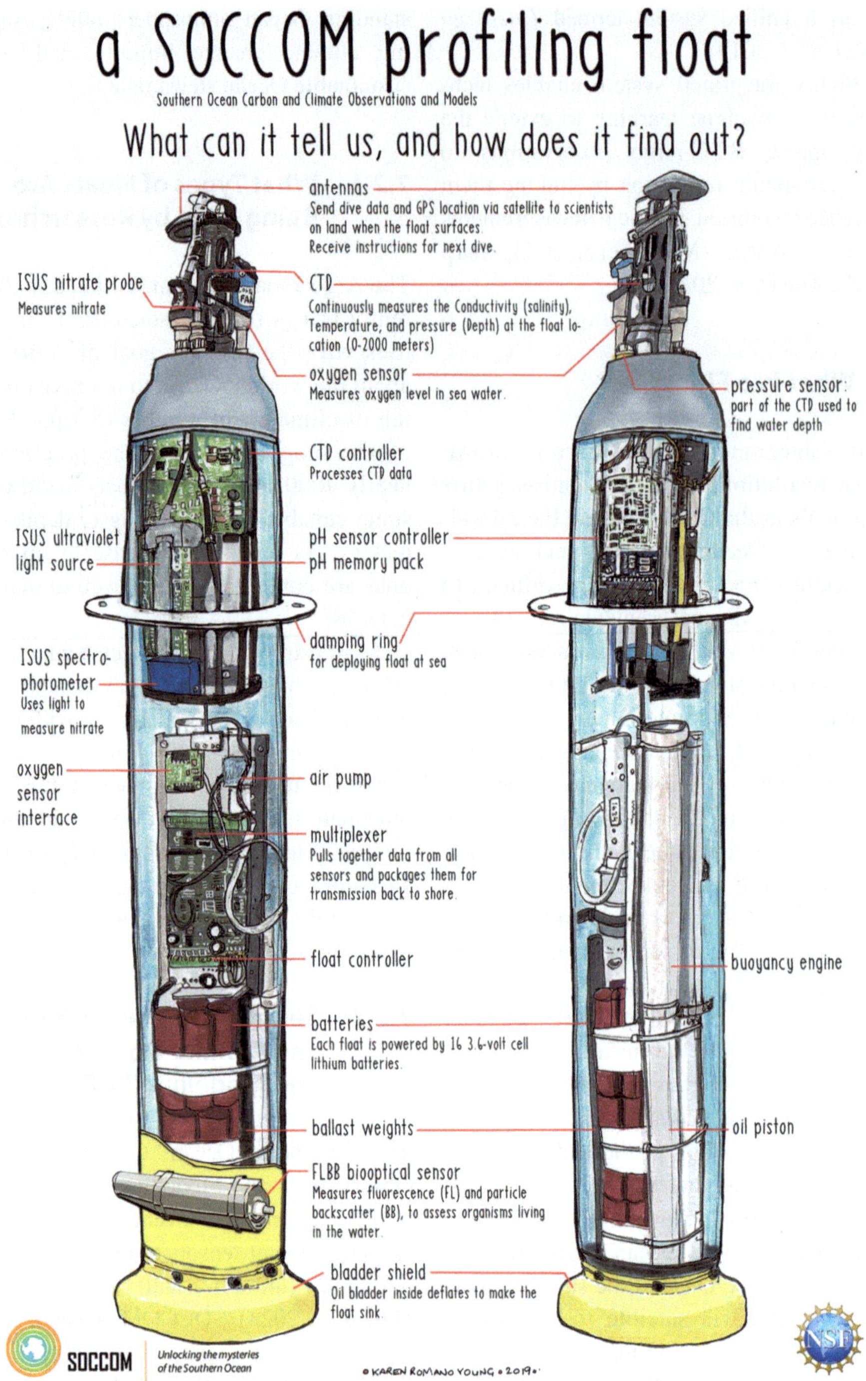

Fig. 7.1 SOCCOM float with parts labeled (Credit: K. R. Young, 2019)

biogeochemical measurements in the Southern Ocean by 10–30 times, with the greatest impact during the Southern Hemisphere winter when observations are typically scarce. All data collected are freely available to the public.

Over 100 SOCCOM floats are currently operating in the Southern Ocean, conducting millions of biogeochemical measurements, including in regions and during seasons that were previously unsampled. What sets SOCCOM apart is its inte-

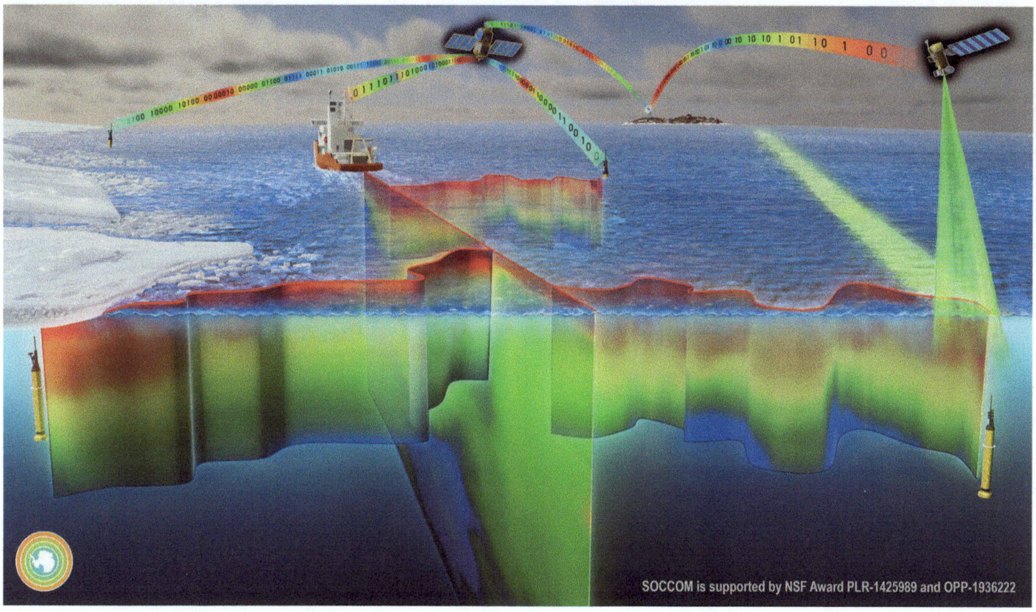

Fig. 7.2 Illustration of the SOCCOM float-based observing system and its intersection with ship and satellite observations. See the resources section for a link to the animation

gration of an innovative observational program with robust modeling efforts, as well as its collaboration with professional outreach partners to broaden the program's impact.

7.2.3 Global Ocean-BioGeoChemical Array (GO-BGC)

A critical part of the BGC-Argo program is the NSF-funded Global Ocean (GO-BGC) array that is developing and deploying over 500 BGC floats between 2020 and 2026. The GO-BGC program is a consortium of institutions including the Monterey Bay Aquarium Research Institute (MBARI), the University of Washington (UW), Scripps Institution of Oceanography (SIO), Woods Hole Oceanographic Institution (WHOI), and Princeton University.

Currently, the GO-BGC float array consists of three primary float types: BGC-APEX (developed by UW/MBARI), BGC-Navis (created by UW and WHOI), and the BGC-SOLO II/S2A (a collaboration between SIO and WHOI) (see Fig. 7.3).

The BGC-APEX is a well-tested, reliable platform that has been successfully deployed in large numbers through the SOCCOM and GO-BGC projects for over a decade. Most BGC-APEX floats are equipped with either five- or six-sensor configurations, measuring eight to nine variables or 12–13 variables, respectively. The six core Argo variables measured include Chlorophyll a fluorescence, oxygen, nitrate, pH, and suspended particles, along with temperature, salinity, and pressure.

The BGC-Navis is a commercially available float developed by SeaBird Scientific. It features a five-sensor configuration; however, the recently developed BGC-NavisV2 includes a six-sensor configuration with significantly improved performance. Additionally, the BGC-SOLO II, developed at SIO, has been licensed to Marine Robotic Vehicles (MRV) for commercialization as the BGC-S2A. These floats are also equipped with six sensors.

The use of three float types is a risk mitigation strategy, in case problems are identified with a particular float type/manufacturer, or a model is discontinued. Diversifying float types throughout the Core-Argo network has been a key focus of

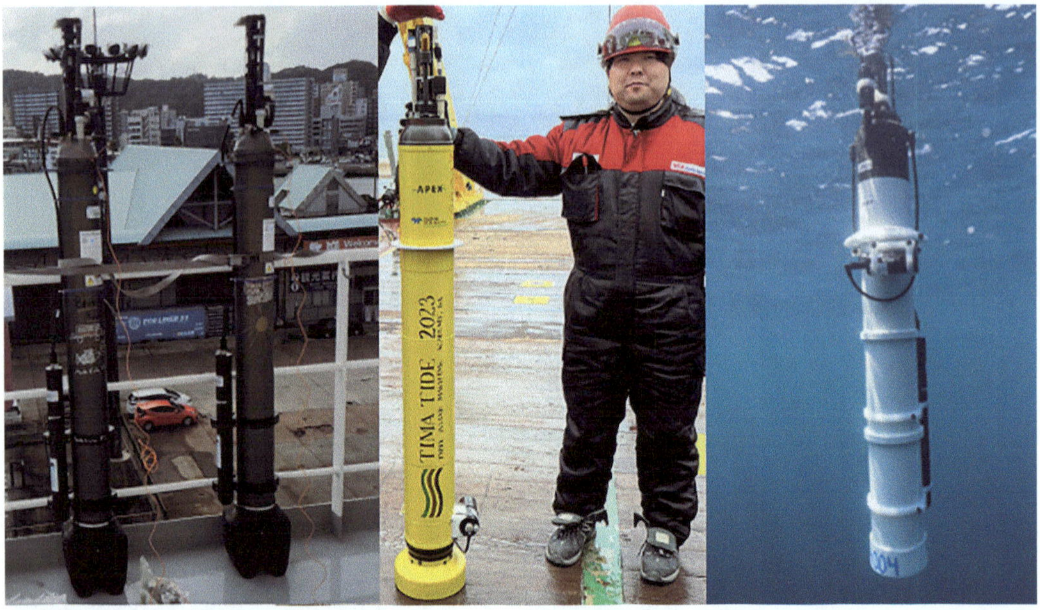

Fig. 7.3 Images of the three primary GO-BGC float types currently being deployed. From left to right: Navis V2 floats undergoing dockside checks before the research vessel departs (*Photo by A. Meyer, 2023*); an Apex float ready for launch on board the IBRV Araon (*Photo by J. L. Magnusson, 2023*); and a Solo II float launched at sea (Photo by Scripps Institution of Oceanography, UC San Diego, 2024)

the program. In 2026, the GO-BGC team plans to build 68 Apex/50 Navis/59 SOLO floats.

The synergy between BGC-Argo and Core-Argo provides a robust framework for understanding the Ocean's chemical and biological processes. This real-time data stream not only advances scientific research but also serves as a powerful educational tool through programs like Adopt-a-Float, enabling students and educators to engage with cutting-edge oceanography in their own classrooms.

7.2.4 Adopt-A-Float Program

The Adopt-a-Float initiative builds on international programs and the earlier SOCCOM program. Importantly, there is no cost or obligation for participants who adopt a float. As of 2025, there have been adoptions from all 50 states, along with American Samoa, Puerto Rico, and 13 other countries (UK, Canada, Germany, Japan, Malaysia, Ireland, Australia, Scotland, Vietnam, Chile, Costa Rica, Saudi Arabia, Iceland). As data collected by the BGC floats deepens our understanding of Ocean biogeochemical processes, it also becomes a powerful educational tool, enabling students to engage with the same real-time data used by scientists. To facilitate data analyses and classroom implementation, lesson plans developed by educators through the MBARI EARTH program are available on the GO-BGC educator website (refer to the resources section of this chapter for more information).

The program serves as an effective engagement strategy for classrooms, clubs, or entire school communities. Participants name their float, create a drawing or logo for the float, and access its data, along with data from other floats, through a specially designed access table for adopted floats. For example, St. Joseph's Academy, an all-girls, Catholic school in suburban St. Louis, Missouri, where Marine Science is offered every other year as a semester-elective course, adopted its first float in the fall of 2018. Although the float deployment did not match up with their Marine Science class, a dozen students still submitted designs for *Angelfish* (the school mascot is the Angels). *Angelfish* made 24 successful dives before ceasing communications in April of 2019. To ensure con-

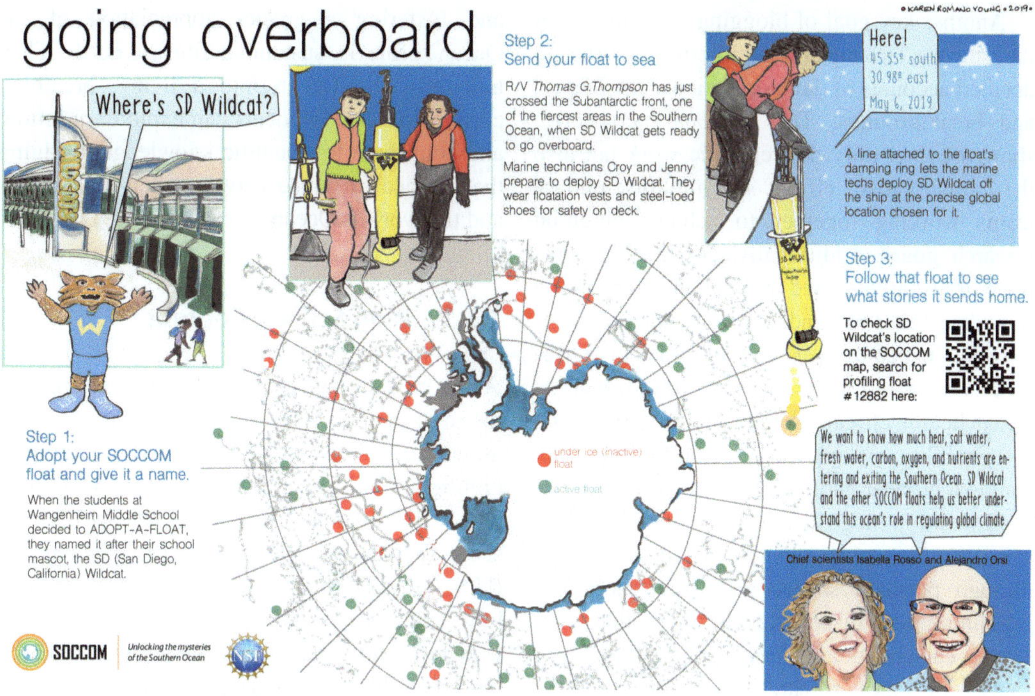

Fig. 7.4 *"Going Overboard"*—Following a SOCCOM Float from School to Sea (Illustration by K. R. Young, 2019)

tinuous access to float data for classroom use, St. Joseph's Academy has adopted three additional floats over the past three years. *Angel Shark* was deployed in February 2022 in the South Atlantic, *Angel Ice Cream Float* was launched in January 2023 in the Southern Ocean, and *Sea Angel* (named after the planktonic organism) was launched in November 2023 in the Pacific Ocean, providing the school with floats in three different Ocean basins. After completing 44 successful cycles, *Angel Ice Cream Float* was under ice from February to December 2024. After it resurfaced, it immediately began uploading profiles that students will examine during the spring 2025 semester and beyond. Fig. 7.4 describes the Adopt-a-Float process, as students follow the journey of their float from naming to deployment and beyond. In this example, students at Wangenheim Middle School named their SOCCOM float after their school mascot, the *SD Wildcat*. The graphic highlights each step of engagement—from naming and launching to tracking and interpreting float data—as a way to connect classrooms with ongoing Southern Ocean research.

7.3 Research Expedition Blogs

Satellite communications technology onboard scientific research vessels has developed significantly over the past few decades, resulting in expeditions that have unprecedented ability to remain connected to the internet, even in the most remote locations. As a result, the science teams onboard most research cruises post on-going Expedition blogs during the duration of their activities at sea. These blogs are diverse, aiming to engage, educate, and document scientific endeavors in an accessible way. One of their primary objectives is education and outreach, as blogs help connect complex scientific research to students, teachers, and civil society, including parents and community members. By breaking down intricate concepts, they make real-world science more understandable, motivating students to explore careers in Science, Technology, Engineering and Mathematics (STEM). For educators, these blogs offer tangible examples that can be integrated into lessons, linking classroom learning with real-world applications.

Another key goal of blogging is to document the science driving research cruises, offering detailed accounts of daily operations, challenges, and breakthroughs. This personal perspective showcases not only the scientific work but also the collaborative efforts of multidisciplinary teams working together to achieve common research goals. Additionally, cruise blogs promote transparency and public engagement by providing real-time updates, allowing audiences to follow expeditions as they unfold. This raises awareness about critical global issues such as Ocean health, climate change, and environmental sustainability, making the research findings relevant to a broader audience.

Expedition blogs also build a compelling narrative that humanizes science, making it relatable through personal stories and experiences. By explaining key discoveries in ways that non-experts can understand, they highlight the significance of the research and its broader environmental implications. Cruise blogs encourage participation by promoting initiatives like the Adopt-a-Float program, inviting readers, especially teachers and their students, to engage directly with oceanographic research. These efforts inspire curiosity, foster active involvement, and create meaningful connections between audiences and ongoing scientific exploration.

This dynamic approach fosters deeper engagement, as students are no longer passive recipients of textbook information but active participants in ongoing research activities. By tracking the movement of GO-BGC floats, analyzing data from different depths, and understanding the physical and chemical properties of the Ocean, students gain hands-on, inquiry-based learning experiences. This bridge between the Ocean and the classroom enhances their understanding of global environmental challenges, cultivates curiosity, and encourages critical thinking. It also instills a sense of connection and responsibility toward the Ocean, empowering students to become informed advocates and stewards of marine environments.

By leveraging real-time science, cruise blogs create educational pathways that align with existing learning goals, promoting STEM education and fostering a greater appreciation of the Ocean's role in sustaining life on Earth. They transform the Ocean into a living, breathing classroom where students can explore, question, and contribute to scientific knowledge, bridging the physical distance between their classrooms and the world's Ocean.

7.3.1 Blog Science Content

While cruise blogs primarily document ongoing research activities, they also capture various aspects of life at sea. The following examples highlight activities from the 2024 52-day GO-SHIP research cruise along the A13.5 transect aboard the *R/V Marcus G. Langseth* (U.S. GO-SHIP 2024). These examples show how teachers and students can engage with activities on board, making oceanographic research more tangible and engaging. By integrating real-time science into classrooms through cruise blogs, teachers bridge the gap between the Ocean and the classroom, transforming abstract scientific research into an interactive, engaging, and accessible learning experience. Classrooms following the day-to-day progress of oceanographic expeditions gain a front-row seat to scientific discovery. This connection not only demystifies the complexities of the Ocean but also makes the scientific process tangible, allowing students to witness data collection, the challenges faced by researchers, and the real-world applications of oceanography.

Two of the A13.5 blogs described GO-BGC float deployments, which are vital for collecting data on the Ocean's temperature, salinity, and oxygen levels as they drift with Ocean currents (Kennedy 2024a; Kennedy et al. 2024b). The blog entries provided teachers with an authentic opportunity to introduce students to real-world technology, explore how floats operate, and gain valuable insights into how scientists collect data from remote and often inaccessible parts of the Ocean. Floats became more than just tools, they transformed into compelling stories of scientific exploration, revealing the mysteries of vast, unseen environments. Students followed their

Fig. 7.5 Kristian Furnes (A13.5 science team member), along with Todd Jensvold and Aaron Martin (LDEO) from the R/V *Marcus G. Langseth*, deploy ***Crubotics Floatosynthesis***, the Argo float adopted by Pensacola Catholic High School (Photos by T. J. Kennedy, UT Tyler, 2024)

float from the moment it was deployed and continue to monitor its journey as it collects valuable data (Fig. 7.5 shows the deployment of the *Crubotics Floatosynthesis*).

Students participating in the Adopt-A-Float program also receive updates on their float's location and collected data, fostering a connection to a global research initiative. This makes abstract scientific concepts more concrete, as students engage with real data from their adopted float and explore its implications for broader scientific questions. The experience can be seamlessly integrated into classroom projects, where students analyze data, make predictions, and create reports or presentations based on their findings. By examining data from their floats, students can compare conditions at different depths, gaining a deeper understanding of the physical and chemical changes that occur as the float descends into the Ocean.

This kind of inquiry-based learning allows students to think like scientists, asking their own questions and using real data to find answers. It inspires curiosity about the Ocean's unseen world. Students can explore questions such as: What happens to water as it gets deeper? Why does the temperature change with depth? How is the deep Ocean connected to the surface? These questions spark discussions about global Ocean circulation, the water cycle, and the impact of human activities on marine ecosystems, among other topics.

A common activity occurring during research cruises deploying CTD (Conductivity, Temperature, and Depth) instruments is to experiment with the effects of pressure by submerging Styrofoam cups (made from a type of plastic called polystyrene foam) during deep-sea deployments. During the A13.5 expedition, cups representing each Adopt-a-Float school monitoring the cruise were decorated with school names and logos and then sent to a depth just 10 meters above the Ocean floor (see Figs. 7.6 and 7.7). A blog post documented their journey to the depths of the Ocean, sparking classroom discussions (Kennedy 2024b). Since extreme pressure changes can be difficult to conceptualize, this activity offered a concrete demonstration of the immense forces present in the deep Ocean, pressures that each school's float would also experience.

As the cups descend with the CTD to depths where pressure can reach hundreds of times the atmospheric pressure at sea level, the air inside the polystyrene structures is compressed, causing the cups to shrink dramatically. Upon return to the ship, the once-normal-sized cups emerge as tiny, intricately decorated souvenirs of the deep sea. This striking demonstration makes the abstract concept of pressure hands-on and memorable for students, allowing them to touch and observe it firsthand when their cups return to their classroom.

By providing students with their own cup to hold, examine, and compare to its original size, they

Fig. 7.6 Adopt-a-Float school cups undergo pressure experiment during CTD deployment along GO-SHIP's A13.5 transect. From left to right: Kristy McTaggart, NOAA's Pacific Marine Environmental Laboratory (PMEL) and Christian Saiz, University of Miami Cooperative Institute for Marine and Atmospheric Sciences (CIMAS) attach the mesh bag of Styrofoam cups to the CTD rosette. The center picture shows the retrieval of the school's shrunken cups by Teresa Kennedy, University of Texas at Tyler. Anna Bruno, University of Washington Cooperative Institute for Climate, Ocean, and Ecosystem Studies (UW/CICOES), and Teresa Kennedy compare pre-and post cup sizes. (Photos by Z. Erickson, NOAA PMEL, 2024)

Fig. 7.7 Shrunken cups. From left to right: Cups before and after descent on GO-SHIP A13.5 *(Photos by J. Anderson, Earth and Space Research (ESR), 2024)*; and shrunken cups made by the science team for 11 Adopt-a-Float school teams (Photo by T. J. Kennedy, 2024)

develop a tangible connection to the challenges and realities of deep-sea exploration. Their shrunken cup serves as a physical representation of the Ocean's power, offering a lesson that transcends textbook explanations. This activity also fosters a sense of ownership and excitement about scientific discovery, as students see the real-world effects of oceanographic research on something they designed, effectively bridging the gap between the classroom and the deep Ocean (see Fig. 7.8).

Research cruises provide powerful educational experiences that bring Ocean science directly into classrooms. Whether through tracking a float, exploring data from different depths, or engaging students in hands-on projects, these resources make global research accessible and engaging. It's not just about learning facts; it's about connecting students to the world around them and showing them that they can play a part in discovering what's still unknown about the Ocean.

The activities described, occurring through the Adopt-a-Float program and the interactive nature of expedition blogs, play a crucial role in advancing Ocean Literacy by connecting the public directly to the science and exploration of the Ocean. Ocean Literacy goes beyond basic knowledge, helping people understand the Ocean's influence on our lives and how human actions affect its health. These activities provide an invaluable hands-on approach to learning, allowing students and teachers to engage with real-time scientific data, monitor oceanographic instruments like floats, and follow the journeys of research expeditions. This level of involvement helps demystify the complex science of the Ocean, making it more approachable and relevant to everyday life.

Fig. 7.8 Classroom engagement. Left to right: Students decorating their cups in the classroom and their final products prior to being sent out to sea (Photos by K. M. Lodes, 2024)

Furthermore, these initiatives align closely with the objectives of the United Nations (UN) Decade of Ocean Science for Sustainable Development (Ocean Decade), which seeks to foster a global movement for Ocean stewardship, involving students as well as broader civil society (Kennedy 2024c; Kennedy et al. 2024a). The Ocean Decade aims to generate actionable knowledge and solutions to support the sustainable development of the Ocean, promoting public engagement, education, and awareness as key drivers of change. Programs like Adopt-a-Float directly contribute to these goals by immersing students in real-world scientific efforts, empowering them to think critically about environmental issues, and encouraging them to be part of the solution.

By integrating these activities into classrooms, educators help cultivate the next generation of Ocean stewards. Students learn not only about the physical and chemical properties of the Ocean but also about broader environmental challenges like climate change, Ocean acidification, and biodiversity loss. These lessons contribute to the Ocean Decade's vision of a clean, healthy, and resilient Ocean by fostering a sense of responsibility and connection to the Ocean among learners at all levels. Through these programs, students not only acquire knowledge but also develop the skills and awareness necessary to address future Ocean-related challenges, supporting the Ocean Decade's mission to transform how we understand and manage the Ocean for the benefit of humanity and the planet.

Ultimately, these activities provide an invaluable platform for educators and students to engage with the Ocean in meaningful ways. They promote a deeper understanding of the interconnectedness between human activities and Ocean health, while also fulfilling the Ocean Decade goal of creating a more Ocean-literate and engaged global community. As students analyze real-time data, observe the Ocean's dynamic systems, and participate in conversations about its future, they contribute to the global movement toward sustainable Ocean stewardship.

7.4 GO-BGC floats and Ocean Literacy

There are seven Ocean Literacy essential principles included in the Ocean Literacy Scope and Sequence for grades K–12 (National Marine Educators Association 2010, 2024), and the biogeochemical float program directly applies to all of them. The floats are deployed and drift around the world's Ocean enabling students to clearly see that the ***Earth has one big Ocean with many features*** (Principle #1).

Although a Google search of "Ocean features" yields countless images of structural features such as continental shelves, abyssal plains, and trenches, biogeochemical features are equally important. Using float data available on the GO-BGC website, students can develop basic graphing skills while exploring the biogeochemical features of the Ocean. For example, students can experiment with adjusting the axes of a graph using the same dataset and create color-coded graphs for each dive (see Fig. 7.9).

Another example of how ARGO floats gather data to support the vast Ocean with its many features comes from St. Joseph's Academy. The real-time data provided by the floats allows students to validate their discussions on how certain physical parameters of the Ocean change with depth. Students select different floats and use the MBARI data visualization program, AdoptAFloatViz 6.0, to create graphs showing temperature, salinity, dissolved oxygen and pH relative to depth. Refer to the MBARI float tables in the resources section for more details. Figure 7.10 (on the left-hand side) provides an example of student work showing a comparison of temperature, salinity, dissolved oxygen, and pH. At a general level, students compare the overall trends in Ocean water with depth discussed in class lectures and predict whether three Ocean basins (Southern, South Atlantic, or

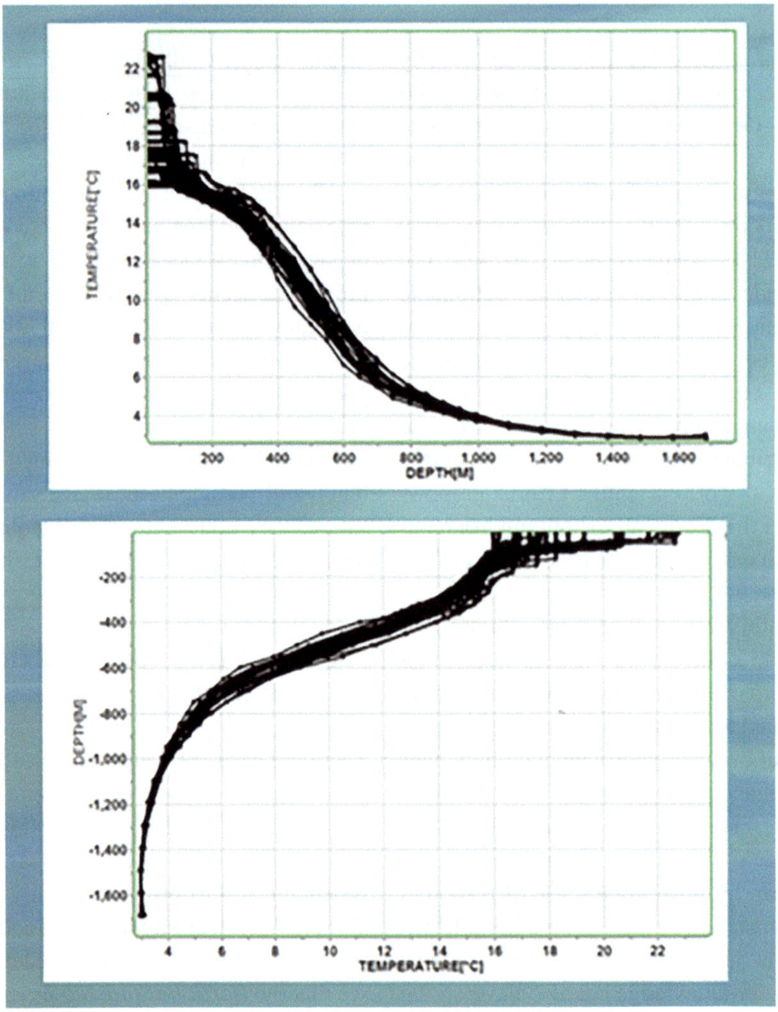

Fig. 7.9 Using GO-BGC float data to explore basic graphing techniques and the impact of axes choices on graph construction

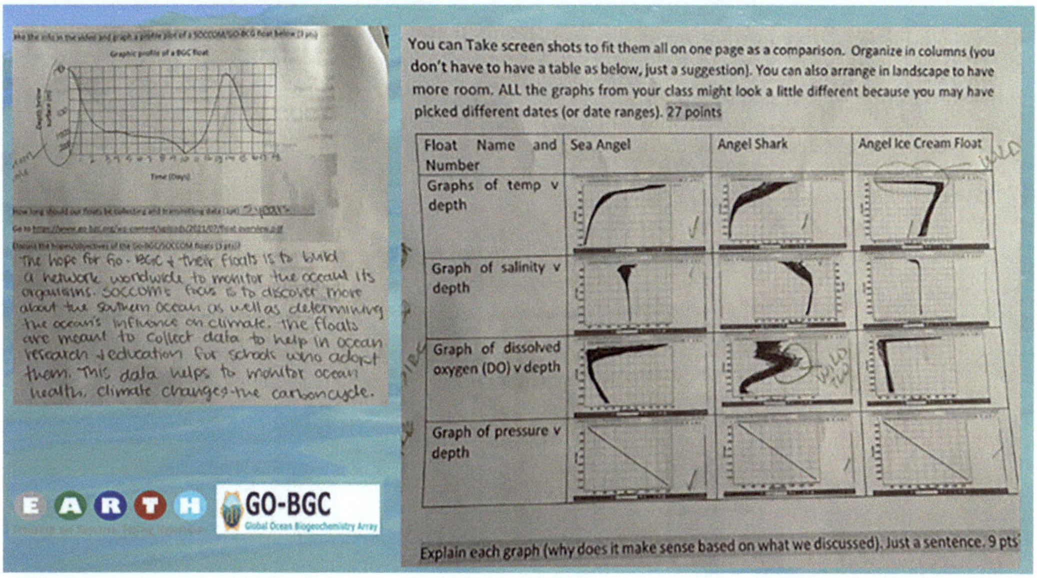

Fig. 7.10 Example of student work showcasing both the graphing aspects of the lesson and the deeper understanding students have developed about Ocean processes

Central Pacific) will exhibit similar or different trends. Figure 7.10 (right-hand side) illustrates an example of student work.

The Ocean is largely unexplored (Principle #7), and despite extensive international effort, there are still regions of the Ocean where no measurements exist. Data from biogeochemical floats are helping us to gain a deeper understanding of how ***The Ocean is a major influence on weather and climate*** (Principle #3) and reveals that ***The Ocean and humans are inextricably interconnected*** (Principle #6). The expedition blogs on the GO-BGC website highlight Principle #6, while new lesson plans, such as *Hurricane Detectives: Diving into the Ocean's Secrets* (featured in the Resources section under MBARI Lesson Plans), are being added to the demonstrate Principle #3 using float data. Additionally, the MBARI lesson *Diving into the Ocean's Secrets* enables students to investigate the connection between deep-water hurricanes and strong storms with changes in Ocean features, such as upwelling of temperature and nutrients. By engaging with real-world phenomena, students enhance their scientific literacy (Kennedy and Fields 2023, 2024).

Argo floats have been deployed for over 20 years, primarily providing temperature and salinity data. In contrast, biogeochemical floats expand on this by measuring additional parameters such as pH, nitrate, chlorophyll, suspended particles, light, and derived variables like dissolved inorganic carbon, pCO2, and total alkalinity. These additional measurements deepen our understanding of how ***The Ocean makes Earth habitable*** (Principle #4), how ***The Ocean supports a great diversity of life and ecosystems*** (Principle #5), and how ***The Ocean and life in the Ocean shape the features of EARTH*** (Principle #2).

Data from ARGO floats can be integrated into ecology curricula to demonstrate the relationship between biotic and abiotic factors. Students begin by selecting a float profiling in a region of the Ocean that interests them. Next, they research organisms found in that general area. Finally, they analyze a parameter from the float (as shown in Fig. 7.11) and match each organism to the depth where it would be able to survive. With the availability of backscatter and chlorophyll data, connections can now be made between primary productivity and algal blooms.

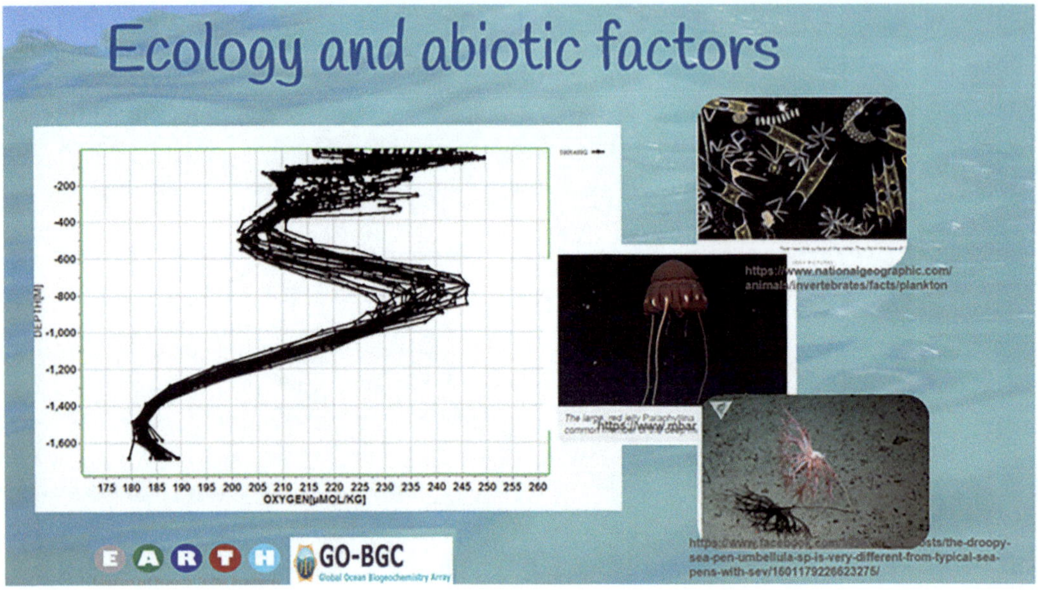

Fig. 7.11 Slide from an educator presentation during the EARTH workshop, illustrating how students utilized data from BGC floats to explore the relationship between Ocean life and conditions in the upper 2000 meters of the water column

Table 7.1 Sample BGC Float Lesson Plans

Lesson Title	Short description	Link to lesson
The Ocean Does the Robot	In this introductory lesson, students will become familiar with accessing and reading float data. Working in teams, students will use the Argo Fleet Monitoring website to access information about a float of their choice, analyze and interpret data, and communicate findings in a creative way. This lesson can be used as a springboard for asking more complex questions and using float data to find answers.	https://www.mbari.org/lesson-plan/the-ocean-does-the-robot/
Our Ocean is Trending	This lesson focuses on what students can learn about the Ocean's oxygen and chlorophyll levels by using Argo and GO-BGC, and SOCCOM Floats. Students learn how to access data from GO-BGC floats in order to analyze and compare properties such as dissolved nitrates, oxygen and chlorophyll levels. Through observing the diversity of factors that are found throughout the Ocean basins, students will be able to correlate how the Ocean supports such a diversity of life and ecosystems.	https://www.mbari.org/lesson-plan/our-ocean-is-trending/

(continued)

Recent research highlights Ocean Literacy as essential for promoting Ocean sustainability and influencing human behavior (Shellock et al. 2024). Using data from biogeochemical floats offers an excellent opportunity to teach the Ocean Literacy Principles through real-time data. Schools that adopt a float can even incorporate data collected from their own deployed float, creating a direct connection to ongoing scientific research.

Many additional lessons have been developed to enable educators and their students to access and explore the available data at various levels. Table 7.1 below provides a list of available lesson plans designed to work with BGC float data. Grade levels are not specified, allowing educators to determine which lessons best suit their students.

Table 7.1 (continued)

Lesson Title	Short description	Link to lesson
Phytoplankton + Productivity 5E	This lesson ties Ocean/float data to the concept of primary productivity by looking for connections between chlorophyll-a and dissolved oxygen.	https://www.mbari.org/lesson-plan/phytoplankton-productivity-5e/
Ocean Data GOES from Space ARGO the sea	In this lesson, students compare and contrast satellite data to float data. As a result, students understand the importance of collecting data from multiple sources and how limiting factors of each data set causes gaps in their interpretation of the data. The first satellite was launched in 1962 and collected three types of data, with NOAA launching the first Ocean-specific satellites in 1970. Floats were first built in 1980, and ARGO launches started in 1999, which collected Ocean data.	https://www.mbari.org/lesson-plan/ocean-data-goes-from-space-argo-the-sea/
Ocean Checkup—How Do You Measure the Health of the Ocean	This lesson focuses on Argo floats, monitoring Ocean health, dissolved oxygen, and chlorophyll a. Students compare dissolved oxygen and chlorophyll a data collected at three different locations by Argo floats. Students discuss the relationship between Chlorophyll A (phytoplankton) and oxygen in the Ocean, and brainstorm how the information would be useful in determining Ocean health.	https://www.mbari.org/lesson-plan/ocean-checkup-how-do-you-measure-the-health-of-the-ocean/
Graphing Temperature	In this lesson, students construct, analyze, and interpret real data from a profiling float in the Ocean. Students explore Ocean temperature trends and climate change developing critical thinking, geographical, mathematical, graphing, analyzing, and interpretation skills.	https://www.mbari.org/lesson-plan/graphing-temperature/
Graphing pH	This lesson explores the topic of Ocean acidification. Students construct, analyze, and interpret a pH graph, developing analytical thinking skills.	https://www.mbari.org/lesson-plan/graphing-ph/
Taking a Dip in the Ocean Data Pool	This lesson is an introduction to the importance of float data, how data are used to gather and communicate information about our day-to-day environment and why we (as students/teachers/scientists/community members) should care. Using temperature data from Ocean floats, this lesson helps to launch the conversation of how and why data are used to understand our changing environment, serving as the gateway to understanding how data communicates information about our everyday environment.	https://www.mbari.org/lesson-plan/taking a dip in the-ocean-data-pool/
LEGO Build a Float	Students walk through components of a float, looking at the process the float goes through to be ready to go out to sea. They create LEGO models of their float.	https://www.mbari.org/lesson-plan/lego-build-a-float/
Taylor on Tour	Students use chlorophyll a data from vertical profiling floats to understand and predict the distribution of Pacific albacore tuna.	https://www.mbari.org/lesson-plan/taylor-on-tour/
Hurricane Detectives: Diving into the Ocean's Secrets	Students study hurricanes by integrating various elements such as tracking, upwellings, Chlorophyll-a, and GO-BGC floats to gather and analyze data. In this lesson, students study hurricanes by integrating various elements such as tracking, upwellings, Chlorophyll-a, and the use of GO-BGC floats to gather and analyze data. Students use real-world data and understand how these elements can be used to analyze hurricanes in different oceanic zones as well as different Ocean basins.	https://www.mbari.org/lesson-plan/hurricane-detectives-diving-into-the-oceans-secrets/

7.5 Conclusion

The U.S. Global Ocean-BioGeoChemical (BGC) profiling array, operating in synergy with the global One-Argo system, represents a pivotal advancement in oceanographic research and public engagement. Providing real-time, open-access data on the Ocean's chemical and biological processes, the BGC array not only contributes to long-term scientific understanding but also plays a critical role in fostering Ocean Literacy, aligning with the goals of UN Decade of Ocean Science for Sustainable Development Challenge 10. This challenge calls for developing and fostering an Ocean-literate society capable of making informed decisions to ensure sustainable stewardship of the Ocean for future generations.

The involvement of educators is essential in advancing this mission. Through programs like GO-BGC Adopt-a-Float, teachers can bring the Ocean into their classrooms, providing students with unique, hands-on learning experiences that go beyond traditional teaching methods. Working with real-time data from BGC floats enables students to develop critical thinking skills, deepen their understanding of the Ocean's role in regulating climate and ecosystems, and engage with global environmental challenges in a meaningful way. These experiences not only enrich their academic journey but also empower them to become the next generation of informed Ocean advocates and stewards.

As educators integrate float data and marine science into their curricula, they bridge the gap between theoretical knowledge and practical, real-world science, inspiring curiosity and responsibility toward the Ocean. Participation in the global BGC program provides teachers with access to lesson plans and tools that enable students to track float data, analyze biogeochemical trends, and contribute to global Ocean conservation efforts. This engagement directly supports the goals of Ocean Decade Challenge 10, nurturing a more informed and engaged public that is prepared to protect and sustain the health of the Ocean for future generations. Educators who actively engage in programs like Adopt-a-Float not only equip students with the tools to understand the Ocean but also help foster the next generation of Ocean stewards and informed citizens. Continued participation in these global efforts is crucial to ensuring a sustainable future for our Ocean and the planet given the role these students will play in shaping future policies and innovations.

7.5.1 Summary

Key concepts explored in this chapter include:

- The U.S. BioGeoChemical (BGC-Argo) profiling array is poised to transform our scientific understanding of biogeochemical cycles by providing a sustained source of high-quality data on the Ocean's chemistry and biology.
- The deployment and impact of BGC-Argo floats in synergy with the Core-Argo and Deep-Argo systems provides the world with a comprehensive global Ocean dataset.
- The combined array has educational potential through programs like the Global Ocean BioGeoChemical Array (GO-BGC) Adopt-a-Float.
- These initiatives advance Ocean Literacy by connecting classrooms and communities with real-time scientific float data as well as cruise details from expedition blogs, fostering hands-on, inquiry-driven learning experiences.
- As a critical component of global Ocean monitoring, the BGC-Argo profiling array contributes to long-term scientific research and empowers the next generation of Ocean stewards through its educational outreach.
- GO-BGC's goals align closely with Ocean Decade Challenge 10, which seeks to foster an Ocean-literate society capable of making informed decisions to ensure sustainable Ocean stewardship.
- Educators can join the GO-BGC program and access lesson plans designed to integrate float data and marine science into their classrooms.

7.5.2 Recommended Resources

The following resources provide further insights and support on this topic:

- The Argo Float program: https://argo.ucsd.edu/
- Adopt a Float: https://www.go-bgc.org/outreach/adopt-a-float
- Adopted Float Data Visualization: https://www.mbari.org/data/adopt-a-float-data/ and https://www.go-bgc.org/outreach/adopted-floats-tables
- BGC-SOLO II developed at SIO and licensed to MRV: https://www.mrvsys.com/
- Biogeochemical Argo: https://biogeochemical-argo.org/
- Float Status: https://argo.ucsd.edu/about/status/
- Float Types: https://argo.ucsd.edu/how-do-floats-work/float-types/
- GO-SHIP: http://www.go-ship.org/
  - GO-SHIP Evolve: https://oceandecade.org/actions/go-ship-evolve/
- GO-SHIP A13.5: https://usgoship.ucsd.edu/2024/03/23/weekly-reports-from-a13-5/
  - R/V *Marcus G. Langseth*: https://lamont.columbia.edu/research-divisions/marine-large-programs/office-marine-operations/langseth
- Monterey Bay Aquarium Research Institute (MBARI) EARTH Workshops: https://www.mbari.org/education/education-research-testing-hypotheses-earth/about-earth/
- Monterey Bay Aquarium Research Institute (MBARI) Lesson Plans: https://www.mbari.org/education/education-research-testing-hypotheses-earth/lesson-plans/
- One Argo Program: https://argo.ucsd.edu/oneargo/
- Seabird Scientific: https://www.seabird.com/
- Southern Ocean Carbon and Climate Observations and Modeling (SOCCOM): https://soccom.princeton.edu/
  - SOCCOM observing systems animation: https://www.youtube.com/watch?v=30XfCTb6ja0
  - SOCCOM Biogeochemical Floats animation: https://www.youtube.com/watch?v=tVoOq93hDuU

Acknowledgements Funding for this work has been provided by NSF Award 1946578 to GO-BGC and NSF Awards (PLR-1425989, OPP-1936222, and OPP-2332379) along with supplemental funds from NASA, NOAA, and the International Argo Program to SOCCOM. Support for GIM comes from the David and Lucile Packard Foundation.

TJK thanks the GO-SHIP A13.5 Chief Scientist Dr. Zach Erickson (NOAA Pacific Marine Environmental Laboratory—PMEL) and Co-Chief Scientist Dr. Jesse Anderson (Earth and Space Research—ESR), as well as oceanographer Kristy McTaggart (NOAA PMEL) and the entire science team, for their leadership and scientific guidance. Special thanks also to the crew of the R/V *Marcus G. Langseth*, especially Captain Breckenridge Crum, Todd Jensvold, Aaron Martin, and Koray Ergun from Lamont-Doherty Earth Observatory (LDEO), for their support during the 2024 South Atlantic research cruise. It was an honor to be part of the dedicated A13.5 science team and to work alongside the exceptional ship crew, whose expertise and collaboration made this expedition such a meaningful and memorable experience. TJK also gratefully acknowledges George Matsumoto and Jennifer Magnusson (MBARI) for their generous assistance in uploading and maintaining the expedition blog posts during the cruise. Their support ensured that daily updates from the ship reached classrooms and communities around the world, expanding the reach and impact of this Ocean learning experience.

References

BGC-Argo Canada (n.d.) Adopt-a-float Canada. https://adopt-a-float.ocean.dal.ca/

Biogeochemical-Argo (n.d.) An extension of the Argo program to include biogeochemical observations. https://biogeochemical-argo.org/

Biogeochemical-Argo Planning Group (2016) The scientific rationale, design and implementation plan for a Biogeochemical-Argo float array. https://biogeochemical-argo.org/cloud/document/relevant-reports/BGC-Argo_Science_Implementation_Plan.pdf

Ifremer (n.d.) Adopt-a-Float. https://www.ifremer.fr/fr/la-mediation-scientifique-l-ifremer/nos-offres-pedagogiques/adoptafloat

Kennedy TJ (2024a) GO-BGC floats. Global Ocean Biogeochemistry Array. Monterey Bay Aquarium Research Institute. https://www.go-bgc.org/expedition/south-atlantic-2024/go-bgc-floats

Kennedy TJ (2024b) Ocean depths: Exploring extreme pressure changes. Global Ocean Biogeochemistry Array. Monterey Bay Aquarium Research Institute. https://www.go-bgc.org/expedition/south-atlantic-2024/ocean-depths

Kennedy TJ (2024c) Advancing Ocean sustainability for climate-resilient change. In: Hufnagel L (ed) Globalization and sustainability: ecological, social and cultural perspectives (Chapter 2). IntechOpen, pp 7–30. https://doi.org/10.5772/intechopen.1005309

Kennedy TJ, Fields DL (2023) Augmenting upper and early learning STEM lessons through phenomenon-based learning. J High Educ Theory Pract 23(18):170–182. https://articlegateway.com/index.php/JHETP/article/view/6630

Kennedy TJ, Fields DL (2024) Preparing students for global qualifications through phenomenon-based learning projects. Int House J 53. https://ihworld.com/ih-journal/issues/issue-53/phenomenon-based-learning/

Kennedy TJ, Dufourneaud O, Castillo MP, Deane S, Durand J, Giaccone T, Jonda S, Mascarenhas T, Mtonga C, Ng K, Palazot S, Paul S, Salazar ER, Santisteban R, Shuwen X, Silva A, Solis Rivera V, van't Land H, Zhou Y (2024a) Ocean Decade manual for non-governmental organizations (NGOs). UNESCO-IOC, Paris. [The Ocean Decade Series, 54]. https://unesdoc.unesco.org/ark:/48223/pf0000392198

Kennedy TJ, Sandborn D, Anderson J (2024b) GO-BGC floats Part 2: What do the data tell us? Global Ocean Biogeochemistry Array. Monterey Bay Aquarium Research Institute. https://www.go-bgc.org/expedition/south-atlantic-2024/go-bgc-floats-part-2

Mignot A, Claustre H, Cossarini G, D'Ortenzio F, Gutnecht E, Lamouroux J, Lazzari P, Perruche C, Salon S, Sauzède R, Taillandier V, Teruzzi A (2023) Using machine learning and Biogeochemical-Argo (BGC-Argo) floats to assess biogeochemical models and optimize observing system design. Biogeosciences 20(7):1405–1422. https://bg.copernicus.org/articles/20/1405/2023/

Morris T, Scanderbeg M, West-Mack D, Gourcuff C, Poffa N, Udaya Bhaskar TVS, Hanstein C, Diggs S, Talley L, Turpin V, Liu Z, Owens B (2024) Best Practices for Core Argo floats – part 1: Getting started and data considerations. Front Mar Sci 11:1358042. https://doi.org/10.3389/fmars.2024.1358042

National Marine Educators Association (2010) Ocean Literacy scope and sequence for grades K–12. *Ocean Literacy Campaign Special Report #3.* https://www.marine-ed.org/ocean-literacy/special-report

National Marine Educators Association (2024) Ocean Literacy: The essential principles and fundamental concepts of Ocean science for learners of all ages (Version 3.2). http://marine-ed.org/ocean-literacy/guide

Princeton University (2024) SOCCOM: Unlocking the mysteries of the Southern Ocean. Southern Ocean Carbon and Climate Observations Modeling. https://soccom.princeton.edu/

Riser S, Freeland HJ, Roemmich D, Wijffels S, Troisi A, Belbéoch M, Gilbert D, Xu J, Pouliquen S, Thresher A, Le Traon PY, Maze G, Klein B, Ravichandran M, Grant F, Poulain PM, Suga T, Lim B, Steri A et al (2016) Fifteen years of Ocean observations with the global Argo array. Nat Clim Chang 6(2):145–153. https://www.aoml.noaa.gov/phod/docs/nclimate2872.pdf

Roemmich D, Alford MH, Claustre H, Johnson K, King B, Moum J, Oke P, Owens WB, Pouliquen S, Purkey S, Scanderbeg M, Suga T, Wijffels S, Ziberman N, Bakker D, Baringer M, Belbeoch M, Bittig HC, Boss E et al (2019) On the future of Argo: A global, full-depth, multi-disciplinary array. Front Mar Sci 6. https://doi.org/10.3389/fmars.2019.00439

Scripps Institution of Oceanography (n.d.) What is argo? https://argo.ucsd.edu/

Sharp JD, Fassbender AJ, Carter BR, Johnson GC, Schultz C, Dunne JP (2023) GOBAI-O_2: Temporally and spatially resolved fields of Ocean interior dissolved oxygen over nearly 2 decades. Earth Syst Sci Data 15:4481–4518. https://doi.org/10.5194/essd-15-4481-2023

Shellock RJ, Fullbrook L, McKinley E, Cvitanovic C, Kelly R, Martin V (2024) The nature and use of Ocean Literacy in achieving sustainable Ocean futures: a systemic map. Ocean Coast Manag 257:1–13. https://doi.org/10.1016/j.ocecoaman.2024.107325

U.S. GO-SHIP (2024) Congrats A13.5 team on a successful cruise! Read weekly reports, blog. https://usgoship.ucsd.edu/2024/03/23/weekly-reports-from-a13-5/

Xue C, Wang Z, Yue L, Niu C (2024) A global four-dimensional gridded dataset of Ocean dissolved oxygen concentration retrieval from Argo profiles. Geosci Data J 00:1–15. https://doi.org/10.1002/gdj3.251

Author Contributions and Biographical Profiles

[1]*George I. Matsumoto: Co-conceptualization; Writing—Original Draft, Review, Editing.

[2]Katie M. Lodes: Writing—Original Draft, Review and Editing.

[3]Teresa J. Kennedy: Co-conceptualization; Writing—Original Draft, Review, Editing.

[4]Jennifer L. Magnusson: Writing—Review and Editing.

*Corresponding Author at mage@mbari.org

George I. Matsumoto, PhD, is the Senior Education and Research Specialist at the Monterey Bay Aquarium Research Institute (MBARI) in California, USA. Previously, he worked at Flinders University of South Australia as a Marine Biology Lecturer and Hopkins Marine Station as an NSF PostDoctoral Scholar. His PhD from UCLA focused on gelatinous zooplankton using SCUBA, human occupied vehicles, and remotely operated vehicles around the world to collect specimens and observations. He has been awarded the QuikScience Ocean Leadership Award for commitment to Ocean Education and recognized as an Association for the Sciences of Limnology and Oceanography (ASLO) fellow. At MBARI, he coordinates the seminar and internship program as well as the NSF funded Adopt-A-Float program.

Teresa J. Kennedy, PhD, is a Professor at the University of Texas at Tyler, USA and former President of the International Council of Associations for Science Education (ICASE). She holds a master's degree in oceanography from the University of Rhode Island and a PhD in STEM education from the University of Idaho, where she also earned a master's in Spanish and two undergraduate degrees. She teaches oceanography and meteorology in the College of Arts and Sciences, and pedagogy courses at the undergraduate level and school improvement courses for master's and doctoral students in the School of Education. Teresa has spent over 90 days at sea on research cruises, is a certified PADI Divemaster, and conducted freefall lifeboat tests in the Netherlands. Her work integrates scientific research, education, and policy, advancing global efforts in Ocean Literacy, sustainability, and STEM education.

Katie M. Lodes is in her 35th year teaching science. Currently at St. Joseph's Academy, an all-girls, Catholic high school located in suburban St. Louis, Missouri, USA, Katie's passion is to bring the people and process of science to her students and colleagues. Katie earned a B.S. in both Secondary Education and Animal Science from the University of Missouri at Columbia. She earned her M.A. in Biology at Washington University, St. Louis. She was named the 2024 National Marine Educators Association Teacher of the Year, awarded a 2023 Grosvenor Teacher Fellowship, and received a 2017 Presidential Award for Excellence in Mathematics and Science Teaching.

Jennifer L. Magnusson is a marine biologist (MS, University of Alaska, Fairbanks, 1998), science communicator, and graphic designer. Jennifer has been going to sea as both a researcher and shipboard educator since 1996 and has recently completed two expeditions as the onboard float and outreach coordinator for the GO-BGC program. Working from her home in Victoria, BC, she combines her desire to learn about and explore the Ocean, her passion for Ocean education and stewardship, and her talent for creating visual resources and lessons that make Ocean education exciting and engaging for all audiences. She is an Independent Contractor for three non-profit organizations—the Monterey Bay Aquarium Research Institute (MBARI), the National Marine Educators Association (NMEA), and the Northwest Aquatic & Marine Educators (NAME)—focusing on outreach, social media, website development, and science communication.

Once Is Not Enough: Connecting Students of Color to their Local Waterways

8

Meghan E. Marrero and Kristen V. Napolitano

Abstract

Ocean Literacy Principle #6 states: "The Ocean and humans are inextricably interconnected." Concept 6G within this principle emphasizes that "we are all responsible for caring for the Ocean" (NOAA, Ocean literacy: the essential principles and fundamental concepts of Ocean sciences for learners of all ages. NOAA, Washington, DC, 2020, p. 11). Yet, marine education and access to Ocean Literacy (OL) opportunities remain limited, at best, for Black and Brown and/or low-income communities. Children in these communities, as a result, may feel little sense of connection to the Ocean and even less sense of agency when it comes to enacting change to improve Ocean health. Thus, if we wish for *everyone* to become Ocean literate, we must purposefully reach audiences who have been historically marginalized, including low-income students and those of color. This mixed-method study explores how students in predominantly Black, Brown, and low-income urban communities developed shifting perspectives of their local waterways and gained a deeper connection and care for their Ocean. Findings suggest OL opportunities need to: discuss the importance of local and distant waterways, provide opportunities to make deeper connections to local waterways, and develop a call to action that empowers students to think critically about their local waterways. In addition, researchers call attention to the *lack* of change after just one round of OL opportunities, recommending that OL become a consistent and on-going part of students' education to ensure sustained shifts in OL, contributing to a diversified Ocean workforce and marine education field.

Keywords

Ocean Literacy (OL) · Environmental justice · Environmental education · Urban environments · Ocean justice · Ocean equity · Ocean education · Marine education · Urban environmental education · New York City

8.1 Introduction

Ocean Literacy Principle #6 states: "The Ocean and humans are inextricably interconnected." Concept 6G within this principle emphasizes that "we are all responsible for caring for the Ocean" (NOAA 2020, p. 11). Yet, Worm et al. (2021) note that "Engagement in marine science has

M. E. Marrero (✉)
Mercy University, Dobbs Ferry, NY, USA
e-mail: mmarrero3@mercy.edu

K. V. Napolitano
Stony Brook University, Stony Brook, NY, USA
e-mail: kristen.napolitano@stonybrook.edu

T. J. Kennedy (ed.), *Ocean Literacy: The Foundation for the Success of the Ocean Decade, Volume I*, Sustainable Development Goals Series, https://doi.org/10.1007/978-3-031-98145-6_8

historically been the privilege of a small number of people with access to higher education, specialized equipment and research funding" (p. 1). A critical component of Ocean Literacy (OL) is being able to "make informed and responsible decisions about the Ocean and its resources" (NOAA 2020, p. 2). While content knowledge is important, studies show that individuals' perceptions of the Ocean are critical to stronger societal relationships with the sea as well as policy aimed at improving Ocean health (Jefferson et al. 2021; McKinley and Burdon 2020). Thus, if we wish for *everyone* to become Ocean literate, we must purposefully reach audiences who have been historically marginalized, including low-income students and those of color.

We must ensure that we are taking an equity-based approach to improving OL, understanding that environmental racism has excluded communities living on or near coastlines, and yet they are disproportionately affected by environmental problems. Equity and justice are critical to Ocean sustainability (Bennett 2022). This study worked directly with urban, primarily Black and Brown students, to examine their beliefs about the Ocean before and after participation in Ocean Literacy-focused opportunities.

8.2 Literature Review and Conceptual Framework

8.2.1 Engagement of Minority Students in Environmental Education

Research shows that minority students and their parents have interest in environmental fields and environmental education, as well as deep concern around environmental issues (Bruyere et al. 2012; Taylor 2019), yet minorities are not well-represented in the environmental workforce (Haynes and Jacobson 2015; Taylor 2018). One barrier to entry into the environmental workforce is simple lack of exposure to nature, outdoor recreation, and environmental careers (Behl et al. 2021; Haynes and Jacobson 2015; Moreno et al. (2020) found that minority and urban students pursuing such careers had previously built a connection to nature. Thus, engagement with local urban environments, including waterways, must start early and include access for underrepresented groups. Additionally, exposure to relatable STEM professionals who reflect students' identities, e.g., gender identity and ethnicity, can promote a sense of belonging that may support students' identities as having a place in STEM fields and increase their chances of pursuing higher education and careers in these areas (Aish et al. 2017; Valla and Williams 2012). Recruiting a broader Ocean workforce requires that we target minority and low-income students in middle and high school, improving students' OL and exposing them to diverse career pathways within Ocean science, Ocean technology, marine education, outreach, and related fields.

While not every student will enter a career in Ocean science, OL is a critical component of scientific literacy (Strang et al. 2007). Myriad local and global environmental issues affect our Ocean and local waterways, and in order to effect behavior change, such as engaging in pro-environmental decisions, people must become emotionally engaged, feeling that their choices and behaviors make a difference (McKinley and Fletcher 2012; Stoll-Kleemann 2019). The goal of this study was to examine students' ideas and beliefs about the Ocean and its importance, before and after participating in opportunities aimed at connecting them with their local waterways and exposing them to Ocean-related career paths.

8.2.2 Conceptual Framework: Ocean Literacy

Ocean Literacy is traditionally defined as "an understanding of the Ocean's influence on you, and your influence on the Ocean." Additionally, an Ocean-literate person:

- understands the Essential Principles and Fundamental Concepts about the Ocean
- can communicate about the Ocean in a meaningful way; and
- is able to make informed and responsible decisions regarding the Ocean and its resources (NOAA 2020).

The Essential Principles and Fundamental Concepts about the Ocean (EP & FCs) outline ideas typical adults should know and understand about the functioning of the Ocean, as well as its role in weather and climate, biodiversity, exploration, culture, human history, and more. Seven Essential Principles overarch 45 fundamental concepts, which represent the "content knowledge" of what people should know about the Ocean. A variety of researchers have focuses on measuring this content knowledge (Fauville et al. 2019; Greely 2008; Guest et al. 2015; Markos et al. 2017; Tsai and Chang 2019). However, this line of inquiry focuses on just one part of what makes someone Ocean literate, leaving out other critical elements, such as connectedness, stewardship, etc. that lead to pro-environmental behaviors (Heimlich and Ardoin 2008; Hungerford and Volk 1990).

Broadening the Definition of Ocean Literacy. In response, OL research has expanded. For example, Glithero and Zandvliet (2021) developed a survey that looked at Ocean perceptions, values, and actions. After analysis, the authors grouped participants as either "Ocean engaged," or "general public," finding that the "general public" respondents did not view the Ocean as connected to their daily lives, although both groups reported that they would be willing to make personal behavioral changes to improve the health of the Ocean (p. 11). This reinforced the idea that awareness of the Ocean is not enough to compel the public to protect or share with others about Ocean health, thereby limiting their Ocean Literacy.

These ideas are further nuanced as researchers have recently developed a more multifaceted view of OL, which leans on a variety of drivers of behavior (Brennan et al. 2019; McKinley and Burdon 2020; Stoll-Kleemann 2019). McKinley et al. (2023) proposed ten dimensions of OL: knowledge; awareness; attitude; behavior; activism; communication; emotional connections; access and experience; adaptive capacity; and trust and transparency. Additionally, a white paper addressing Challenge 10 of the UN Ocean Decade, *Restoring Society's Relationship with the Ocean*, published by UNESCO and written by a collaborative, worldwide group of OL experts, explains that "Ocean Literacy is an outcome—a society that understands, values, and cares for the Ocean" (Glithero et al. 2024, p. 10). Ocean Literacy is critical to meeting the UN Sustainable Development Goals, and critical for the Ocean's future sustainability (Chen and Tsai 2016).

Fostering these nuanced elements of OL in young children may be an important step towards restoring and sustaining public connection to waterways. Studies have found that children frequently engage in pro-environmental behaviors based on what they learn and can influence their families to do the same (Hartley et al. 2015; Uzzell 2000). As the authors of this study examined students' beliefs and OL over the course of participation in a variety of opportunities, OL became the conceptual framework guiding data collection and interpretation.

Ocean Literacy in Greater New York City Area. OL is critical in local contexts as well. This study took place in New York City (NYC) and another local urban area, Yonkers, NY. Of the five boroughs of NYC, the Bronx, Brooklyn, Manhattan, Queens, and Staten Island, two (Manhattan and Staten Island) are islands. Brooklyn and Queens are geographically located on Long Island. The only NYC borough connected to the mainland, the Bronx, is a peninsula. Major waterways bordering the five boroughs include the Atlantic Ocean, New York Bay, Long Island Sound, and the East River. Yonkers is located directly north of the Bronx on its west side, along the Hudson River. New York City and its surrounding areas have been strongly connected to local waterways throughout their history, yet many urban New Yorkers are unaware of the waters' importance and are not particularly connected with nature (Achard 2022; O'Neil et al. 2020).

Many local issues related to water affect New Yorkers' lives. For instance, the runoff from urban streets is collected in combined sewer outflows and sent to water treatment plants. When there is heavy rainfall, however, the treatment plants consistently overflow, spilling sewage and other contaminants into waterways, rendering them unsafe for recreation (Cohen 2011). New York's metro area is also very susceptible to roadway flooding and storm surge (Solecki 2012). These disproportionately affect low-income, high-minority neighborhoods, raising concerns about how climate change will continue to impact these communities (Herreros-Cantis and McPhearson 2021; Reckien et al. 2018). Additionally, the lack of familiarity or safe introductory experiences to the Ocean can be dangerous. Achard (2022) studied this lack of access to safe recreational Ocean experiences in Rockaway, NY, a barrier beach community and home to several of New York City's public beaches, highlighting concerning cultural and racial barriers to water access. Improving OL in urban communities is an environmental and social justice issue.

8.3 Methods

8.3.1 Context and Research Questions

Nine teachers, teaching at five traditional public and public charter schools (three middle schools serving students ages 11–14 and two high schools serving students ages 14–18) in New York City and Yonkers, NY that served a student population of whom at least 40% were from low-income families, were recruited to participate in a series of marine education opportunities with their students. The research team also partnered with a nonprofit organization dedicated to empowering women and girls of African descent. This nonprofit organization supported eight Black girls, ages 12–14, to attended some of the opportunities. The team planned a series of five opportunities, including two webinars (76 students), an annual community estuary event (154 students), a marine education conference (25 students), and a science-based field trip (64 students), all aimed at improving students' OL and connection to their local waterways.

Students and their teachers participated in the two webinars featuring marine scientists. Seventy-six students from the partner schools attended one or both webinars. Following the webinars, 154 students attended "It's My Estuary Day" (IMED) held on Coney Island, an urban coastal neighborhood located on a barrier island in Brooklyn, New York. This annual event allowed students to participate in several opportunities including a beach cleanup, seining and species identification, ROV demonstrations, engagement with local environmental organizations, and kayaking.

The students and teachers were also invited to participate in the National Marine Educators Association (NMEA) Student Conference. Twenty-five students attended. They listened to a keynote speaker share his journey as a Black man in STEM and chose from a variety of Ocean-focused workshops. Finally, sixty-four students took a field trip to another local nonprofit, located on the banks of the Hudson River. The students learned about the Hudson River Estuary and conducted hands-on water quality measures, comparing them to sensor-collected data and analyzing changes over time in scientific data sets. The original intention was to involve the same students in all opportunities. While a few students attended two or three events, most only attended one of the opportunities. One of the challenges for future organizers/presenters to address in planning such a program is ensuring students' accessibility to and commitments for the entire program.

8.3.2 Demographics

Figures 8.1 and 8.2 illustrate the self-reported race and ethnicity of the 147 students whose responses were included in the study. Some stu-

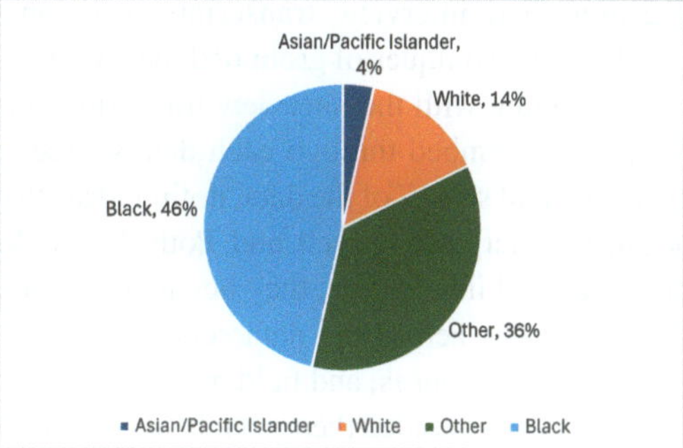

Fig. 8.1 Self-reported Race of Participating Students

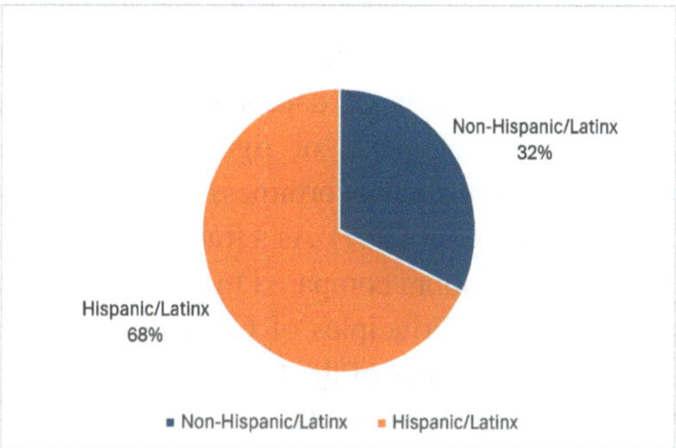

Fig. 8.2 Self-reported Ethnicity of Participating Students

Table 8.1 Percentage Economically Disadvantaged Students Attending Each of the Five Schools

Participating School #	School Level	Location	% Economically Disadvantaged
1	MS	NYC	63%
2	MS	Yonkers	82%
3	MS	Yonkers	86%
4	HS	Yonkers	86%
5	HS	NYC	86%

dents' data were not included for various reasons, including lack of both pre- and post-survey responses. Of the participating students, 46.4% identified as Black, 35.7% other, 14.3% White, and 3.8% Asian/Pacific Islander. 67.9% of students identified as Hispanic/ Latinx, with 32.1% identifying as Non-Hispanic/Latinx (Figs. 8.1 and 8.2).

Table 8.1 shows the percentage of economically disadvantaged students attending each of the five schools. These data were not requested in the surveys, therefore the overall school demographics provide insight into the student body.

8.3.3 Research Questions

The researchers used mixed methods to answer the following research questions regarding minority and low-income students participating in the grant opportunities:

1. How does participation in Ocean Literacy-focused opportunities affect students' beliefs about the importance of the Ocean?
2. How does participation in Ocean Literacy-focused opportunities affect students' ideas about local waterways and their relevance in their lives?
3. What do students take away from participating in Ocean Literacy-focused opportunities?

The researchers employed a mixed methods study design (Creswell et al. 2003) to use qualitative approaches to get a deeper, more comprehensive understanding of participants' shifts in perspectives after participation in the opportunities.

8.3.4 Data Sources

Prior to and after participating in opportunities, students completed a survey on Ocean beliefs adapted from Bidwell (2017) and attached at the end of this chapter as Appendix A. Bidwell's survey was originally used to examine community members' values and beliefs about the Ocean as they pertained to the development of an offshore wind farm in the northeastern United States. The instrument is grounded in values-norms-beliefs (VBN) theory which "provides a model of environmentally significant behavior, in which foundational values influence environmental worldviews and other beliefs, which in turn influence personal norms and a wide range of behavior" (p. 101). The survey contained demographic questions and examined students' beliefs about the importance of waterways, arranged around the latent variables of *appreciative, non-use, and consumptive* reactions. The survey consisted of 12 Likert-style items in which students indicated the level of agreement (1—strongly disagree to 7—strongly agree) with statements about the Ocean, e.g., "The Ocean is a place of beauty," "The Ocean is a place of wildness."

Additionally, students were asked to respond to several open-ended questions about the importance of waterways to their families and communities. After participating in their final activity, students completed the post-survey, which was identical to the pre-survey, plus two questions: "How are waterways important in general?" and "How has learning more about the Ocean affected how you feel about waterways and their importance?"

Following opportunities, focus groups were conducted with three groups of 4–10 students and one group of three teachers. All focus group interviews were semi-structured and ranged from 20–45 minutes, conducted via video conference by this chapter's authors. Student focus group interviews were designed to explore students' beliefs about the relevance of waterways to them and their communities, as well as to learn what they took away from the opportunities. Students were asked questions, such as "Before participating in these events, how did you feel about your connection to the Ocean? How do you feel now?" and "In what ways do you affect your local waterways?" The teacher focus group interview sought to get teacher perspectives on the students' experiences. Teachers were asked questions, such as "Did you notice any shifts in students' interests/connections to the Ocean?" and "How could we support helping students feel more connected [to the Ocean]?"

Interviews were recorded and transcribed for analysis. One researcher also recorded field notes following the webinars, It's My Estuary Day, and student conference events to be used for triangulation of data and provide additional context for findings.

8.3.5 Data Analysis

The Likert survey data were analyzed using an unpaired t-test to examine changes in pre/post responses. Qualitative data from the open-ended questions and interview transcripts were analyzed using techniques of grounded theory analysis. Beginning with the interview transcripts, the researchers combed through each data source to get a "general sense" of the data, noting ideas that began to emerge (Creswell and Poth 2024). As they analyzed line by line, they began to identify common ideas across the data sets (interviews, open-ended responses, and field notes), using the techniques of grounded theory analysis (Charmaz 2006), to highlight and chunk common ideas.

Finally, codes were collapsed into emergent themes. The authors took several passes through the data and participated in peer debriefing discussions to crystallize emergent themes (Charmaz 2006). Triangulation of data sources, and peer debriefing are elements of rigor in qualitative research, assuring trustworthiness of the findings (Guba and Lincoln 1989). As a final pass through the data, the authors compared the ideas with the seven Essential Principles of Ocean Literacy to examine which areas of the framework most resonated with the participants.

8.4 Findings

Of the 12 Likert-type items, statistically significant pre/post changes ($p < 0.05$) were only identified for one item: "The Ocean is a place for human culture." The average value for this item increased from 3.78 on the pre-survey to 4.48 on the post-survey. The results for all 12 items are illustrated in Table 8.2.

Analysis of the qualitative data revealed three emergent themes: *Misconceptions about Drinking Water, Viewing Waterways in New Ways* and *Increased Caring about the Ocean.*

8.4.1 Theme 1: Misconceptions About Drinking Water

One question in the pre-survey asked, "How are waterways (like rivers or the Ocean) important to your community?" The most common responses were around drinking water, e.g., "they provide us water," "source of water," "because we drink from them," "they are our water supply." None of

Table 8.2 Survey Items Initial/Post Means

Question	Content	Initial	Post
1	The Ocean is a place of beauty	5.64	5.92
2	The Ocean is a place of recreation	5.25	5.18
3	The Ocean is a source of relaxation	5.6	5.93
4	The Ocean is a place of inspiration	5.52	5.86
5	The Ocean is a home for wildlife	6.54	6.1
6	The Ocean is a place of wildness	5.92	5.61
7	The Ocean is a place for learning	5.65	5.89
8	The Ocean is a place for exploration	5.96	5.75
9	The Ocean is a place of human culture	3.78	4.48
10	The Ocean is a source of food	5.07	5.32
11	The Ocean is a place of commerce	4.39	4.92
12	The Ocean is a source of energy	5.46	5.66

the interventions focused on drinking water, and unsurprisingly, the post survey results revealed unchanged responses. Students said, for instance, "water to drink," "it provides tap water," "it supplies our drinking water," "pipe water and shower water," and many other similar responses.

Interestingly, *none* of the communities from which these students hail get their drinking water from local waterways or the Ocean. Both NYC and Yonkers supply municipal water from reservoirs and lakes in mountains more than 100 miles north of the cities. While the topic of drinking water is not explicitly covered with the Fundamental Concepts of Ocean Literacy, the authors propose that drinking water sources should be a component of understanding one's environment.

8.4.2 Theme 2: Viewing Waterways in New Ways

A recurrent theme across data sources was that having access to this series of opportunities did shift students' thinking about waterways. These opportunities allowed students to literally "get their feet wet" in the Hudson River and the Ocean and therefore consider the waterways in ways they had not before. Prior to the opportunities, student responses to open-ended questions most frequently tied waterways' importance to seafood, transportation, habitat for organisms, as places of beauty, and less frequently, for swimming. For instance, when asked why waterways were important to their communities, sample answers included, "I eat a lot of seafood," "food like seafood," "they help transport goods between people," "commercial shipping," "they provide a habitat for fish, birds, and other wildlife," "they provide habitats for species that live there." These comments reflected students' general familiarity with the Ocean, but not their personal or local connections. Even so, there were some students who came into the experience with already established relationships with local waterways and the Ocean. For instance, one student explained,

> The East River is important to my community because personally it makes us more amazed on how amazing nature is. Noticing how bad all the other places in New York is and then seeing how beautiful the river is it makes us think that we need to take care of nature more if we want to see this.

This comment shows that the student viewed waterways as a place of beauty. Another student had a different perspective, stating, "I live in the Bronx, all the waterways are polluted basically so there's no use for them like that." While opposing, the researchers viewed both perspectives as more deeply engaged in OL than those that captured more distant, impersonal views of their local bodies of water.

Post-survey responses reflected many of the same ideas as above, e.g., "me and my family really like fish," "they give a way of transportation." However, the post-responses also demonstrated a broader understanding of the importance of the Ocean and waterways and recognition of their importance, with many fewer "I don't know" or "it doesn't" responses to the questions of how waterways affect their family or communities. More students mentioned recreation,

Ocean exploration, and oxygen, which are all reflected explicitly in the fundamental concepts of OL.

Students were also asked, "How has learning more about the Ocean affected how you feel about the Ocean and its importance?"

Sample answers included:

- "I always knew they were important but now I have an understanding of why."
- "Learning about the Ocean made me feel more connected to waterways and made me think they are more important."
- "It made me realize that the Hudson River is not that dirty."
- "I realized that the Ocean provides a lot of things for us."
- "It made its importance ten times more to me."
- "Oceans regulate our climate and absorb carbon dioxide."
- "It made me feel that the Ocean is more important than just fishing, swimming, kayaking. It is a place of beauty and importance."
- "I realized that waterways are crucial to our society and without them I would not have access to the goods I have now."
- "It made the Ocean feel much more complex."

These responses captured their growing sense of connectedness to their local Ocean and waterways offering a glimpse into what was most important to them including viewing them as cornerstones in climate, wildlife, and society.

The shifts in student beliefs about the Ocean were also evident in the interviews. In a focus group, one student noted that after going on the field trip, they better understood the Hudson River as an ecosystem:

> I just feel I am more conscious of the life that's in lots of rivers specifically after going on the trip and being around it more all of these fish that lived in there and smaller animals these little things that are in there that made a lot more sense to the ecosystem as a whole.

It should be noted that this student's school was in Yonkers and that the Hudson River was visible from classroom windows. So, although the students saw the river on a daily basis, they had a new appreciation for the river as a complex ecosystem.

After participating in IMED, another student explained how she gained a new perspective on the water through kayaking:

> I never did kayaking in my life and I went kayaking for the first time there. I learned kayaking was pretty cool to achieve something like a new skill… but the reason why I loved it so much is that it was peaceful for me. Then also on top of that, it was just cool being in the water. I liked it so much that I didn't even realize how deep I went in because when I looked back everybody was so small on the beach. I went even further and further and it was just cool to be in the water and look at the environment.

The teachers felt that all the opportunities provided new activities and viewpoints for their students. During the focus group, the three teachers discussed the shark webinar, in which two of their classes had participated.

> There was a shark webinar and I didn't think my kids were gonna be into it because they were in 8th grade but they were totally into it!… I guess because they don't really get exposure like that you know and they have this fascination with sharks so they really enjoyed the webinar. They participated, they were telling me to put stuff in the chat and questions to ask.

All three teachers noted that for some of their students, it was the first time ever going to a beach when they attended IMED.

> Some of my kids, they're 12th graders, and for some of them the first time to go to a beach. They are urban city kids." Another teacher agreed and provided additional cultural context, saying, "The ones that actually go to the beach are the ones that are from the Caribbean, like the Jamaicans, Puerto Ricans, and Dominicans. Those are the ones that go to the beach, mostly in their home country. But the ones that are born here, it's the first time they've been to a beach. They don't even realize that New York has a beach.

All the teachers stressed the importance of providing access and exposure to their students, so they can learn new things and see the world in new ways.

8.4.3 Theme 3: Increased Caring About the Ocean

During a good portion of the teacher focus group, the educators discussed how their students were impacted by participating in the different opportunities. For instance, one teacher noted that the students who participated in the beach cleanup at IMED saw firsthand the importance of recycling. They became very upset that their school does not recycle because the building would have to pay to recycle materials. The teacher explained that the students began a recycling program to try and quantify how many plastic bottles were being thrown out at their school each day, noting that after IMED, they became more intent on recycling after that trip, and on like 'don't litter,' 'don't be a litterbug.'… They were saying stuff like 'if we don't recycle that's going to wind up in the Ocean!'…So I saw more of an increase in just wanting to take better care of the Ocean.

In their post survey, when asked how learning about the Ocean affected how they feel about waterways, students gave many answers coded under this theme, in comparison to the pre-survey. Responses included:

- "It makes you more caring about your environment."
- "It made me more empathic towards sea animals."
- "It makes me want to keep the Oceans more safe."
- "Cleaning up the beach and seeing the plastic and the straws made me more motivated to help keep our Oceans clean."
- "It's given me a deeper care about the Ocean and an internal reminder not to pollute, litter, or act reckless in the Ocean."
- "It affected me positively because I know to care for the Oceans and cleaning up the beach and seeing the straws, plastic motivated me to help keep our Oceans clean."

In the focus groups, the interviewer asked if students had made any changes to their behavior based on their experiences. Some students said they had not, but others discussed specific examples of ways in which they have made changes. Table 8.3 below is an excerpt from one focus group of students ages 14–18.

While Student 3 questioned their ability to effect change and impact the Ocean, they conceded that small things do matter, echoing the comments of their classmates in terms of personal behavior around garbage, littering, and plastics use.

Together, these illustrated for the research team a trending shift towards deeper consciousness towards and responsibility for the health of their local waterways. This sense of caring transcended their recognition that Oceans and waterways are places of beauty, ecosystems, and food sources, to think critically about large scale issues causing harm and about themselves as Ocean caretakers despite these greater problems.

Table 8.3 Excerpt from Student Focus Group

Interviewer:	What about you guys; have you changed your behavior at all? It is okay to say no, by the way.
Student 1:	I don't know but I'm more conscious around the way that I dispose of my trash in my house at least recycling water bottles. But some people just dump their things out on the side and I am more conscious of that and if I see something that's a trash or whatever I might pick it up and put it in and just be conscious.
Student 2:	Yes, like I might be more aware about trash being there and just staying there. Even if it's a small thing I would feel bad if I just left it out there. I try to avoid littering and I just make sure I am disposing of things the right way.
Student 3:	No, I haven't done anything. It is because if you look at it on a larger scale I am one person so sometimes you are thinking. "Does anything I do really affect… But sometimes small things matter so it's about adapting that mindset and thinking well what I do does matter because I am a person living near the water.

8.5 Conclusion

8.5.1 Once is Not Enough

While the quantitative data did not illustrate changes in beliefs about the Ocean, the qualitative data revealed that the experiences had an impact on these learners in terms of how they viewed Ocean and waterways and thought about their connections to them. Since most students had very little exposure to OL activities prior to the opportunities described above, it is unsurprising that their beliefs were not significantly altered through one to two opportunities. It would be interesting to see if their beliefs change appreciably after more sustained experiences, and if so, how significantly.

In addition to more exposure, the breadth and quality of their OL experiences matters. For instance, the findings above show that even after several opportunities, children did not understand where their drinking water comes from and, therefore, what other ways the Ocean and local waterways connect to their lives. Furthermore, students could have benefited from specific calls to action regarding the Ocean to help mitigate the overwhelming feeling that the issues plaguing waterways are too large for them to make any significant change. As with any justice-focused education, change-oriented OL requires engaging students in thinking critically about themselves and their communities, identifying opportunities for change, and strengthening their sense of agency as it relates to Ocean health.

With that said, the students still felt that the opportunities they engaged in were useful and made them see the Ocean in new ways, including as places for recreation and as a habitat for a wide variety or organisms. All these examples are elements of OL, reflecting the seven Essential Principles of Ocean Literacy:

1. Earth has one big Ocean with many features.
2. The Ocean and life in the Ocean shape the features of the Earth.
3. The Ocean is a major influence on weather and climate.
4. The Ocean makes Earth habitable.
5. The Ocean supports a great diversity of life and ecosystems.
6. The Ocean and humans are inextricably interconnected.
7. The Ocean is largely unexplored.

Furthermore, some students began to develop self-awareness towards their and their communities' impacts on the Ocean, getting closer to an understanding of the interconnectedness among themselves, their communities, and the Ocean. McKinley et al. (2023) explain, "The access, exposure and experience individuals have to the Ocean, whatever that might look like, can shape their attachment and connection, their desire to learn more and ultimately their Ocean Literacy" (p. 5). While this study argues that once is not enough to create fully Ocean literate children, little ripples ultimately become big waves.

8.5.2 Inclusivity for Sustainable Development

The findings also underscore the critical importance of providing opportunities for access and exposure for *all learners*. Teachers and students alike noted that many of the participants had little prior experience with their local waterways and knew little about water bodies such as the Hudson River and Atlantic Ocean. To have an Ocean literate society, access to learning about and experiences with waterways cannot be limited to a privileged few. The student voices represented in this study speak to this point. While all the children in this study were geographically situated among some of the busiest and most highly impacted waterways in the region, their understanding of the significance of those was limited. The lack of access to marine education, compounded by sociohistorical environmental racism, may be to blame. Increasing the number and quality of Ocean Literacy opportunities for *all* students may promote a greater sense of agency in protecting waterways and also increase diversity in local Ocean workforce development. Ultimately, to meet UN Sustainable Development Goals, particularly *11: Sustainable Cities and*

Communities, and *13: Climate Action,* societies will need to ensure that all community members are given opportunities to become Ocean literate.

Previous studies have illustrated that although it is not a direct link from access and exposure to positive environmental behaviors, these opportunities lead to emotional connections and changes in mindset. Ultimately these shifts can lead to behavioral change. The qualitative data provide a glimpse into this idea; as students reflected on their experiences, they realized that they had altered their perceptions, even slightly. They also began to care more about the Ocean and its inhabitants, which may ultimately lead toward positive behavioral changes. At the same time, the quantitative data demonstrate that one to three opportunities are not enough to result in a change in beliefs, underscoring the importance of providing *all* students with multiple opportunities to experience and learn about their waterways.

8.5.3 Summary

Key concepts explored in this chapter include:

- Students, even those in schools or communities situated among major waterways, may not have a strong sociocultural connection to those bodies of water.
- Students need support through multiple, or continuous, Ocean Literacy-focused opportunities to help them recognize and articulate their connection to the Ocean and waterways.
- Students and adults need clarification on the importance of local and distant waterways, including understanding the origin of their drinking water and food supply.
- Students may develop closer connections to their local waterways through hands-on, engaging activities, such as seining, kayaking, and beach clean ups.
- Traditional quantitative survey instruments may not adequately capture small yet important shifts in perceptions, underscoring the importance of qualitative research in this area.

8.5.4 Recommended Resources

The following resources provide further insights and support on this topic:

- Environmental Justice NYC (https://climate.cityofnewyork.us/topic/environmental-justice/) provides many resources, including mapping tools and actions to take to understand and promote urban environmental justice
- The National Marine Educators Association's Ocean Literacy site (https://www.marine-ed.org/Ocean-literacy/overview) has many resources to support educators and others in increasing Ocean literacy in their communities.

Acknowledgements This publication was supported by an agreement with Cornell University, under Prime Agreement CM04068 from the New York Department of Environmental Conservation. Any opinions, findings, and conclusions or recommendations expressed in this publication are those of the author(s) and do not necessarily reflect the views of Cornell University nor the opinions, interpretations or policy of New York State.

Appendix A

Ocean Beliefs Survey (Adapted, Bidwell 2017)
*Pre-survey only
**Post-survey only

1. Which of the following best describes your race?*
   (a) Asian/Pacific Islander
   (b) Black
   (c) Native American/Alaska Native
   (d) White
   (e) Other
2. Which of the following best describes your ethnicity?*
   (a) Hispanic/Latinx
   (b) Non-Hispanic/Latinx
3. What are the last four digits of your cell phone number? If you don't have a cell phone, you can use your parent's number.

4. Which school do you attend? (Options excluded here to preserve anonymity)
5. Which activities did you participate in? (Options excluded here to preserve anonymity)**

   For the following questions, choose the number that represents how much you agree with each statement about the Ocean (1 = Very Strongly Disagree – 7 = Very Strongly Agree)
6. The Ocean is a place of beauty.
7. The Ocean is a place for recreation (like swimming, fishing, or sailing).
8. The Ocean is a source of relaxation.
9. The Ocean is a place of inspiration.
10. The Ocean is a home for wildlife.
11. The Ocean is a place of wildness.
12. The Ocean is a place for learning.
13. The Ocean is a place for exploration.
14. The Ocean is a place of human culture.
15. The Ocean is a source of food.
16. The Ocean is a place of commerce, like shipping and commercial fishing.
17. The Ocean is a source of energy.

    Open-Ended Responses
18. How are waterways (like rivers or the Ocean) important to your community?
19. How are waterways (like rivers or the Ocean) important to your family?
20. How are waterways (like rivers or the Ocean) important in general?
21. How has learning more about the Ocean affected how you feel about waterways and their importance?

References

Achard M (2022) Rockaway, NY: A community diving headfirst into closing the water and Ocean Literacy gap. Senior Comprehensive Project, Occidental College

Aish N, Asare P, Miskioğlu EE (2017) People like me increasing likelihood of success for underrepresented minorities in STEM by providing realistic and relatable role models. In: 2017 IEEE Frontiers in Education Conference (FIE). IEEE, pp 1–4

Behl M, Cooper S, Garza C, Kolesar SE, Legg S, Lewis JC, White L, Jones B (2021) Changing the culture of coastal, Ocean, and marine sciences. Oceanography 34(3):53–60. https://doi.org/10.5670/Oceanog.2021.307

Bennett NJ (2022) Mainstreaming equity and justice in the Ocean. Front Mar Sci 9:873572. https://doi.org/10.3389/fmars.2022.873572

Bidwell D (2017) Ocean beliefs and support for an offshore wind energy project. Ocean Coast Manag 146:99–108. https://doi.org/10.1016/j.ocecoaman.2017.06.012

Brennan C, Ashley M, Molloy O (2019) A system dynamics approach to increasing Ocean Literacy. Front Mar Sci 6. https://doi.org/10.3389/fmars.2019.00360

Bruyere BL, Wesson M, Teel T (2012) Incorporating environmental education into an urban after-school program in New York City. Int J Environ Sci Educ 7(2):327–341

Charmaz K (2006) Constructing grounded theory: A practical guide through qualitative analysis. Sage

Chen C, Tsai C (2016) Marine environmental awareness among university students in Taiwan: A potential signal for sustainability of the Oceans. Environ Educ Res 22(7): 958–977. https://doi.org/10.1080/13504622.2015.1054266

Cohen SA (2011) Sustainable New York City: A work in progress. https://spm.ei.columbia.edu/sites/default/files/content/Publications/Sustainable%20New%20York%20City%20A%20Work%20in%20Progress.pdf

Creswell JW, Poth CN (2024) Qualitative inquiry and research design: Choosing among five approaches. Sage Publications

Creswell JW, Plano Clark VL, Gutmann ML, Hanson WE (2003) Advanced mixed methods research designs. In: Tashakkori A, Teddlie C (eds) Handbook of mixed methods in social and behavioral research. Sage, pp 209–240

Fauville G, Strang C, Cannady M, Chen Y (2019) Development of the International Ocean Literacy Survey: Measuring knowledge across the world. Environ Educ Res 25(2): 238–263. https://doi.org/10.1080/13504622.2018.1440381

Glithero LD, Zandvliet DB (2021) Evaluating Ocean perceptions and Ocean values: The Canadian Ocean Literacy survey. Can J Environ Educ 24(1):216–232

Glithero LD, Bridge N, Hart N, Mann-Lang J, McPhie R, Paul K,... Curtin F (2024) Ocean Decade Vision 2030 White Papers-Challenge 10: Restoring Society's Relationship with the Ocean. https://unesdoc.unesco.org/ark:/48223/pf0000390126

Greely T (2008) Ocean Literacy and reasoning about Ocean issues: The influence of content, experience and morality [Doctoral dissertation, USF Tampa]. https://digitalcommons.usf.edu/etd/271

Guba E, Lincoln Y (1989) Fourth generation evaluation. Sage

Guest H, Lotze HK, Wallace D (2015) Youth and the sea: Ocean Literacy in Nova Scotia Canada Marine

Policy 58: 98–107. https://doi.org/10.1016/j.marpol.2015.04.007

Hartley BL, Thompson RC, Pahl S (2015) Marine litter education boosts children's understanding and self-reported actions. Mar Pollut Bull 90(1–2): 209–217. https://doi.org/10.1016/j.marpolbul.2014.10.049

Haynes NA, Jacobson S (2015) Barriers and perceptions of natural resource careers by minority students. J Environ Educ 46(3):166–182. https://doi.org/10.1080/00958964.2015.1011595

Heimlich JE, Ardoin NM (2008) Understanding behavior to understand behavior change: A literature review. Environ Educ Res 14(3): 215–237. https://doi.org/10.1080/13504620802148881

Herreros-Cantis P, McPhearson T (2021) Mapping supply of and demand for ecosystem services to assess environmental justice in New York City. Ecol Appl 31(6):e02390. https://doi.org/10.1002/eap.2390

Hungerford HR, Volk TL (1990) Changing learner behavior through environmental education. J Environ Educ 21(3): 8–21. https://doi.org/10.1080/00958964.1990.10753743

Jefferson R, McKinley E, Griffin H, Nimmo A, Fletcher S (2021) Public perceptions of the Ocean: Lessons for marine conservation from a global research review. Front Mar Sci 8:711245. https://doi.org/10.3389/fmars.2021.711245

Markos A, Boubonari T, Mogias A, Kevrekidis T (2017) Measuring Ocean Literacy in pre-service teachers: Psychometric properties of the Greek version of the survey of Ocean Literacy and experience (SOLE). Environ Educ Res 23(2): 231–251. https://doi.org/10.1080/13504622.2015.1126807

McKinley E, Burdon D (2020) Understanding Ocean Literacy and Ocean climate-related behaviour change in the UK-work package 1. Evidence synthesis. Daryl Burdon Ltd., Hull. https://darylburdon.co.UK

McKinley E, Fletcher S (2012) Improving marine environmental health through marine citizenship: A call for debate. Mar Policy 36(3):839–843. https://doi.org/10.1016/j.marpol.2011.11.001

McKinley E, Burdon D, Shellock RJ (2023) The evolution of Ocean Literacy: A new framework for the United Nations Ocean Decade and beyond. Mar Pollut Bull 186:114467. https://doi.org/10.1016/j.marpolbul.2022.114467

Moreno B, Crandall C, Monroe MC (2020) Factors influencing minority and urban students' interest in natural resources. J For 118(4):373–384. https://doi.org/10.1093/jofore/fvaa008

National Oceanic and Atmospheric Administration (NOAA) (2020) Ocean Literacy: The essential principles and fundamental concepts of Ocean sciences for learners of all ages. NOAA, Washington, DC

O'Neil JM, Newton RJ, Bone EK, Birney LB, Green AE, Merrick B, Goodwin-Segal T, Moore G, Fraioli A (2020) Using urban harbors for experiential, environmental literacy: Case studies of New York and Chesapeake Bay. Reg Stud Mar Sci 33:100886. https://doi.org/10.1016/j.rsma.2019.100886

Reckien D, Lwasa S, Satterthwaite D, McEvoy D, Creutzig F, Montgomery M, Schensul D, Balk D, Khan I (2018) Equity, environmental justice, and urban climate change. In: Rosenzweig C, Solecki W, Romero-Lankao P, Mehrotra S, Dhakal S, Ali Ibrahim S (eds) Climate change and cities: Second assessment report of the urban climate change research network. Cambridge University Press, pp 173–224

Solecki W (2012) Urban environmental challenges and climate change action in New York City. Environ Urban 24(2):557–573

Stoll-Kleemann S (2019) Feasible options for behavior change toward more effective Ocean Literacy: A systematic review. Front Mar Sci 6:273. https://doi.org/10.3389/fmars.2019.00273

Taylor DE (2018) Racial and ethnic differences in the students' readiness, identity, perceptions of institutional diversity, and desire to join the environmental workforce. J Environ Stud Sci 8:152–168. https://doi.org/10.1007/s13412-017-0447-4

Taylor D (2019) Can the environmental movement attract and maintain the support of minorities? In: Race and the incidence of environmental hazards. Routledge, pp 28–54

Tsai L, Chang C (2019) Measuring Ocean Literacy of high school students: Psychometric properties of a Chinese version of the Ocean Literacy scale. Environ Educ Res 25(2): 264–279. https://doi.org/10.1080/13504622.2018.1542487

Strang C, DeCharon A, Schoedinger S (2007) Can you be science literate without being Ocean literate? Current: J Mar Educ 23(1): 7–9.

Uzzell DL (2000). The psycho-spatial dimension of global environmental problems. J Environ Psychol 20(4): 307–318. https://doi.org/10.1006/jevp.2000.0175

Valla JM, Williams W (2012) Increasing achievement and higher-education representation of under-represented groups in science, technology, engineering, and mathematics fields: A review of current K-12 intervention programs. J Women Minorities Sci Eng 18(1) https://doi.org/10.1615/JWomenMinorScienEng.2012002908

Worm B, Elliff C, Fonseca JG, Gell FR, Serra-Gonçalves C, Helder NK, Murray K, Peckham H, Prelovec L, Sink K (2021) Making Ocean Literacy inclusive and accessible. Ethics Sci Environ Polit 21:1–9. https://doi.org/10.3354/esep00196

Author Contributions and Biographical Profiles

[1]*Meghan Marrero: Conceptualization; Writing—Original Draft, Review, and Editing.

[2]Kristen Napolitano: Writing—Drafts, Review, and Editing.

*Corresponding Author at mmarrero3@mercy.edu

Meghan E. Marrero EdD, is a Professor of Secondary Science Education at Mercy University in New York, USA and co-Director of the Mercy Center for STEM Education, working primarily to improve access to STEM education for diverse learners. Her research centers on Ocean Literacy of students and teachers, as well as STEM teacher education. Meghan was a Fulbright Scholar to Ireland in 2018, where she focused on teaching and research around family learning in science and engineering for early childhood students and their families. A former high school science teacher in New York City, Meghan has been involved with several national and international Ocean Literacy initiatives and currently serves as the United States National Coordinator for the USA Blue Schools. She served as President of the New York State Marine Education Association (NYSMEA) from 2009–2014, and President of the National Marine Educators Association (NMEA) in 2018–2019. Meghan holds a B.S. in Biological Science from Cornell University, an M.A. and EdD in Science Education from Teachers College, Columbia University, and an advanced certificate in educational leadership from Queens College.

Kristen V. Napolitano is a Lecturer in the Institute for STEM Education at the State University of New York at Stony Brook, USA. She holds her EdD in Science Education with a specialization in Teacher Education from Teachers College, Columbia University. Prior to her work at Stony Brook University, Kristen was Senior Researcher at the Mercy University Center for STEM Education where she taught graduate level courses in secondary STEM education and led Wipro Reimagined, a collaborative grant-funded project for teacher leadership. Since starting at Stony Brook University in September 2024, Kristen now teaches STEM teacher education courses for the Ph.D. and MAT programs. Her research interests are in STEM teacher leadership, professional development, and community-based STEM education.

Shaping Scientific Literacy: How Underserved Students Conceptualize Marine Scientists and Themselves

9

Sarah Nuss, Anna Caputo, Lisa Lawrence, Greta Olsen Gardiner, Bethany Smith, and Celia Cackowski

Abstract

Developing Ocean Literacy for all citizens is an important component of protecting and stewarding the Ocean and its resources. Not all communities have access to marine science education opportunities, and barriers to developing Ocean Literacy (OL) within these communities must be considered. By engaging with all communities, we can diversify the field of marine science and also meet the complex challenges that threaten our global Ocean. This study explores the impact of a summer enrichment program aimed at students from low socioeconomic communities. Using Chambers (Science Education 67:255, 1983) Draw a Scientist Test as an example, we developed a tool specific to marine science and collected 4 years of longitudinal data from program participants. We ascertained existing biases students had of marine scientists at work, their attitudes toward scientists, and their ability to describe scientific work. We also assessed how the students viewed themselves as scientists. This study provides insight on the importance of working with underserved populations in STEM, and how longitudinal data can showcase changes in students' attitudes and perspectives.

Keywords

Stereotypes of scientists · Underrepresented in STEM · Marine science · Ocean Literacy (OL) · Self-identity · Draw-a-scientist

S. Nuss (✉) · A. Caputo · L. Lawrence · B. Smith · C. Cackowski
Virginia Institute of Marine Science, Gloucester Point, VA, USA
e-mail: mcguire@vims.edu; ayers@vims.edu; bsmith@vims.edu; ccackowski@vims.edu

G. O. Gardiner
Phillips 66, Ferndale, Washington, USA

9.1 Introduction

To protect the Ocean, we must understand it. Developing Ocean Literacy (OL) in citizens is one way to increase the public's awareness of the Ocean, and to encourage all citizens to be responsible stewards of the Ocean and its resources. The unbelievable vastness of the Ocean, the potential for mysteries in unexplored depths of water, and multitudes of remarkable sea creatures makes marine science intriguing to many students (Plankis and Marrero 2010). Marine science's implicit connection to foundational science disciplines, including chemistry, biology, and geology, make it an ideal starting point to generate interest in science.

In ensuring that all citizens understand the values and services of the Ocean for human well-being, culture, and sustainable development (Ocean Decade Challenge 10), we cannot overlook any population, and in fact may need to target specific popu-

T. J. Kennedy (ed.), *Ocean Literacy: The Foundation for the Success of the Ocean Decade, Volume I*, Sustainable Development Goals Series, https://doi.org/10.1007/978-3-031-98145-6_9

lations. Studies have found that BIPOC, girls, and students living below the poverty line are less likely to have positive attitudes towards science as they grow into adolescence (Staus et al. 2020). Therefore, the unique engagement opportunity that marine science offers can be crucial to keeping students enthusiastic about exploring science and related technology, engineering, and math (STEM) careers.

9.2 Lack of Diversity in Marine Science

BIPOC communities, those living below the poverty line, and people who do not have access to marine science opportunities have been at the forefront of systemic inequity in developing their OL (Graham et al. 2022). The lack of BIPOC students and students living below the poverty line pursuing STEM careers means these communities that have already been at the forefront of systemic inequities are also less likely to reap the benefits of a STEM career (CEOSE 2018). Boosting the number of underrepresented students who pursue STEM can help address the shortage of qualified scientists and engineers while propelling those students forward with the financial benefits and job security of a STEM career. Additionally, marine science is among the least diverse science fields in the United States with only 2% of marine science doctoral degrees being awarded to Black scholars (BIMS in Review 2022). The barriers to developing OL within these communities must be considered, as well as solutions to better reach these audiences to impact their relationship with and understanding of the Ocean. With diverse perspectives entering the STEM workforce, it is likely that potential solutions and innovations to the problems facing our Ocean would increase, thus increasing capacity for creative solutions to the challenges of pollution, sustainability, climate change, and resilience currently facing our global Ocean. As we amplify the voices of these communities, our capacity to meet the complex challenges that threaten our global Ocean increases.

Informal learning experiences offer a pathway for underrepresented students to become interested in STEM and see it as a future career. Opportunities offered by science centers, workshops, and camps allow students to gain a deeper understanding of science than schools can provide (National Research Council 2009). Participation in the hands-on activities and career exploration provided by informal education can cultivate a fascination with STEM and grow into a scientific career (Weber 2011). Low-income and Black students are equally or more likely to have an interest in STEM careers than their respective high-income and white counterparts (Lichtenberger and George-Jackson 2013), but are less likely to think they will be able attain their ideal job (St Clair and Benjamin 2011). The significant achievement gap for Black and low-income students (Reardon 2011), a lack of knowledge of STEM careers and how to achieve them (Staus et al. 2020), and a student's view that he or she could not be a scientist (Kricorian et al. 2020) keeps many disadvantaged students from seeking a STEM degree, but informal learning can help to break down these barriers.

9.3 Participants

For this study, we focus on a 2-week summer enrichment program that has been offered to high-ability students from low-income families for nearly a decade. To be eligible for this enrichment program, students must score in the 90th percentile on standardized tests or have a school recommendation accompanied by evidence of performance. Additionally, the student's family income must not exceed 185% of the U.S. HHS Poverty Guidelines. The program strives to propel students into a successful educational and professional future. The program largely serves students who are underrepresented on multiple fronts, including but not limited to gender identity, income, and race or ethnicity, with approximately 72% of participants belonging to underrepresented racial or ethnic groups. Four years ago (2021), this program expanded its offerings to include a marine science component for students entering their first 2 years of high school (approximately 13–15 years old). The students preparing to enter their first year of high school (subsequently referred to as Year 1 students) participated in a 1-week introductory marine science camp (and spent their second week of camp in another STEM discipline), and those entering their second year of high school (Year 2 students) spent 2 weeks in an advanced marine science camp. Participants in this study were students who attended either the introductory or advanced marine science class from 2021 through 2024 (see Table 9.1).

The marine science camps had intended learning outcomes that included an introduction to the field, engagement with professionals working in marine science, development of students' awareness of careers in marine science, and development of students' awareness of academic paths leading to a career in marine science, with the advanced marine science program (Year 2) including significant time spent learning research methods and completing and presenting an independent research project. This informal educational experience used marine science topics and exposure to scientific methods as a bridge towards interest and curiosity in science (Table 9.1).

Table 9.1 Program Themes and Aligned Sustainable Development Goals (SDG)

Programmatic Themes	United Nations SDGs Targeted
Year 1 Programming (1-week, Introductory Marine Science)	
Water Quality	4, 10, 13
MacroInvertebrates	4, 14
Nature Journaling	4
Biodiversity (seining)	4, 13, 14
Dissection (Squid)	4, 14
Plankton	4, 14
Canoing	4
Human Impacts on the Marine Environment	4, 8, 10, 13, 14, 17
Year 2 Programming (2-weeks, Advanced Marine Science)	
Research Methods	4, 8, 10
Estuarine Habitats & Organisms	4, 13, 14
Dissection (Shark)	4, 14
Ichthyology Collections	4, 14
Biodiversity (seining, fishing)	4, 13, 14
Maritime History & Marine Archaeology	4, 8, 10
Canoing	4
Aquaculture	4, 8, 13
Plankton & Invertebrates	4, 14
Ocean Exploration	4, 8, 10, 13, 14

Note: Both Year 1 and Year 2 curriculums include games, engineering design challenges, field data collection, marine career exploration, infographic/scientific poster design, and presentations

Across the 4 years considered in this study, 131 participants completed pre- and post-tests for a given year and had parental consent to be included in the research data. Table 9.2 displays the number of students per year, age level, and demographic information.

From the total 131 participants, 22 attended both years of camp and completed all four pre- and post-tests. These participants will be called "long-term" students or campers.

9.4 Methodology

The objective of this study was to determine what impacts, if any, the program had on students' attitudes and perceptions of science and science careers, including their personal identification as a scientist. For our study, we conducted pre- and post-test sampling of all students with marine science content questions and questions about their general interest in and attitudes toward science. To assess students' perceptions of marine scientists, we used the Draw a Scientist Tool (DAST) created by David Chambers in 1983, a test implemented to learn at what age the well-known stereotypic image of the scientist first appeared. In a traditional DAST, students are asked to imagine and draw a scientist at work, and checklists and rubrics are used to measure the degree that students identify with or have misconceptions about scientists (Chambers 1983; Finson et al. 1995). To meet the specific goals of

Table 9.2 Participant Demographics for 2021–2024

	2021	2022	2023	2024	TOTALS
Total # of Participants	25	24	43	39	131
Grade Level					
Year 1	17 (68%)	14 (58%)	30 (70%)	27 (69%)	88 (67%)
Year 2	8 (32%)	10 (42%)	13 (30%)	12 (31%)	43 (33%)
Gender					
Female	13 (52%)	10 (42%)	19 (44%)	23 (59%)	65 (50%)
Male	9 (36%)	13 (54%)	22 (51%)	15 (38%)	59 (45%)
Other or prefer not to say	3 (12%)	1 (4%)	2 (5%)	1 (3%)	7 (5%)
Ethnicity					
Asian or Pacific Islander	1 (4%)	2 (8%)	2 (5%)	2 (5%)	7 (5%)
Black or African American	16 (64%)	14 (58%)	21 (49%)	13 (33%)	64 (49%)
Hispanic or Latino	3 (12%)	2 (8%)	3 (7%)	3 (8%)	11 (8%)
Multiracial or Biracial	0 (0%)	0 (0%)	5 (12%)	5 (13%)	10 (8%)
White or Caucasian	5 (20%)	6 (25%)	12 (28%)	16 (41%)	39 (30%)

Note: Includes only those with parental consent that completed both a pre-test and post-test for the given year

the marine science component of the program, students were explicitly asked to draw a marine scientist in what we are calling the Draw-A-Marine-Scientist Test (DAMST).

Students were given a blank piece of paper. A prompt was read that asked students to imagine they were going to visit a marine scientist at a place where the marine scientist was working. Students were asked to draw the scientist busy at work. We asked students to draw and color as many details as possible, including what the scientist looked like, what they are holding, what else is in the area, and where they are working. As an opportunity to indicate race or ethnicity, students were asked to use colored pencils with their drawings. After 10 min of drawing time, students were prompted to add three words that came to mind when looking at their scientist. The test was administered at the beginning of instructional time on the first day and then repeated on the final day of the camp. In addition to the DAMST, students completed a short survey with content and attitude questions, as well as questions regarding their future.

For the DAMST portion of the study, a rubric which assessed drawings for things such as facial expression, multiple scientists present, ethnicity, gender, location of scientists, specific tools included, and whether the drawing indicated a 'mad scientist' archetype, was used. The research questions included: 1) Based on gender and racial markers, to what degree do students self-identify with scientists? 2) Does this change between pre- and post-tests? and, 3) Does specific knowledge of the work marine scientists do increase at the conclusion of the program? For analysis of the pre- and post-test content knowledge questions, we evaluated individual tests, looking for trends by age level and year of participation. For self-identification as a scientist, we calculated the percentage of students that responded "True" to the statement "You are a scientist" across all 4 years of camp. We analyzed these data by age level, gender identity and race/ethnicity.

9.4.1 Theoretical Framework

Situated learning theory was used as the theoretical framework for this study, as the researchers focused on the potential impact of hands-on contextual learning on the student in a real-world community of practice. Situated learning theory, originally developed by Jean Lave and Etienne Wenger, questions how learning occurs and places more emphasis on the social aspect of learning and experiences in communities where all are learning collectively (Smith 2009). An example of these communities of practice are summer enrichment programs, where students come together with a common curiosity and develop their knowledge in the field through their interactions with each other (Smith 2009). Situated learning theory is appropriate for this study in many ways, including the focus on the experience of the individual student, the expansion and connection to prior knowledge they bring to the camp experience, and how the student develops into a part of the marine science community (Kurt 2021).

Curriculum development for the marine science component of the program also followed a situated learning model. Instructors served more as facilitators than lecturers, creating a reflective, collaborative, and discursive classroom, and providing real-world activities and research in which students participated, including problem-based learning (Kurt 2021). As much as possible, students were actively involved in field experiences, research projects, and investigations that highlighted authentic issues and scientific data in the field of marine science.

9.5 Results

Within the DAMST data, several themes emerged about students' conceptualizations of marine scientists. In the drawings and text descriptors of the tests, we ascertained existing biases of marine scientists at work, students' positive perspectives when it comes to scientists, the impact of programs on the students' ability to describe scientific work, as well as how the student views themself as a scientist.

9.5.1 Stereotypes

When asked to draw a marine scientist for the pre-test, students typically illustrated either marine biologists working with animals, or they reverted to stereotypical scientists such as those found in Chamber's study (a person wearing a lab coat holding a test tube or microscope; 1983) but sometimes adding a coastal setting or a lab with an aquarium. This focus on marine biology eliminated the variety of disciplines represented in marine science and demonstrated the students' lack of awareness of this diversity. This changed in the post-test where students included a greater variety of field research in various capacities and less reliance on the stereotypical scientist in their drawings (Fig. 9.1).

Despite some changes in the scope of work scientists were depicted completing, the post-tests continued to heavily represent the work of marine biologists. This may be reflective of the program, as instructors' backgrounds heavily represented biological sciences. The post-test images did reflect more specifics from the program such as dissections, seining, and water quality testing.

The stereotype of scientists being "mad" or "crazy" was not, in general, represented in the students' drawings. Overall, the demeanor of marine scientists depicted positive traits and descriptions. Even on pre-tests, the illustrated scientists were generally happy (facial expressions). Additionally, students often included positive traits in their three-word summaries. The majority of those positive traits were focused on intellect (e.g., smart or intelligent), but other positive words were included such as brave, interesting, and creative. While these words were affirming to hear, we also recognize that they could be applied to any career, and are not specific to marine science, and students may have relied on these general positive words more in the pre-test.

Fig. 9.1 Stereotypical Changes from Pre- to Post-Test. Note: Pre-test (on left) showing stereotypical scientist, and post-test (on right) showing a specific fish seining exercise that students participated in during camp

9.5.2 Knowledge and Confidence Gain

In addition to viewing scientists in a positive light, the student participants also demonstrated an increase in knowledge of and ability to accurately describe the work of marine scientists. In the pre-tests, we observed the use of general descriptive words, either about people or commentary on their own drawing. For example, students used words such as happy and helpful, and stick figures to describe their scientist. In the post-tests, we observed less of these general descriptive words. Instead, students were able to use terms related to science and marine science. An example of this was their ability to list specific tools used in marine science research (e.g., net, microscope).

We also observed differences in students' knowledge of and ability to accurately depict the work of scientists in their drawings. Even with a low percentage of students drawing "mad scientists" overall, we still observed a decrease of this archetype (e.g., lab coats, test tubes, crazy hair) in post-tests in both 2022 and 2023 camp data. A lower representation of mad scientists suggests that most of our participants do not see scientists in the way Chambers (1983) found in his original study, and for those in our study that did view scientists this way, the enrichment program is helping to dismantle that stereotype.

Students were also asked marine science content questions that covered the camp curriculum on the pre- and post-tests to measure any knowledge gained. Unlike the attitude questions, the Ocean science content questions were different between the two camps as their content was different. The questions also changed each year to reflect changes in the curriculum. With the exception of 2023 Year 1 students (which showed no gain or loss), the results of the marine science content assessments showed a gain in content knowledge for all students, though the extent of these gains varied across years and ages levels (Table 9.3). For example, Year 1 students showed considerable gain in 2021 but only moderate gain in 2022, while Year 2 students' gains were

Table 9.3 Participants' Pre- and Post-Test Content Knowledge Average Scores per Age Level for 2021–2024

	Year 1 Participants (~Age 14)			Year 2 Participants (~Age 15)		
Year	Pre-Test	Post-Test	% Change	Pre-Test	Post-Test	% Change
2021	47%	63%	+16%	46%	51%	+5%
2022	59%	65%	+6%	75%	90%	+15%
2023	55%	55%	0%	53%	66%	+13%
2024	57%	80%	+23%	47%	69%	+12%

opposite. However, both groups of students showed the highest gain in content knowledge in 2024 with 23% for Year 1 students and 22% for Year 2 students. The considerable increases for all students in 2024 may suggest that formative changes to the program in terms of curriculum, teaching strategies, and assessments are having a positive impact on the efficacy of the program.

9.5.3 Human Dimensions

In addition to changes in the conceptualizations of marine scientists' work we recognized in the DAMST, we also observed some changes in the racial and gender diversity of marine scientists, and increased inclusion of collaboration among scientists. In general, the majority of gender and race depictions were unclear or ambiguous which may reflect the researchers' hesitancy to interpret gender and race in the drawings to avoid inaccurate assumptions. Gender and race/ethnicity did not seem to differ significantly from pre- to post-drawings, however, this was difficult to ascertain given it was completely reliant on the researchers' interpretation of the drawings. Given this, the number of females represented in the post-tests did increase slightly in three out of the 4 years, however the unclear category was the highest across all years and pre- and post-tests.

Another example of changes within the human dimension category included students drawing more examples of collaboration in the post-tests, as seen in Fig. 9.2 below.

Students who illustrated multiple scientists in their drawings may be representing their new understanding of the collaborative nature of science and research in general. This was reinforced during instruction, as each camp had a team of instructors, and the students also interacted with scientists who were part of research labs and teams.

9.5.4 Self as Scientist

One way we were able to ascertain gender identity and ethnicity was through the students drawing themselves as the scientist in their images (e.g., Fig. 9.1). One of the most significant findings for the research team, as informal educators, was that many students drew themselves in the post-test, indicating that they see themselves as scientists and view the work they completed in the enrichment program as "doing science." In Fig. 9.3, when asked to draw a marine scientist, one student drew a mirror and used the words "educated, hardworking, and creative" to, essentially, describe themself.

Additionally, in the written pre/post-test, students were asked to respond with true or false to the following statement "You are a scientist" (Table 9.4). For Year 1 students, 44% agreed with the statement on their pre-test. By the end of their 1-week camp, this increased to 65%. This increase of more than 20% suggests that even a brief, 1-week exposure to marine science activities can enhance students' self identification as a scientist.

For Year 2 students, we see students start camp with a higher percentage of agreement with the statement than students at the start of Year 1, but lower than Year 1 students at the end of camp. This suggests that in the year absence between programming, their confidence in themselves as scientists decreases. But, by the end of Year 2, students' 2-week camp, the number agreeing increases to 65%, the same proportion as at the end of Year 1. This finding suggests that repeated exposure to short-term STEM interventions may be critical to maintaining the self-perception gains made during the initial short-term exposure.

When looking at the same question analyzed by campers' gender identity (Fig. 9.4) or race/ethnicity (Fig. 9.5), we see additional trends in campers' self-identity as a scientist.

Fig. 9.2 Pre- to Post-Test Comparison. Note: Pre-test (on left) showing scientist in a generic lab in an aquarium, while post-test (on right) shows collaboration in the field to accomplish a fish tagging

Fig. 9.3 Students' Perception of Self as Scientist. Note: A student participating in the 2024 camp drew a mirror (suggesting their reflection) when asked to draw a marine scientist

Table 9.4 Participant Agreement With "You are a scientist" Statement

Year Class	Pre-Test	Post-Test	% Change
Year 1 Students (88 Total Students)	44%	65%	+21%
Year 2 Students (43 Total Students)	56%	65%	+9%
Long-Term Students (22 Total Students)	41%	77%	+36%

Note. For long-term students, the pre-test is the test at the beginning of their first year and post-test is the test at the end of their final year

Students who identify as male are least likely (40%) to identify as a scientist at the start of camp, compared to their peers that identified as female or preferred not to say, 48% and 50% respectively. However, by the end of camp, both male- and female-identifying students had a similar percent increase in their self-identity as a scientist (17–18%). Students who selected "other or preferred not to say" had a much higher increase in scientific identity (33%) at the end of camp. The notable increase in these students identifying as scientists may be an area of future study where we separate "other" from "prefer not to say" in the answer options.

Examining students' scientific identity based on their race or ethnicity reveals additional trends.

On their pre-tests, students were asked to identify the race/ethnicity that best described them, resulting in five groupings of students—Asian, Black, Hispanic, White, and multiracial. Upon the start of camp, Hispanic students were least likely to identify as a scientist (33%), while Asian students were most likely (67%). Additionally, there was no difference between Black and White students' self-identification

Fig. 9.4 Participant Agreement with "You are a scientist" Statement by Gender Identity

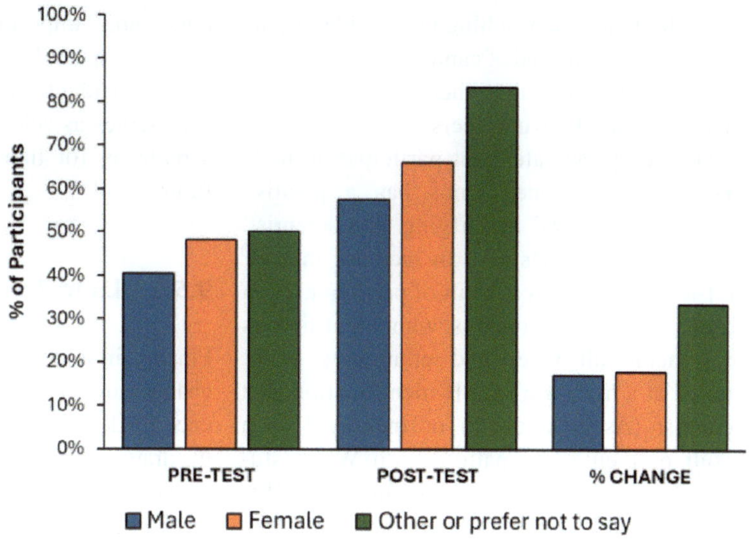

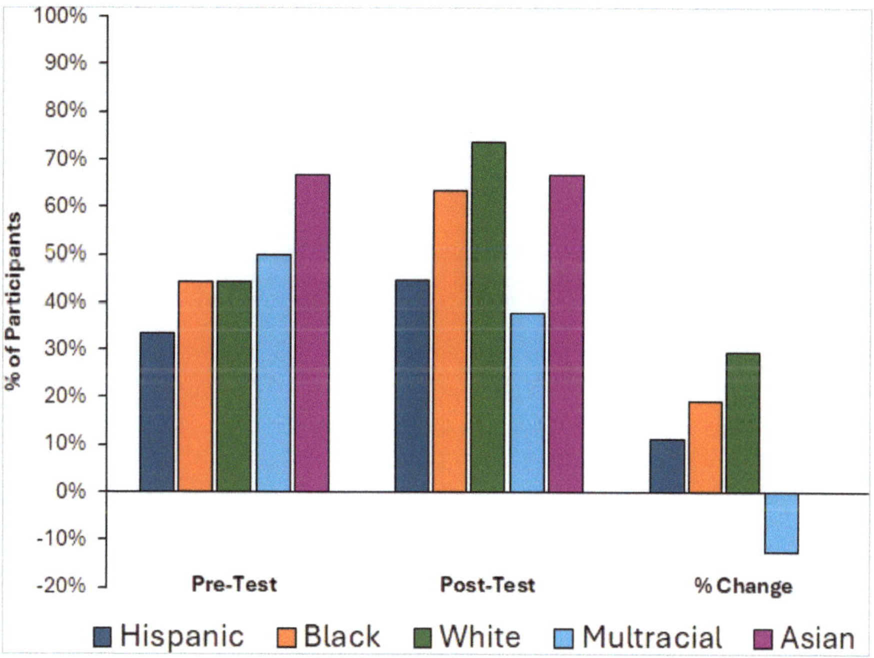

Fig. 9.5 Participant Agreement with "You are a Scientist" Statement by Race/Ethnicity

(44%) and only a slight difference for multiracial students (50%).

At the end of camp, students again indicated whether they agreed with the statement "You are a scientist." Asian students, the group with the highest starting rate, showed no change in their responses at the end of camp. Hispanic students, the group with the lowest starting rate, ended camp increasing their scientific identity, but at a lower rate (11%) than their Black or White peers. At the start of camp, Black, White, and multiracial students all had mid-range levels (44–50%) of scientific identity but their post-camp level varied considerably. Multiracial students were the only group to decrease (−13%) in their self-identification as a scientist. White students had the highest increase resulting in 74% identifying as a scientist at the end of camp, while Black students increased less than their White peers but more than their Hispanic peers.

Our study indicates that participation in 1- and 2-week science camps has a positive impact on students' identifying as a scientist. All gender identities saw an increase in their self-identification as a result of participating in camp. Except for students who identified as multiracial, all races and ethnicities either started at a high rate of self-identification as a scientist (Asian students) or increased as a result of their participation, with White students having the highest rate of increase. This suggests that STEM enrichment programs like ours are important for underrepresented groups, especially Black, Hispanic, and non-cisgender-identifying people, to begin to see themselves as scientists and potentially create a pathway for them to pursue science in the future.

9.5.5 Long-Term Effects

Finally, an important finding was the significant changes in perception and attitudes toward science for long-term students. This is reflected in the drawings of marine scientists when compared across both years of participation for long-term students. Figure 9.6 shows an example of pre-

Fig. 9.6 One Student's Pre- and Post-Tests Over Two Years of Participation. Pre-test 2022 on top left; Post-test 2022 on top right; Pre-test 2023 on bottom left; post-test 2023 on bottom right. Note: Pre-test 2022 (on top left) shows a generally negative depiction of a marine scientist, which shifts for the post-test 2022 (on top right) into an image of actual scientists that the camper met during the week, working on fish gut content analysis. For year 2, the pre-test 2023 (bottom left) returns to a solitary scientist with an aquarium and dissection but the image is positive (e.g. smile alongside positive words). The 2023 post-test (bottom right) shows a combination of the scientist from the year before, alongside other research the student saw throughout the camp

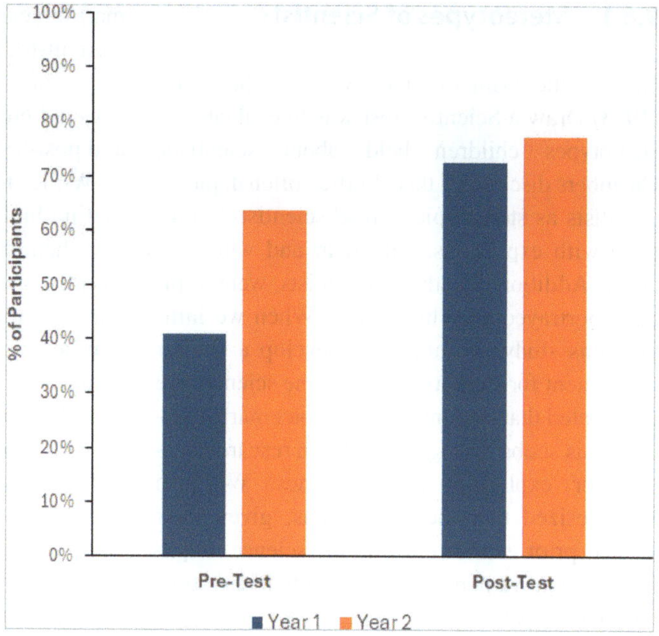

Fig. 9.7 Long-Term Participants' Agreement with "You are a scientist" Statement

and post-tests from the same student, across both years, suggesting that students retained attitudes and conceptualizations of marine scientists from year to year.

These images also suggest the importance of longer interactions with students having a greater impact on change from pre- to post-test. We observed an increase in the knowledge of scientists' work, ability to accurately portray the work of scientists, and general positivity from the pre- to post-DAMST each year. Often, there was a drop from post-drawing to the following year's pre-drawing, but the students rebounded by the time of their second post-drawing. This suggests that students retained some attitudes and conceptualizations of scientists from year to year, but repeated exposures for students returning in Year 2 further dismantled the mad scientist archetype, increased positive perceptions of marine scientists, and demonstrated a deeper understanding of collaborative science.

We also saw this reflected within the written tests. For the 22 long-term students, we observed changes in their responses to whether they saw themselves as scientists. Figure 9.7 displays the longitudinal impact on the students' perception of themselves as a scientist.

Among the 22 long-term students, 41% agreed with the statement, "You are a Scientist" at the start of their first year of camp, with a substantial increase to 77% by the end of their final year of camp. This was 12% higher than the post-test results for all Year 2 students combined.

9.6 Conclusion

This study aimed to investigate the impacts of a summer enrichment program on students' attitudes and perceptions of science careers, and marine science in particular, as well as their personal identification as a scientist. The results illustrated five main themes: stereotypes of scientists, knowledge and confidence gains within participants, human dimensions of scientists, self-identity as a scientist, and longitudinal impacts. Two of those themes, stereotypes of scientists and longitudinal impacts, will be discussed further here as they have direct implications for informal marine educators.

9.6.1 Stereotypes of Scientists

One of the primary objectives of Chambers' (1983) Draw a Scientist Test was to evaluate the stereotypes children held about scientists. Chambers discovered that children often depicted scientists as stereotypical 'mad scientists,' complete with explosions, wild hair, and white lab coats. Additionally, these scientists were typically portrayed as White males. When we initiated this study, we aimed to develop a similar assessment for stereotypes in marine science. We anticipated that students would depict marine scientists as scuba diving, working on research vessels, or exploring in submarines. We also hypothesized that the participants, given their limited prior exposure to marine science, might harbor negative perceptions of the field as a possible career pathway.

However, as our results indicated, we did not observe these stereotypical images of marine scientists, nor did we find any negative connotations from students regarding marine science. The absence of these stereotypical depictions, especially in the pre-tests, was surprising and raised questions about whether perceptions of scientists have shifted in the 40 years since Chambers' study. It also prompted us to consider whether other studies have examined changes in perceptions of scientists using a pre- and post-test methodology.

For informal marine educators, one recommendation is to critically examine the biases inherent in their curricula. After completing this study, we realized that our entire instructional team had backgrounds in marine biology or biological sciences. This inevitably influenced the design of our curriculum and learning experiences, potentially shaping students' perceptions. Furthermore, the guest scientists who interacted with our students may have impacted how they perceived the work of marine scientists. For instance, if many guest speakers were marine biologists, students might be more inclined to depict that type of work in their drawings. Additionally, the lack of gender identity and ethnic diversity within our instructional team may have influenced students' conceptualizations of marine scientists. If students were drawing on their instructors as representations of scientists, the team's backgrounds, interests, gender identity, and ethnicities could have shaped their pre- and post-test responses.

While many student drawings focused on marine biology, rather than on other disciplines like chemistry, geology, or physics, this isn't necessarily negative. We see the value in using marine biology as a gateway to the broader marine sciences, particularly for our Year 1 students. Although it's essential to teach the diverse aspects of marine science, biological connections may be necessary to first engage students with an unfamiliar topic. Moreover, despite some students continuing to emphasize biological concepts in their post-tests, others incorporated elements of physical science, such as water quality—a major focus of our camp. This suggests that with a well-rounded curriculum, students can be exposed to various marine science disciplines and may integrate them into their understanding of the field.

9.6.2 Longitudinal Impacts

This research study examined a program in which students participated over four consecutive summers, with the option to take marine science in the final 2 years. The program's design, offering continual engagement from early adolescence through the middle of high school, underscores the importance of repeated exposure to STEM fields (Alexander 2021). By the time students enter our Year 1 and Year 2 marine science programs, they have already engaged in 2 years of other STEM-related camps. Unsurprisingly, this has resulted in a high initial interest in STEM among our students, as reflected in the pre-tests, where 77% of long-term students reported an interest in a STEM career at the beginning of their time with us.

What stood out, however, were the changes observed from pre- to post-test each year and the impact of a second year of participation. Data from 22 long-term students were included in this analysis. These students (17% of the total study population) participated in both years and completed all pre- and post-tests. Even within a program designed for repeated exposure, collecting longitudinal data posed significant challenges. What happens to these students' perceptions of marine scientists—and more importantly, their self-concept as scientists—after a year or more has passed since the program ended? This is where targeted marine education programs can offer solutions.

With marine science opportunities already limited for many students (BIMS in Review 2022), providing repeated exposures can be challenging. Nonetheless, we propose not only offering opportunities at each grade level, but also ensuring these opportunities continue as students mature. One solution within our own program is an extension of the camp for students in their final 2 years of high school. This initiative arose from a commitment to repeated exposure, allowing a subset of our motivated Year 2 students to return to the research institute, work with graduate student mentors, and conduct a 2-week research project in marine science. For marine education providers, we suggest tracking students' progress over time, acknowledging that this requires considerable time and resources and that it may be complicated to track students participating in a variety of programs. However, the insights we gained into students' growth have been invaluable.

9.6.3 Limitations

While this study offers valuable insights into students' perspectives of marine scientists and themselves as scientists, it is important to acknowledge certain limitations that may affect the interpretation and generalization of the results. These limitations are inherent to the research design, methodology, and context, and addressing them can help guide future research efforts in this field. First, the longitudinal study sample size was limited due to difficulties obtaining parental consent as well as students missing one or more pre/post-tests due to departing early because of illness or family reasons. These limitations were further complicated by our first year of programming being delivered virtually due to the COVID-19 pandemic. However, the trends in the reduced sample size, long-term longitudinal data are similar to the trends seen in the data from the larger sample, lending validity to the long-term longitudinal data despite the reduced sample size.

Additionally, there were challenges with the implementation of data collection including student apathy to the pre/post-test framework as well as disruptions in data collection due to differences in staffing and the way testing was administered. In the most recent years of the study, data collection methods were more standardized to limit student testing apathy as well as to deliver testing prompts with a more standardized process. For the Draw a Scientist portion of the study, we also found that there was some misinterpretation or apathy when students were asked to write three words that described their scientist. Some students wrote adjectives that described the drawing itself (e.g. stick figure, person, circle head, etc.…) or wrote that their drawing was of a scientist rather than the qualities of their imagined marine scientist. To reduce the frequency of this trend we moved the 3-word prompt before the drawing and recommend clarifying the wording of this part of the prompt in the future.

Lastly, we recognize that there are implicit biases related to the interpretation and coding of gender identity and ethnicity in student drawings. Students were not explicitly prompted to indicate gender identity or ethnicity in their drawings and had limited availability of colored pencils/markers. Additionally, exposure to marine scientists of varying ethnicities during camp may have been limited due to a lack of diversity at the study institution, as well as in the field of marine science as a whole. Researchers were careful to not infer gender identity or ethnicity where it was not explicitly stated or obvious, therefore our results could be an underrepresentation of the racial and gender diversity intended by students.

9.6.4 Future Research

Efforts to increase diversity in STEM fields can help bring different perspectives that are desperately needed to innovate solutions to current and future problems. The results from this study suggest the important role of informal education programs in the formation of individual participants' science identity by impacting their knowledge and attitudes towards science and the role science plays in their own future. These results could assist program educators, as well as other informal educators, in reaching these underrepresented audiences by using marine science as a "hook" to engage students in STEM. These results also demonstrate the importance of longitudinal data, when appropriate, to track the impacts of such programs.

Armed with data from the first 4 years of the program, instructors and organizers hope to broaden the existing curriculum and offer additional opportunities in marine science for participants. This could be done by connecting participants to internships, courses, and other marine science programs both on campus and at other research institutions.

9.6.5 Summary

Key concepts explored in this chapter include:

- The field of marine science lacks diversity and a focus on reaching all populations is key to meeting the Ocean Decade goals.
- Students generally have a positive viewpoint of marine scientists, but may lack the knowledge and understanding to specifically describe the work that they do.
- After enrichment programs, students were better prepared to discuss marine scientists' work.
- Long-term participation (2 successive years) proved impactful for changing students' perspective of themselves as a scientist.
- When possible, longitudinal data should be generated to assess impacts of marine education efforts.

9.6.6 Recommended Resources

The following resources provide further insights and support on this topic:

Books

- Campbell, G., Dened, R., & Morrison, C. (Eds.). (2000). *Access denied: Race, ethnicity, and the scientific enterprise.* Oxford University Press. https://doi.org/10.1093/oso/9780195107746.001.0001
- Holmegaard, H. T., & Archer, L. (Eds.) (2022). *Science Identities: Theory, method and research.* Springer. https://doi.org/10.1007/978-3-031-17642-5
- Graham, J., Caceres, C., & Santos de Azevedo Menna, D. (Eds.). (2022). *Minorities in Shark Sciences: Diverse Voices in Shark Research.* CRC Press. https://doi.org/10.1201/9781003260370

Articles

- Ambrosino, C. M., & Rivera, M. A. J. (2022). A longitudinal analysis of developing marine science identity in a place-based, undergraduate research experience. *International Journal of STEM Education, 9*(1). https://doi.org/10.1186/s40594-022-00386-4
- Barkan, J., John, J., Drenkard, E., & Talley, D. (2023). Ocean Discovery Institute's Model for Empowering Underrepresented Students in STEM: Community-Based, Continuous Belief. *Oceanography, 36*(4). https://doi.org/10.5670/Oceanog.2024.117
- Çolakoğlu J., Steegh, A., & Parchmann, I. (2023). Reimagining informal STEM learning opportunities to foster STEM identity devel-

opment in underserved learners. *Frontiers in Education, 8*:1082747. https://doi.org/10.3389/feduc.2023.1082747

- Dewitt, J., Osborne, J., Archer, L., Dillon, J., Willis, B., & Wong, B. (2013). Young children's aspirations in science: The unequivocal, the uncertain and the unthinkable. *International Journal of Science Education, 35*(6), 1037–1063. https://doi.org/10.1080/09500693.2011.608197
- Farland-Smith, D. (2012). Development and field test of the modified draw-a-scientist test and the draw-a-scientist rubric. *School Science and Mathematics*, *112*(2), 109–116. https://doi.org/10.1111/j.1949-8594.2011.00124.x
- Lewis, S., Holloway, A., & Yarincik, K. (2023). Assessing Diversity in US Ocean Science Institutions: Insights from Fifteen Years (2007–2021) of OSER Data. *Oceanography, 36*(4). https://doi.org/10.5670/Oceanog.2024.134
- Zeigler, M. C. (2015). Engaging audiences underrepresented in STEM fields, presented at Public Libraries & STEM: A National Conference on Current Trends and Future Directions, Denver, 2015. https://www.lpi.usra.edu/education/stemlibraryconference/events/Zeigler.pdf

Websites

- Black in Marine Science (BIMS). (2024). *Home*. https://www.blackinmarinescience.org
- Minorities in Shark Science (MISS). (2024). *Home*. https://www.misselasmo.org
- Minorities in Aquaculture (MIA). (2024). *Home*. https://www.mianpo.org
- Black Girls Dive Foundation. (2024). *Home*. https://blackgirlsdivefoundation.org

Acknowledgements We would like to express our sincere gratitude to the dedicated staff who provided invaluable program and instructional support, as well as to the students whose participation made this study possible. We also extend our appreciation to our editor for her thoughtful guidance, and to our business administrator for her invaluable assistance. Their contributions have been instrumental in bringing this work to fruition.

References

Alexander AB (2021) Exploring student attitudes and outcomes toward STEM careers after repeated participation in STEM outreach. (Publication No. 28847826) [Doctoral dissertation, North Carolina State University]. ProQuest Dissertations Publishing

Black in Marine Science (BIMS) (2022) BIMS in review 2020-2022. https://www.blackinmarinescience.org/bimsreview.html

Chambers DW (1983) Stereotypic images of the scientist: The draw-a-scientist test. Sci Educ 67(2):255–265. https://doi.org/10.1002/sce.3730670213

Committee on Equal Opportunities in Science and Engineering (CEOSE) (2018) Biennial report to Congress 2017–2018: Investing in diverse community voices. https://www.nsf.gov/od/oia/activities/ceose/reports/CEOSE_ReportToCongress_RP_FVmp_508.pdf

Finson K, Beaver J, Cramond B (1995) Development and field test of a checklist for the draw-a-scientist test. Sch Sci Math 95(4):195–205. https://doi.org/10.1111/j.1949-8594.1995.tb15762.x

Graham J, Hodsdon G, Busse A, Crosby MP (2022) BIPOC voices in Ocean sciences: A qualitative exploration of factors impacting career retention. J Geosci Educ 71(3):369–387. https://doi.org/10.1080/10899995.2022.2052553

Kricorian K, Seu M, Lopez D, Ureta E, Equils O (2020) Factors influencing participation of underrepresented students in STEM fields: Matched mentors and mindsets. Int J STEM Educ 7(16). https://doi.org/10.1186/s40594-020-00219-2

Kurt S (2021) Situated learning theory. Educational Technology. https://educationaltechnology.net/situated-learning-theory

Lichtenberger E, George-Jackson C (2013) Predicting high school students' interest in majoring in a STEM field: Insight into high school students' postsecondary plans. J Career Tech Educ 28(1):19–38. https://doi.org/10.21061/jcte.v28i1.571

National Research Council (2009) Learning science in informal environments: People, places, and pursuits. In: Learning science in informal environments. National Academies Press. https://doi.org/10.17226/12190

Plankis B, Marrero M (2010) Recent Ocean Literacy research in United States public schools: Results and implications. Int J Environ Educ 1(1) https://files.eric.ed.gov/fulltext/EJ1055372.pdf

Reardon SF (2011) The widening academic achievement gap between the rich and the poor: New evidence and possible explanations. In: Duncan G, Murnane R (eds) Whither opportunity? Rising inequality, schools, and children's life chances. Russell Sage Foundation, pp 91–116. http://www.jstor.org/stable/10.7758/9781610447515

Smith MK (2009) Jean Lave, Etienne Wenger and communities of practice. The Encyclopedia of Pedagogy and Informal Education. https://infed.org/mobi/jean-lave-etienne-wenger-and-communities-of-practice/

St Clair R, Benjamin A (2011) Performing desires: The dilemma of aspirations and educational attainment. Br Educ Res J 37(3):501–517. https://doi.org/10.1080/01411926.2010.481358

Staus NL, Falk JH, Penuel W, Dierking L, Wyld J, Bailey D (2020) Interested, disinterested, or neutral: Exploring STEM interest profiles and pathways in a low-income urban community. Eurasia J Math Sci Technol Educ 16(6). https://doi.org/10.29333/EJMSTE/7927

Weber K (2011) Gender and grade level differences in interest, perceived personal capacity, and involvement in technology and engineering-related activities. (Publication No. 3478117) [Doctoral dissertation, Fielding Graduate University]. ProQuest Dissertations Publishing

Author Contributions and Biographical Profiles

[1]*Sarah Nuss: Co-conceptualization, Writing—Original Draft, Review, and Editing, Methodology, Project Administration, Supervision.

[2]Anna Caputo: Co-conceptualization, Writing—Original Draft, Methodology—Data Collection and Analysis.

[3]Lisa Lawrence: Writing—Original Draft, Review, and Editing, Methodology—Data Collection and Analysis, Project Administration.

[4]Greta Olsen Gardiner: Writing—Original Draft, Project Administration.

[5]Bethany Smith: Writing—Original Draft, Project Administration.

[6]Celia Cackowski: Writing—Original Draft, Project Administration.

*Corresponding Author at mcguire@vims.edu

Sarah Nuss, PhD, is the Director of the Virginia Institute of Marine Science (VIMS) Office of Outreach and Engagement, as well as the Education Coordinator for the Chesapeake Bay National Estuarine Research Reserve in Virginia (CBNERR-VA). Sarah has worked with VIMS and CBNERR-VA for the last 20 years where she and her team specialize in education projects such as the VIMS summer camp program, experiential field experiences for students, and teacher and graduate student professional development. Sarah's research focuses on the influence of pre-service teacher preparation on teachers' inclusion of environmental education, impact of experiential marine science on students and teachers, and pathways leading to a career in Ocean science.

Anna Caputo is a former Marine Education Specialist at the Chesapeake Bay National Estuarine Research Reserve in Virginia. She received her M.A. from the Virginia Institute of Marine Science in 2023 and her B.A. in Environmental Studies from Green Mountain College in 2016. Both of her degrees were focused on environmental education. As a marine science educator Anna has worked at two other National Estuarine Research Reserves in Rhode Island and New Jersey. Her professional interests involve connecting people with marine science and the natural world by providing hands-on learning opportunities that combine field experiences, research, and nature journaling.

Lisa Lawrence is the Associate Director of Broader Impacts and Strategic Initiatives for the VIMS Office of Outreach and Engagement. She earned a master's degree in fisheries from VIMS during which time she became involved in education and outreach, igniting her passion for and launching her career in marine education. Lisa's work focuses on development of resources and professional development for teachers, guiding graduate students in science communication, and collaborating with researchers to translate their work for public impact. Lisa is a recipient of the Mid-Atlantic Marine Education Association's Informal Educator Award and the National Marine Educators Association's James Centerino Award.

Bethany Smith is the Associate Director of Formal Education for the VIMS Office of Outreach and Engagement. After completing her master's degree in biological oceanography at the College of William and Mary's Virginia Institute of Marine Science, Bethany spent 11-years as a classroom teacher, inspiring the next generation of Ocean science professionals. Since joining VIMS in 2020, Bethany's work has focused on the intersection of Ocean science research and K–12 education. Bethany's programs bridge the gap between Ocean science research and applied learning for students and educators.

Greta Olsen Gardiner is a 2020–2022 NOAA Ernest F. Hollings Scholar. During her internship with the Chesapeake Bay National Estuarine Research Reserve in Virginia, Greta collaborated on projects with the education team at the Virginia Institute of Marine Science. Her research focused on assessing the impact of informal marine science education on underserved communities and developing educational resources about Ocean acidification. After earning a B.S. in Chemical Engineering from the University of Kansas in 2022, Greta now works as a process engineer at an oil refinery.

Celia Cackowski serves as the Digital Resources Coordinator for the VIMS Office of Outreach and Engagement. She holds a master's in biology from Miami University and an executive certificate in science communication from Duke. Before joining VIMS in 2016, Celia spent 2 years as a National Science Foundation Fellow and 8 years at the University of Rhode Island as an outreach scientist for both the Census of Marine Life and the Inner Space Center. Her current projects promote Ocean Literacy and workforce development.

Investigating the Longer-term Student Outcomes of a Two-day Ocean-focused Program for Underserved Youth in Cape Town, South Africa

10

B. Troy Frensley, Hannah Eaton, Claire Carretti, Katherine Wallis, Amalia De Abreu, and Peter Marshall

Abstract

We present the findings from a mixed-methods evaluation focused on measuring longer-term student outcomes from the Ocean Guardians Workshop (OGW). The OGW is a two-day, Ocean education program for 11–13-year-old underserved students in western Cape Town, South Africa. The OGW incorporates beach clean-ups, ecosystem explorations, snorkeling, mindfulness approaches, and more. We used pre-, post-, and follow-up surveys developed with the organization to measure outcomes related to Ocean Learning, Attitudes Towards the Ocean, and Ocean Stewardship Behaviors for 266 students that attended the OGW. Our findings suggest positive short-term and longer-term gains across many of these student outcomes. Particularly noteworthy outcomes include positive changes in student's attitudes towards the Ocean, which persisted months after the program, along with over 70% of students reporting performing one or more Ocean stewardship behaviors in the four months after the OGW. This study helps to fill a gap in measuring diverse, and longer-term, student outcomes. In this chapter, we share details on the collaborative nature of this evaluation study, findings from these efforts, and future directions.

Keywords

Ocean Literacy (OL) · Program evaluation · Ocean education · Environmental education · Environmental literacy · Evaluation capacity

10.1 Introduction

Environmental and Ocean education programs for youth often engage students in immersive, hands-on experiences directly in the natural environment that provide unique learning opportunities. For many programs like these, increasing

B. T. Frensley (✉) · H. Eaton
University of North Carolina, Wilmington, NC, USA
e-mail: frensleyb@uncw.edu; eatonh@uncw.edu

C. Carretti · A. De Abreu · P. Marshall
I AM WATER: Ocean Conservation, Cape Town, South Africa
e-mail: claire@iamwaterfoundation.org; amalia@iamwaterfoundation.org; peter@iamwaterfoundation.org

K. Wallis
Blue Alliance Marine Protected Areas, Cape Town, South Africa
e-mail: KatherineW@bluealliance.earth

T. J. Kennedy (ed.), *Ocean Literacy: The Foundation for the Success of the Ocean Decade, Volume I*, Sustainable Development Goals Series, https://doi.org/10.1007/978-3-031-98145-6_10

students' Ocean Literacy (OL) or elements of environmental literacy can be aspirational goals. OL, in the broadest sense, focuses on increasing student understanding of the socio-ecological relationships that humans have with the Ocean (E.g., the impact of the Ocean on humans and the impact humans can have on the Ocean; McKinley et al. 2023). OL is then broken down into the seven essential principles with one or more of these principles often being the focus of Ocean education programs for youth (Halversen et al. 2021). Environmental literacy, which was first described in the international Tblisi Declaration (UNESCO 1977), focuses on increasing environmental knowledge, positive attitudes and dispositions towards the environment, skills needed to address environmental issues, and ultimately behaviors that make a positive difference.

10.1.1 Evaluating Ocean Education Programs for Youth

A program evaluation, more specifically a summative evaluation, can be used to measure outcomes after participating in an environmental or Ocean education program to help determine if programmatic goals are being met. Reviews of decades of evaluation studies on environmental and Ocean education programs for youth suggest a wide range of outcomes can be achieved including increased knowledge, more positive environmental attitudes, increased skills, stronger behavioral intentions, and even behaviors (Ardoin et al. 2018; Stern et al. 2014; Thomas et al. 2018). As such, well-designed program evaluations can provide evidence on the strengths and areas for improvement for a specific program and these data can facilitate meaningful reflection, learning, and programmatic changes to promote more positive results in the future (Stern et al. 2022).

Despite these benefits, program evaluations are uncommon due to the technical skills, resources, and time required to conduct an evaluation. This leaves many organizations either partnering with external evaluators, if they can secure funding to do so, or not conducting any formal evaluation at all (Anderson et al. 2022). Developing the outcomes and corresponding data collection measures can be a complicated process in evaluation. In the Ocean education field specifically, there has been a "knowledge-centric" (McKinley et al. 2023) focus to many student outcomes measured in evaluation studies. For example, the International Ocean Literacy Survey (Fauville et al. 2018) was developed and statistically validated to assess the Ocean knowledge of 15–17-year-olds and has been used around the world. Changes in knowledge are important to understand but there is not a linear relationship between knowledge and behaviors. If an ultimate goal of an Ocean education program is behavior change (E.g., actions related to Ocean conservation), other outcomes such as measuring changes in attitudes or behavioral intentions, are also important to incorporate into evaluations of Ocean Literacy (OL) programs (Ashley et al. 2019; McKinley et al. 2023).

Additionally, most program evaluations focus only on short-term student outcomes immediately after a program because it is extremely difficult to reach student participant's weeks, months, or even years later to collect follow-up data. As such, very few studies measuring longer-term student outcomes exist and as a result, we know very little about how program outcomes persist over time (Stern et al. 2014). For example, we can measure behavioral intentions after an Ocean education program, but intentions alone may not lead to behaviors (Ajzen 1985). Student behaviors can only be measured using a follow-up approach when they have had time to perform relevant behaviors. This study seeks to address these gaps by sharing a mixed-methods evaluation study using pre-, post-, and delayed follow-up surveys to measure short- and longer-term student outcomes associated with an Ocean education program called the Ocean Guardians Workshop (OGW).

10.2 The Ocean Guardians Workshop

Ten critical challenges have been identified in the United Nations Decade of Ocean Science for Sustainable Development (UNESCO 2021) that are pivotal for ensuring the protection of our Oceans. These ten challenges, along with the seventeen United Nations Sustainable Development Goals (SDGs; 2015), offer aspirational targets for those striving to foster a healthier Ocean and a more sustainable future. I AM WATER: Ocean Conservation is a nonprofit organization located in Muizenberg, Cape Town, South Africa, that develops transformative human-Ocean experiences designed to meet some of these Ocean Decade challenges and SDGs. Their flagship experience, and the focus of this evaluation study, is a youth program called the Ocean Guardians Workshop (OGW), that is closely tied to SDGs 4 and 14, as with Ocean Decade Challenges 1, 2, 3, 9, and 10 (see Table 10.1).

The OGW is a two-day immersive Ocean education program reaching 11–13-year-old underserved youth attending schools in Western Cape Town, South Africa (SDG 4, see Table 10.1). Throughout these two days, students engage in a variety of hands-on activities beginning each day with mindfulness techniques focused on breathing and stress management followed by stations focused on ecosystem explorations, snorkeling, and service-learning projects (a detailed description of each day's activities can be found below). Student groups of 36 individuals, or less, travel by bus from their schools to different Marine Protected Area (MPA) sites (SDG 14, see Table 10.1) or these field-based programs (Ocean Decade Challenge 2, see Table 10.1). There are four potential MPA sites used in the OGW: St. James (1), Millers Point (2), Glencairn (3), and Windmill Beach (4; numbers correspond with locations on the map, see Fig. 10.1). On the first day, students visit either St. James or Glencairn because these sites contain calm, concrete swimming areas for students to learn to snorkel and build their confidence. Windmill Beach or Miller's Point are the options for day two sites because they contain sheltered kelp forests and rocky shorelines that allow the students to venture farther out into the Ocean and explore.

One of the unique aspects of the OGW is that the educators, called coaches, are from many of the same communities as the student participants. These coaches are not required to have a professional background in education, fisheries, or other relevant fields. The only requirement is that they are passionate about the Ocean. New coaches undergo a rigorous training program and are put on teams with more experienced coaches to provide support in the field when delivering the OGW. The coaches range in age from 18 to 35 and reflect the diversity of the student population and communities served. As such, these coaches are uniquely equipped to connect with the participating students, understand their perspectives and challenges, and ignite a passion for Ocean protection. Students may view these coaches as role models due to these strong cultural and com-

Table 10.1 OGW Links to Ocean Decade Challenges and SDGs

SDG 4 is to "ensure inclusive and equitable quality education and promote lifelong learning opportunities for all" (United Nations 2015)
SDG 14 is to "conserve and sustainably use the Oceans, seas, and marine resources for sustainable development" (United Nations 2015)
Ocean Decade Challenge 1 is to "understand and beat marine pollution" (UNESCO 2021)
Ocean Decade Challenge 2 is to "protect and restore ecosystems and biodiversity" (UNESCO 2021)
Ocean Decade Challenge 3 focuses on "sustainably nourishing the global population" (UNESCO 2021) and students learn about the role of MPA's and sustainable fishing
Ocean Decade Challenge 9 is to develop "skills, knowledge, technology, and participation for all" (UNESCO 2021)
Ocean Decade Challenge 10 is to "restore society's relationship with the Ocean" (UNESCO 2021)

Fig. 10.1 Marine Protected Area (MPA) sites used in the OGW (Esri 2023)

munity matches, potentially resulting in more positive experiences and enhanced student outcomes (Stern et al. 2018).

10.2.1 Ocean Guardians Workshop: Day 1

On the first day of the OGW, after putting on their wetsuits and boots, the students begin with yoga-inspired stretching, breathing exercises, and mindfulness activities. Most students have never worn a wetsuit or been snorkeling before and are anxious about the experience, so these mindfulness approaches are intentionally designed to overcome these stressors and help the students be present and open to learning and connecting. The coaches then present key OL concepts, information on local species and ecosystems, and details about the MPA site. Students are subsequently split into smaller groups where they rotate through three stations: ecosystem exploration, snorkeling, and performing a beach cleanup. The ecosystem exploration focuses on increasing participants' knowledge of MPAs, the importance of MPAs as nurseries for marine life, sustainable fishing practices, local species identification, and more (Ocean Decade Challenge 3, see Table 10.1). The beach cleanup focuses on educating participants about the impacts of pollution on marine ecosystems and the students see firsthand the positive benefits of their efforts to clean up the beach (Ocean Decade Challenge 1, see Table 10.1). At the snorkeling station, all equipment is provided, and students are split into even smaller groups (typically four or less) with a coach for each group. After going over water safety, students learn how to wear a mask and breathe out of a snorkel on the beach and then slowly venture out into the shallow, calm swimming area where they hold onto a round float and are guided by their coach as they get comfortable looking under the water (see Fig. 10.2). Breathing, yoga, and mindfulness activities are referenced by the coaches to remind the students how to control their emotions as many see life under the surface of the Ocean for the first time (Ocean Decade Challenge 9, see Table 10.1). Day one ends with lunch and then a sharing circle, where participants reflect on their experiences before returning to school.

Fig. 10.2 Snorkeling Station and exploration for beginning students (Photo credit: I AM WATER, Peter Marshall)

10.2.2 Ocean Guardians Workshop: Day 2

On the second day participants are transported to either Miller's Point or Windmill Beach MPA sites. These locations offer an open beach front, rocky shoreline, and kelp forest with no separate swimming area. After changing into their wetsuits and boots, participants begin their day with yoga, stretching, and breathing exercises, reinforcing the concepts of stress management and mindfulness from the previous day. Then participants are split into three groups and rotate again through ecosystem exploration, snorkeling, and beach cleanup stations. At the ecosystem exploration station, many of the concepts covered on the first day are reinforced and students also learn about charismatic species found in these MPAs, such as the endangered African penguin (*Spheniscus demersus*). The beach cleanup station offers a similar and positive experience as the prior day and students reflect on the impacts of pollution. At the snorkeling station, students and their coaches explore deeper water, including snorkeling along a rocky shoreline and through a kelp forest (see Fig. 10.3). If participants feel confident in their snorkeling ability, they are allowed to dive under the water beside their coach. After lunch, the OGW concludes with each student reflecting on this experience and deciding what they will do in the future to protect the Ocean. Each student then stands in front of the group and announces what they will do to protect the Ocean. After they share this pledge with everyone, they are given an Ocean Guardian certificate recognizing their efforts and pledge to be an Ocean Guardian from now on (see Fig. 10.4).

Fig. 10.3 Snorkeling Station located in deeper water along rocky shoreline and kelp forest (Photo Credit: I AM WATER)

Fig. 10.4 Ocean Guardian Pledge Ceremony (Photo credit: I AM WATER, Danielle Davenport)

10.3 A Participatory Evaluation Approach

We began these efforts in 2021 and modeled our participatory evaluation approach on prior relevant efforts (Powell et al. 2006). We started by working with organization staff to learn as much as we could about the OGW program and their corresponding theory of change. The organization was interested in diverse and aspirational student outcomes and was excited about measuring aspects of environmental literacy, Ocean Literacy (OL), and social-emotional learning in OGW participants. As such, our survey development was informed by existing statistically validated instruments such as the Environmental Education for the Twenty-first Century (EE21; Powell et al. 2019) and the International Ocean Literacy Survey (Fauville et al. 2018). We also worked iteratively with the organization to refine survey items, and a final pre-survey and post-survey were developed and used in a pilot evaluation in 2022.

We then built upon these pilot efforts and added a follow-up survey (4 months after participation) in 2023. We use a mixed-methods approach that includes both closed-ended and open-ended items. The closed-ended items use an 11-point, unipolar Likert-type scale to counter issues regarding positive-skew and lack of variability associated with 'balanced' bipolar Likert-type scales when conducting evaluation research in this context (Powell et al. 2019). These 11-point scales are relatable to the pain scale youth often see when they visit the doctor's office, rating pain from 0 to 10. In this chapter, we will focus only on a selection of longer-term student outcomes from a comprehensive OGW program evaluation in 2023. This selection includes the student outcomes of Ocean Learning, Attitudes Towards the Ocean, and Ocean Stewardship Behaviors, each of which are described below.

10.3.1 Ocean Learning

Seven closed-ended survey items measure 'Ocean Learning' which we define as students' knowledge regarding the interconnectedness and interdependence between humans and the Ocean environments (see Table 10.2 for items; Powell et al. 2019). These knowledge items focus on the overarching aim of Ocean Literacy (OL; Halversen et al. 2021). These items are retrospective in nature, only occurring on the post-survey and follow-up survey, to minimize pre-survey response bias commonly observed with knowledge-based questions (Powell et al. 2019). The survey stem for each of these seven items asks students "*How much did you learn about the following things as a result of this experience*" and the Likert-type scale is focused on measuring self-reported changes in knowledge based on the experience (0 = nothing at all; 5 = a fair amount; 10 = a huge amount). On the follow-up survey, we also asked students an open-ended question, "*What was the most important thing you learned from this experience?*" and students provided written responses.

10.3.2 Attitudes Towards the Ocean

Seven closed-ended items measure students' 'Attitudes Towards the Ocean' at three points in time: before the OGW (pre-survey), immediately after the OGW (post-survey), and 4 months later (follow-up survey). We define 'Attitudes Towards the Ocean' as the students' sensitivity, concern, and positive dispositions towards the Ocean (see Table 10.3 for items; Powell et al. 2019).

On the post-survey, we also asked students "*Because of this workshop, do you feel differently about the Ocean now? If yes, how have your feelings about the Ocean changed?*" and students who answered yes, provided written responses. On the follow-up survey, we asked students an open-ended question, "*Because of this experience, how do you feel about the Ocean now?*" to better understand any longer-term shifts in attitudes. These open-ended items allow for more nuance and individual changes in students' attitudes to be captured.

10.3.3 Ocean Stewardship Behaviors

As previously described, the students finish their second day with the Ocean Guardians ceremony, where they share their pledge to protect the Ocean, which are behavioral intentions, and receive their certificate stating they are now an Ocean Guardian. On the follow-up survey, we asked students an open-ended question, "*Do you still remember your promise to the Ocean on the final day of the workshop? If yes, please write your promise in the space below.*"

We have conceptualized 'Ocean Stewardship Behaviors' as students' motivations and performance of Ocean stewardship-related behaviors (Powell et al. 2019). We measured seven different behavioral intentions related to Ocean stewardship (see Table 10.4 for items) on the pre-survey and post-survey to gauge changes in students' intentions but our focus in this chapter is on whether students actually performed these behaviors during the four months since their OGW, as behaviors are rarely reported. To measure this, the follow-up survey asks students "*Since your workshop, have you done any of the following things?*" and students circled '*Yes*' or '*No.*' A single open-ended item used on the follow-up survey also asks the students *"Is there anything else you want to tell us about what you have done since the workshop? If yes, please write in the space below"* to capture additional details about their Ocean stewardship behaviors.

10.3.4 Building Evaluation Capacity

As already described, the staff from the I AM WATER: Ocean Conservation organization in Cape Town, South Africa played a key role in the development of the evaluation methods. The OGW coaches also administered each survey and staff members entered data used in this evaluation. These data were entered into Microsoft Excel with support from the evaluation team at UNCW. The evaluation team at UNCW conducted all data analyses and reporting. Quantitative data were analyzed using IBM SPSS version 28 and qualitative data were transcribed and thematically coded by the two lead authors to ensure reliability. This evaluation was reviewed and granted exempt status by the University of North Carolina Institutional Review Board (IRB #H24-0023) prior to any data collection.

10.4 Evaluation Results

A total of 13 OGW programs were delivered to 378 youth from eight schools during March 2023 to May 2023. I AM WATER staff and coaches administered pre-surveys immediately before the start of the first day, post-surveys at the end of the second day, and the follow-up surveys were facilitated by I AM WATER staff to these participating students at their schools approximately four months after their OGW program. Individual student pre-, post-, and follow-up surveys were matched and any student completing less than 50% of any survey, or that missed one or more of the surveys, were removed from the sample.

This resulted in a final sample of 266 student pre-, post-, and follow-up surveys used for analyses. This represents a 79% total response rate for all students that participated in the OGW. In this sample of students, 59.6% self-identified at female (40.4% male) with no students self-reporting any other gender identity. For race/ethnicity, 70.3% of students self-reported as being Coloured (mixed racial/ethnic heritage) followed by 28.8% reported as Black, 1.5% as White, and 0.4% as Other.

10.4.1 Ocean Learning Outcomes

Students' mean scores for each of the seven survey post-survey Ocean Learning items were very positive, suggesting the students learned a lot about each of these during the OGW. We performed a paired-samples t-test to explore any changes in mean scores from the post-survey to the follow-up survey four months later (statistically significant changes are noted in bold in Table 10.2). Three items show a statistically significant decrease in mean scores four months later (*How different parts of the environment interact with each other; How my actions affect the Ocean;* and *How climate change is impacting the Ocean*) although the mean scores remain very positive on this Likert-type scale. Conversely, student mean scores for four items show no statistically significant differences four months later, which is extremely exciting and suggests positive learning gains are persisting months after the program. The four items are: *How people's actions are changing the Ocean; How changes in the Ocean can impact my life; How people can benefit from the Ocean;* and *How marine protected areas help conserve the Ocean.*

Table 10.2 Paired samples mean comparisons of Ocean Learning survey items

Stem: How much did you learn about the following things as a result of this experience?	Post-experience		Follow-up				
	Mean	*SD*	*Mean*	*SD*	*t*	*df*	*p*
How different parts of the environment interact with each other.	8.65	1.97	7.96	1.98	−4.27	251	**<0.001**
How people's actions are changing the Ocean.	8.68	2.00	8.76	1.86	.530	254	0.596
How changes in the Ocean can impact my life.	8.67	1.86	8.47	1.81	1.41	236	0.160
How my actions affect the Ocean.	8.66	1.94	8.32	2.04	2.07	246	**0.040**
How people can benefit from the Ocean.	8.56	2.03	8.45	1.94	0.68	252	0.495
How marine protected areas help conserve the Ocean.	8.44	2.12	8.42	1.87	0.11	249	0.916
How climate change is impacting the Ocean.	8.64	1.93	8.30	1.99	2.19	250	**0.029**

Note: Likert-type scale 0–10; 0 = nothing at all; 5 = a fair amount, 10; a huge amount

Four months later, students were asked to recall the most important thing they learned participating in the OGW. We received responses from 255 students and after qualitative coding, three primary themes emerged. The first theme shared by almost half of these students (122) focused on learning about the impacts of pollution on the Ocean with responses such as, "*we need to keep the Ocean clean*" and "*I learned not to throw pages*." This was followed by 78 students noting the importance of protecting wildlife with statements like "*I learned we must not hurt sea animals, and we must not let the Ocean dirty*" and "*To respect and keep animals safe*." Finally, 42 students said the most important thing they learned was about protecting the Ocean with comments like "*…it is important to protect the Ocean*" and "*how to protect the Ocean*."

10.4.2 Attitudes Towards the Ocean

We performed a paired-samples t-test to identify any statistically significant changes in mean scores for Attitudes Towards the Ocean from pre-experience to post-experience and then again from pre-experience to follow-up (statistically significant differences are noted in bold in Table 10.3). All survey items but one show a statistically significant increase in student mean scores from before the OGW (pre-survey) to immediately after (post-survey) suggesting a positive short-term attitude shift because of the OGW. In the longer-term, four items remained statistically significantly higher when compared to the pre-survey (baseline) indicating the positive attitudinal shift seen immediately after the OGW also persisted months later. These four items are: *I have the power to protect the Ocean; I am comfortable when I visit the beach; It is important to eat sustainably caught fish*; and *Marine Protected Areas are important.*

On the post-survey immediately after the OGW, students were asked if they feel differently about the Ocean now. Over 81% of the students (186) responded 'yes,' they did feel different about the Ocean now. We then asked any student that responded yes to explain how their feelings have changed, and two common themes emerged after qualitative coding. The most common theme (53 students) focused on general positive feelings about the Ocean and Ocean conservation with responses such as, "*Yes, I feel differently about taking care of the Ocean and keeping it safe too,*" "*I feel the Ocean is another place I can call home*" and "*I fell [feel] that what I learned is right. I learned that we need to protect the Ocean.*" Many students (46) also noted they were less fearful of the Ocean and shared feelings such as, "*I was scared of the Ocean and now I'm not,*" "*Because I always thought that the Ocean is deep and the moment you step in it, you will drown,*" and "*I am not scared of the Ocean anymore and I feel safe*."

Four months later, we asked the students how they feel about the Ocean now. A total of 257 stu-

Table 10.3 Paired samples mean comparisons of Attitudes Towards the Ocean survey items

Pre-experience vs. immediate post-experience							
	Pre-experience		Post-experience				
Stem: To what extent do you agree with the following statements?	*Mean*	*SD*	*Mean*	*SD*	*t*	*df*	*p*
I feel it is important to take good care of the Ocean.	9.29	1.56	9.30	1.37	0.152	243	0.879
Humans are a part of nature, not separate from it.	8.43	2.23	8.76	1.91	2.97	246	**0.003**
I have the power to protect the Ocean.	8.07	2.51	8.63	1.90	3.16	237	**0.002**
I am comfortable when I visit the beach.	8.18	2.46	8.77	2.01	3.16	237	**0.002**
It is important to eat sustainably caught fish.	6.57	3.13	7.91	2.58	6.34	243	**<0.001**
I care about marine animals.	8.37	2.22	8.92	1.74	3.32	228	**0.001**
Marine Protected Areas are important.	7.99	2.32	8.68	1.83	4.06	245	**<0.001**
Pre-experience vs. follow-up							
Stem: To what extent do you agree with the following statements?	Pre-experience		Follow-up				
	Mean	***SD***	***Mean***	***SD***	*t*	*df*	*p*
I feel it is important to take good care of the Ocean.	9.26	1.57	9.26	1.37	0.152	252	0.943
Humans are a part of nature, not separate from it.	8.33	2.23	8.34	2.02	0.048	246	0.962
I have the power to protect the Ocean.	8.03	2.54	8.58	1.91	3.06	243	**0.002**
I am comfortable when I visit the beach.	8.27	2.43	8.88	1.75	3.70	248	**<0.001**
It is important to eat sustainably caught fish.	6.57	3.06	7.09	2.94	2.46	248	**0.015**
I care about marine animals.	8.39	2.20	8.63	1.94	1.33	237	0.185
Marine Protected Areas are important.	7.97	2.43	8.62	1.76	3.86	254	**<0.001**

Note: Likert-type scale 0–10; 0 = not at all; 5 = somewhat, 10; strongly agree

dents responded to this open-ended question and the same two top themes emerged of general positive feelings about the Ocean (90 students) and overcoming a fear of the Ocean (77 students). The following student responses capture some aspects of these two main themes that persisted over time, "*I believe it is an amazing place to be at,*" "*I feel safe and I enjoy being in the Ocean,*" "*I feel so relaxed and not afraid of [the] water,*" and "*I feel very safe and proud to go under the water and I have courage because of I AM WATER.*"

10.4.3 Ocean Stewardship Behaviors

At the end of day two of the OGW, students make a pledge to do something to protect the Ocean and share that promise to the Ocean aloud during the Ocean Guardians certificate ceremony. This ceremony can be an emotional time for students and OGW coaches as they reflect on their experiences and students share what they are going to do to protect the Ocean moving forward. We wanted to know if four months later, students remembered the promise they made during this Ocean Guardians certificate ceremony. On the follow-up survey, we asked the students and 82.7% responded "yes, they remembered their promise to the Ocean." For any student responding yes, we then asked them to write down their promise. Three main themes emerged after qualitative coding. The most common was focused on behaviors dealing with pollution (102 students) and students wrote responses such as, "*I promised that I will pick up trash and make our Ocean clean,*" and "*Yes I do. I promised to never litter again.*" The second most common theme was Ocean conservation (78 students) with the focus being on general conservation promises such as "*I promised to protect the Ocean.*" The final common theme was focused on protecting wildlife (26 students) with students focused on a range of species, mostly animals, that live in and around the Ocean as one student noted, "*Yes, to protect all of the animals in the Ocean.*"

Table 10.4 summarizes the frequencies of student's reported behaviors for each Ocean

Table 10.4 Frequencies of student self-reported Ocean stewardship behaviors with examples of reported behaviors from the students for each Ocean-stewardship category

Survey Questions Stem: *Since the workshop, have you done any of the following things?*	Follow-up Frequency (sample)	Example of Student Behavior
Help to protect the Ocean.	**Yes: 88.3% (233)** No: 11.7% (31)	*"Since the workshop I have told my family and friends about it and two weeks after we went to Fish Hoek beach and Camps Bay and cleaned it and helped a baby turtle that got lost"*
Make a positive difference in my community.	**Yes: 69.5% (182)** No: 30.5% (80)	*"I have started recycling and now everyone know so now they are now and when were all at the beach, we clean a little"* *"Yes, over the holidays I joined a clean-up community to clean the beach"*
Spend more time outside.	**Yes: 81.9% (213)** No: 18.1% (47)	*"Started to love swimming"* *"After going to I AM WATER, I have been going outside because I used to be a house person"*
Talk with others about ways to protect the Ocean.	**Yes: 79.2% (210)** No: 20.8% (55)	*"Yes, I told my friend how to protect the Ocean"* *"I now spend more time outside and visiting the beach to pick up trash and I am talking with others about ways to protect the Ocean"*
Pick up trash.	**Yes: 88.5% (231)** No: 11.5% (30)	*"I went to the beach and I saw a child throwing papers and I said she must not throw papers away"* *"I am…trying to just keep my world clean"* *"I started recycling"*
Visit a beach.	**Yes: 88.6% (234)** No: 11.4% (30)	*"I visit the Ocean regularly"* *"I have been visiting the Ocean more"*
Learn more about the Ocean.	**Yes: 88.6% (234)** No: 11.4% (30)	*"I have been learning a bit more about the Ocean"*

Stewardship item on the follow-up survey along with some examples of written responses providing details for each behavior. Seventy percent or more of students reported that they have performed at least one Ocean stewardship behavior in the 4 months since their OGW and many students reported doing more than one of the Ocean stewardship behaviors in this battery. Amazingly, over 88% of students reported having performed four Ocean Stewardship behaviors (*Help to protect the Ocean, Pick up trash, Visit a beach, and Learn more about the Ocean).*

10.5 Conclusion

Overall, the results of this evaluation study suggest that the OGW has a positive short-term and longer-term impact on student participants with respect to our measures of Ocean Learning, Attitudes Towards the Ocean, and Ocean Stewardship Behaviors. In evaluation research, it is common to see statistically significant declines in student outcome scores over time (Stern et al. 2014). While we did see slight declines in student mean scores for some items in this study, other items had mean scores that persisted, or even increased, during the months after student participation in the OGW. For example, in the Ocean Learning battery, mean scores for four items remained very positive (*How people's actions are changing the Ocean, How changes in the Ocean can impact my life, How people can benefit from the Ocean;* and *How marine protect areas help conserve the Ocean*) and did not significantly decline in the four months after the OGW. Overall, our findings suggest that students are showing strong self-reported Ocean Learning gains as a result of their participation in the OGW and some of these gains persist well after the workshop is completed, which is an extremely exciting finding.

In the Attitudes Towards the Ocean battery, students reported significant positive short-term changes in mean scores for all but one item. The mean scores for the following four items were also significantly higher four months later when compared to the pre-survey (*I am comfortable when I visit the beach; It is important to eat sustainably caught fish*; *Marine Protected Areas are important,* and *I have the power to protect the Ocean;).* The OGW focuses on shifting student emotions through intentional practices of mindfulness, breathing, and yoga each day along with empowering many students to overcome their fear of the Ocean and snorkeling. Students' written responses commonly reflect that many of them did indeed overcome a fear of the Ocean and developed a strong connection with the Ocean environment (Ocean Decade, Challenge 10), with many visiting a beach again. The coaches emphasize the importance of the MPAs the students visit and sustainably caught fish, which is relevant as many communities these students are from rely on local fish for food. These direct experiences and relevant links made by the coaches seem to have a positive impact on students' attitudes towards these places and their feelings about Ocean conservation (Ocean Decade, Challenge 2).

Students leave the OGW feeling like they have the power to protect the Ocean, and that feeling persists over time. This exciting finding may be due to the Ocean Guardians ceremony at the end of day two. Verbal commitments to take actions to protect the Ocean have been shown to encourage pro-environmental behaviors (Cooper et al. 2024), making it an important aspect of closing the OGW. This heightened self-efficacy may lead to more Ocean stewardship behaviors (Ashley et al. 2019), which we observed in this study. On average, over 70% of students reported performing one or more Ocean stewardship related behavior in the four months after participating in the OGW. Many students completed more than one behavior with the most common behaviors being those that the students have the greatest capacity to perform (e.g., picking up trash, learning more about the Ocean, etc.). It is also worth noting that the coaches perform the beach cleanup activities alongside the students and cheer each student after they share their pledge to protect the Ocean as they receive their Ocean Guardian certificate. By doing so, these coaches who come from the same communities as these students are modeling and emphasizing Ocean stewardship behaviors that may have a lasting impact on the students (Stern et al. 2018). These findings are exciting and contribute to our growing understanding of student behaviors after an environmental or Ocean education program. The student behaviors reported in this study directly support Ocean Decade Challenges 1 and 2 along with the SDG 4 and SDG 14.

This evaluation study is not without its limitations. Like any questionnaire, our survey is prone to social desirability bias and students may respond the way they think others may want rather than how they really feel (Podsakoff et al. 2003). I AM WATER organization staff administered each survey and encouraged students to answer honestly and with their own opinions each time to minimize bias. Additionally, the outcomes and survey items used in this evaluation were designed to be broad and general, rather than specific, to capture more of the diverse and aspirational outcomes the organization is seeking to achieve in the OGW. While we have only included a portion of the outcomes measured in the full OGW evaluation in this chapter, each outcome was designed to be measured quickly and concisely with each survey being completed in 10 min or less. We are not suggesting that our measure of Ocean Learning, for example, is a substitute for a survey focused only on measuring Ocean Literacy (OL) and its associated key principles.

We are currently in the process of building upon this work and leveraging lessons learned for the new OCEAN FOR ALL: Alliance, which is led by I AM WATER: Ocean Conservation. This Alliance brings together Ocean educators from I AM WATER and organizations across South Africa, Ecuador, the Galapagos Islands, and Argentina and the evaluation researchers from this study, to form an evidence-based learning network, a type of community of practice, designed for collaboration, learning, and improvement (see Stern et al. 2022). We encourage future researchers and practitioners to also build upon the study described in this chapter and continue to improve our understanding of Ocean education programs and associated diverse student outcomes.

10.5.1 Summary

Key concepts explored in this chapter include:

- The Ocean Guardians Workshop (OGW) is a unique 2-day program that reaches underserved 11–13-year-old youth in western Cape Town, South Africa. This program, led by the I AM WATER: Ocean Conservation organization combines snorkeling, beach cleanups, mindfulness and breathing/yoga, and ecosystem explorations.
- The coaches (educators) in the OGW are from the same communities as the students which seems to have a positive impact on the students. The coaches are not required to have a background in education or the Ocean and are trained by organization staff to deliver the OGW.
- Evaluators and organization staff co-created the evaluation methods and built the organization's evaluation capacity to administer pre-, post-, and follow-up (four months later) surveys and enter the survey data for analyses.
- Measuring longer-term student outcomes after an Ocean or environmental program are extremely rare, so the findings from this evaluation study help contribute to our understanding of how student outcomes persist over time.
- Overall, the findings indicate the OGW has positive short and longer-term results for the outcomes we measured. Students reported learning a great deal about the socio-ecological connections between humans and the Ocean because of participation, which is a primary goal of Ocean Literacy (OL).
- Students also developed more positive attitudes and a stronger connection with the Ocean after this experience with many overcoming a fear of the Ocean through breathing and mindfulness techniques and snorkeling with their coaches.
- Over 70% of the students in this study reported performing one or more Ocean stewardship behaviors (e.g., picking up trash, educating others, etc.) and open-ended student responses provide examples of these behaviors.

10.5.2 Recommended Resources

The following resources provide further insights and support on this topic:

- I AM WATER: Ocean Conservation: https://www.iamwaterfoundation.org/
- OCEAN FOR ALL: Alliance (OFAA) | IAmWaterFoundation: https://www.iamwaterfoundation.org/oceans-for-all-alliance

Acknowledgements We would like to thank I AM WATER: Ocean Conservation for funding this project and the entire I AM WATER: Ocean Conservation staff and coaches for their support and efforts throughout this work. We would also like to thank the students who took the time to complete these surveys and their teachers who facilitated the classroom visits for the follow-up survey data collection. Finally, we would like to thank the Oceans Finance Company for their support of the new OCEAN FOR ALL: Alliance.

References

Ajzen I (1985) From intentions to actions: A theory of planned behavior. In: Kulh J, Beckman J (eds) Action-control: from cognition to behavior. Springer, Heidelberg, pp 11–39

Anderson KC, Stern MJ, Powell RB (2022) Identifying areas and approaches for improving evaluation processes in environmental education in the United States of America. J Environ Educ 53(5):290–303. https://doi.org/10.1080/00958964.2022.212918

Ardoin NM, Bowers AW, Roth NW, Holthuis N (2018) Environmental education and K-12 student outcomes: A review and analysis of research. J Environ Educ 49(1):1–17. https://doi.org/10.1080/00958964.2017.1366155

Ashley M, Pahl S, Glegg G, Fletcher S (2019) A change in mind: Applying social and behavioral research methods to the assessment of the effectiveness of Ocean Literacy initiatives. Front Mar Sci 6:288

Cooper E, Dolnicar S, Grun B (2024) Understanding how a commitment-based pledge intervention encourages pro-environmental tourism behavior. Tour Manag 104:1–13. https://doi.org/10.1016/j.tourman.2024.104928

Esri (2023) Country boundaries map [Map]. ArcGIS Online. https://www.arcgis.com/index.html

Fauville G, Strang C, Cannady MA, Chen Y-F (2018) Development of the international Ocean Literacy survey: Measuring knowledge across the world. Environ Educ Res 25(2):238–263. https://doi.org/10.1080/13504622.2018.144038

Halversen C, Schoedinger S, Payne D (2021) A handbook for increasing Ocean Literacy: Tools for educators and Ocean Literacy advocates. National Marine Educators Association, College Park

McKinley E, Burdon D, Shellock RJ (2023) The evolution of Ocean Literacy: A new framework for the United Nations Ocean decade and beyond. Mar Pollut Bull 186:114467. https://doi.org/10.1016/j.marpolbul.2022.114467

Podsakoff PM, MacKenzie SB, Lee JY, Podsakoff NP (2003) Common method biases in behavioral research: A critical review of the literature and recommended remedies. J Appl Psychol 88(5):879–903. https://doi.org/10.1037/0021-9010.88.5.879

Powell RB, Stern MJ, Ardoin N (2006) A sustainable evaluation framework and its application. Appl Environ Educ Commun 5(4):231–241. https://doi.org/10.1080/15330150601059290

Powell RB, Stern MJ, Frensley BT, Moore D (2019) Identifying and developing crosscutting environmental education outcomes for adolescents in the twenty-first century (EE21). Environ Educ Res 25(9):1281–1299. https://doi.org/10.1080/13504622.2019.1607259

Stern MJ, Powell RB, Hill D (2014) Environmental education program evaluation in the new millennium: What do we measure and what have we learned? Environ Educ Res 20(5):581–611. https://doi.org/10.1080/13504622.2013.838749

Stern MJ, Powell RB, Frensley BT, Ardoin NM (2018) What difference do role models make? Investigating outcomes at a residential environmental education center. Environ Educ Res 24(6):818–830. https://doi.org/10.1080/13504622.2017.1313391

Stern MJ, Powell RB, Frensley BT (2022) Education and interpretation on public lands: Lessons from research and new directions. J Park Recreat Adm 40(1):228–237. https://doi.org/10.18666/JPRA02021-11003

UNESCO (1977) Final report intergovernmental conference on environmental education. Tblisi, USSR, 14—26, October 1977. UNESCO, Paris

UNESCO, Intergovernmental Oceanographic Commission (2021) Ocean Decade: Challenges. https://Oceandecade.org/challenges/. Accessed 29 Aug 2024

United Nations, Department of Economic and Social Affairs (2015) Transforming our world: The 2030 agenda for sustainable development. https://sdgs.un.org/publications/transforming-our-world-2030-agenda-sustainable-development-17981. Accessed 29 Aug 2024

Author Contributions and Biographical Profiles

[1]*B. Troy Frensley: Co-conceptualization; Writing—Original Draft, Review, and Editing; Methodology—Data Collection and Analysis; Project Administration and Supervision.

[2]Hannah Eaton: Co-conceptualization, Writing—Original Draft, Review, and Editing; Methodology—Data Collection and Analysis; Project Administration and Supervision.

[3]Claire Carretti: Co-conceptualization; Methodology—Data Collection, Supervision.

[4]Katherine Wallis: Co-conceptualization; Methodology—Data Collection, Supervision.

[5]Amalia De Abreu: Co-conceptualization; Methodology—Data Collection, Supervision.

[6]Peter Marshall: Co-conceptualization; Supervision.

*Corresponding Author at frensleyb@uncw.edu

B. Troy Frensley, PhD, is an Associate Professor in the Environmental Sciences Department at the University of North Carolina Wilmington where he teaches courses in environmental education, environmental interpretation in protected areas, environmental nonprofit organizations, and global environmental problems. He also teaches in the Master of Science in Environmental Sciences concentration in Environmental Education and Interpretation. He has coordinated research and evaluation projects with diverse collaborators and stakeholders around the United States and abroad and his work seeks to enhance our understanding of the impacts of environmental and Ocean education, interpretation, and participatory science projects with diverse individuals and communities.

Claire Carretti is a dedicated conservationist with nearly two decades in the non-profit sector, progressing from social work studies and frontline crisis work to a seasoned professional. Over the last 15 years, she has excelled in securing funding for welfare and conservation organizations. As the Development Director for I AM WATER in Cape Town, South Africa, Claire specializes in organizational development, stakeholder management, and strategic planning. A charismatic leader, she excels in conflict resolution, team management, and decision-making, committed to driving performance, integrity, and positive change.

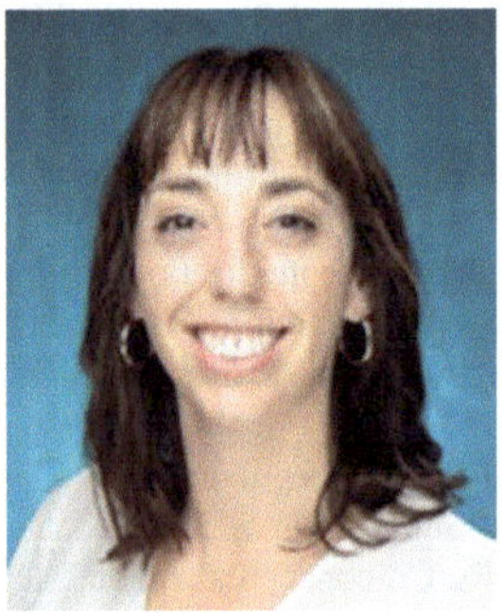

Hannah Eaton is a Lecturer and Lab Coordinator at the University of North Carolina at Wilmington (UNCW). She is an alumna of the University of North Georgia, a Georgia 4-H Camp Counselor alumna, and a participant in the University of Purdue Military Extension Internship. She received her Master of Science in Environmental Studies at UNCW, where she concentrated in Environmental Education and Interpretation. Her interests include environmental education program evaluation, communities of practice, and positive youth development. Hannah has worked with local, state, and military programs to enhance their environmental education efforts while creating welcoming communities for youth.

Katherine Wallis was the Education and Communications Manager at I AM WATER, during the time of this project. She is now the Communications Manager for the Blue Alliance Marine Protected Areas. With a background in Marine Biology, she focuses on using environmental education and communication to connect people to nature. An avid freediver, Katherine loves exploring the chilly kelp forests around Cape Town, South Africa.

Amalia De Abreu is a detail-oriented Operations Coordinator for I AM WATER with a background in marine biology and ecology, passionate about educational outreach and community engagement. She excels in managing logistics, marketing programs to schools, and coordinating the coaching community. Her love for learning, strong organizational skills, and ability to connect with others make her a valuable asset. Amalia thrives in fostering collaboration and achieving organizational goals. In her spare time, she enjoys being in nature, whether on the mountains or in the sea.

Peter Marshall swam on the U.S. National Team for 11 years, breaking 8 World Records and claiming 3 World Champion titles. He swam on a scholarship at Stanford University where he also completed a degree in Economics. Peter has been involved in program design and development for multiple non-profits and was instrumental in developing the Ocean Guardians Workshop program and leading the start of I AM WATER's partnership with UNCW.

11 Ocean Equity: Unveiling Gender Opportunities and Challenges in Brazil's Ocean Literacy Activities

Tatiana Martelli Mazzo, Adriana Lippi, William Remo Pedroso Conti, Andrezza Justino Gozzo Andreotti, Camila Keiko Takahashi, Barbara Lage Ignacio, Marilia Bueno, Ronaldo Adriano Christofoletti, and Leandra R. Gonçalves

Abstract

Amidst global discussions about the science we need for the Ocean we want, promoting Ocean Literacy has become central to expanding the scientific impact and fostering greater social engagement. A key component of these initiatives is the need to prioritize gender equity as a foundational pillar for building a just and sustainable future. Women continue to be underrepresented in Ocean sciences, both globally and in Brazil, facing numerous barriers such as hiring difficulties, limited career progression, harassment, lack of support during motherhood, and lower publication rates. These challenges are deeply rooted in historical and structural biases that still shape the field of Ocean science today. The United Nations Ocean Decade (2021–2030) and the 2030 Agenda have promoted international cooperation and investments to address these obstacles. Public policies must be developed to foster women's participation across all levels of Ocean research and governance. This chapter examines the challenges and opportunities for promoting gender equity within Ocean Literacy initiatives in Brazil, with a particular focus on the Maré de Ciência Program and its partners. The chapter outlines strategic pathways to create a more inclusive and equitable Ocean science community by showcasing best practices and offering a detailed view of female representation.

Keywords

Ocean Literacy (OL) · Gender equity · Gender biases · Ocean Decade · 2030 agenda · Ocean sciences · Female representation · Public policy · Sustainable development · Women in STEM · Science education · Marine conservation · Social inclusion · Scientific collaboration · SDG 5 · SDG 14

T. M. Mazzo (✉) · A. Lippi · W. R. P. Conti · A. J. G. Andreotti · B. L. Ignacio · M. Bueno · R. A. Christofoletti · L. R. Gonçalves
Federal University of São Paulo, Institute of Marine Sciences, Santos, Brazil
e-mail: tatiana.mazzo@unifesp.br; adriana.lippi@unifesp.br; wrpconti@unifesp.br; ajgozzo@unifesp.br; barbara.lage@unifesp.br; marilia.bueno@unifesp.br; christofoletti@unifesp.br; goncalves.leandra@unifesp.br

C. K. Takahashi
Maré de Ciência, Federal University of São Paulo, Santos, Brazil

T. J. Kennedy (ed.), *Ocean Literacy: The Foundation for the Success of the Ocean Decade, Volume I*, Sustainable Development Goals Series, https://doi.org/10.1007/978-3-031-98145-6_11

11.1 Introduction

In this chapter, we explore the issue of gender equity in Ocean Literacy activities carried out in Brazil by the Maré de Ciência Program and its partners, contextualizing this discussion within the global landscape of scientific and socio-environmental initiatives. Among these initiatives are those promoted during the United Nations Decade of Ocean Science for Sustainable Development (Ocean Decade), spanning from 2021 to 2030. The Ocean Decade has placed a spotlight on the critical need for advancements in Ocean science and policy to address global challenges, from climate change to biodiversity loss, and central to these efforts is the role of Ocean Literacy. This chapter targets a diverse audience, including educators, researchers, policymakers, and community leaders involved in formal education, informal learning environments, and community-based initiatives. The focus is on actions that foster gender equity and Ocean Literacy across multiple contexts, with special attention to engaging youth, particularly girls, in Ocean science and sustainability discussions.

Ocean Literacy, a concept that has been gaining momentum in global socio-environmental engagement agendas, has the potential to radically transform society's relationship with the Ocean. By promoting knowledge, awareness, and practices aimed at sustainable Ocean stewardship, Ocean Literacy serves as a bridge between scientific communities and the public, ensuring that Ocean science is not confined to the laboratory but becomes an integral part of societal values and actions. However, for this transformation to be truly inclusive and effective, gender equity must be incorporated as a fundamental pillar of these actions. Without addressing the deep-rooted gender disparities in Ocean sciences, we risk perpetuating a system that excludes valuable perspectives and talents, ultimately limiting the effectiveness of Ocean sustainability efforts.

Despite significant progress in access to education and increased female participation in various scientific fields, the underrepresentation of women in Ocean sciences remains a concerning reality, both globally and in Brazil. Women continue to face barriers that hinder their full participation and advancement in Ocean science careers. These barriers are multifaceted, including historical and structural challenges that have long shaped the field. Women often experience lower representation in leadership roles, fewer opportunities for participation in marine expeditions, and face additional challenges in balancing personal and professional responsibilities, particularly regarding motherhood. These difficulties are exacerbated by a lack of supportive policies in institutions, where gender biases persist.

The underrepresentation of women in Ocean sciences is not just a professional issue but a societal one. The exclusion of women from these critical fields results in a loss of talent and diversity, which are essential for innovation and problem-solving in the face of complex global challenges. The Ocean Decade provides a unique opportunity to address these disparities by promoting gender equity as a central goal of global Ocean initiatives. Efforts to dismantle the barriers that women face in Ocean sciences are not only about achieving fairness but also about enhancing the capacity of Ocean science to generate solutions that benefit all of society as well as the field itself.

This chapter is structured to address these challenges and explore the opportunities for promoting gender equity in Ocean Literacy initiatives in Brazil. Initially, we delve into the historical and contemporary obstacles that perpetuate gender inequality in Ocean science and governance. These obstacles range from the exclusion of women from early scientific expeditions to modern-day issues like lower publication rates for women, fewer opportunities for funding, and gender-based harassment in professional settings. By understanding these challenges, we can identify where change is needed most and how Ocean Literacy can be a vehicle for that change. Next, we examine the role of Ocean Literacy as a powerful tool to overcome these barriers and promote a more inclusive environment within the scientific community.

Ocean Literacy does more than educate the public about Ocean issues; it empowers individuals, especially women and girls, to participate actively in Ocean conservation and science. Through community engagement, education, and advocacy, Ocean Literacy initiatives can help break down the stereotypes and biases that have historically kept women out of the sciences. When gender equity is integrated into these initiatives, the result is a richer, more diverse scientific field that is better equipped to address the global challenges we face.

Practical examples, such as the good practices of the Maré de Ciência Program and its partners, are presented to illustrate how these initiatives can create real and meaningful opportunities for girls and women. These examples demonstrate the tangible impact of gender-focused Ocean Literacy programs, from mentoring young women interested in marine sciences to creating safe spaces for women to lead and innovate in their fields. By fostering an environment of inclusion and empowerment, programs like Maré de Ciência are paving the way for the next generation of female Ocean scientists and leaders. The engagement of participants in our Ocean Literacy initiatives is facilitated through established networks such as the Blue School Brazil Program. Additionally, we leverage the social media platforms of Maré de Ciência and its partners to amplify outreach and encourage widespread participation.

Finally, we discuss how these actions not only contribute to the Ocean Decade and the Sustainable Development Goals (SDGs) but also strengthen the building of a more just, equitable, and prepared society to face global challenges. By addressing gender inequality in Ocean sciences, we create a fairer and more inclusive field as well as enhance our collective capacity to protect the Ocean and ensure a sustainable future for all. Thus, this chapter aims not only to identify the challenges but also to highlight practical and strategic paths to advance gender equity within Ocean Literacy, contributing to a more sustainable and inclusive future for all.

11.2 Challenges of Gender Equity in Science and Ocean Governance

During the advancement of Ocean science during World War II (WWII), there was a significant push in research, equipment development, and institutional support, primarily funded by military investment (Oreskes 2021). In this context, women were present in many projects, often performing essential but invisible roles. They frequently worked as "human computers," a meticulous job involving complex calculations and data analysis. Although these roles were critical to significant discoveries, such as the understanding of thermocline structure and dynamics (Oreskes 2000), they were often viewed as mundane or auxiliary tasks.

Women were there from the beginning, though disguised, as in the case of Jeanne Baret, which illustrates the historical challenge faced by women who wished to engage in scientific expeditions. Disguised as a man, Baret participated in L.A. Bougainville's French expedition, becoming the first woman to circumnavigate the globe. However, the dominant narrative of Ocean science focuses on the achievements of men, obscuring the participation of women like Baret and other pioneers (Cavaleri Gerhardinger et al. 2024). This historical erasure is a clear example of how gender bias has shaped marine sciences, from the earliest expeditions to modern research institutions.

In the last century, many countries implemented policies to promote gender equality in access to education, as part of a global effort to reduce disparities. A recent milestone in this agenda is the 2030 Agenda for Sustainable Development, which establishes gender equality in access to education as one of its core goals (SDG 4—Quality Education). However, reality is still far from this ideal. Globally, women represent about half (45–55%) of graduate and master's levels, but only 28.4% of PhD researchers are women (UNESCO 2021b). This reveals a phenomenon of attrition throughout academic progression, especially in more specialized fields like Ocean sciences.

The Global Ocean Science Report (UNESCO 2021a) highlights that while women represent 38.6% of researchers in this field, disparities between countries are vast. In Brazil, female participation in Ocean science reaches 50%, a figure that suggests progress in terms of access. However, in countries like the Democratic Republic of Congo, women represent only 7% of Ocean scientists, while in Ireland, this number reaches 72%, reflecting significant variations in gender policies and access to science worldwide.

A detailed analysis of some European institutions (Giakoumi et al. 2021) revealed that although women are the majority at the PhD level, their representation drastically declines in senior positions, dropping to a range of 13% to 24%. In the United States, statistics indicate a similar pattern: women make up 51.9% of PhDs, but only 19.8% achieve tenured faculty positions (Lewis et al. 2023). In Brazil, the situation also reveals a sharp decline in career progression: 47.5% of PhDs are women, but they occupy only 28.6% of faculty positions in graduate programs (Lippi 2024). This phenomenon, often called the "leaky pipeline", reflects how women, throughout their academic and professional trajectories, face multiple barriers that hinder their ascent to leadership positions.

The gender disparities in Ocean sciences are possibly more documented than in other Ocean sectors, but there are examples in areas such as in Maritime Industry (Women occupy 5% of maritime leadership roles 2021) and Maritime Affairs (Kitada and Rodríguez-Chaves 2024). These barriers include, but are not limited to, lower citation rates when women are the first authors of publications, lower success rates in job and funding applications, less access to laboratories, samples, and research infrastructure, and a high incidence of harassment and discrimination in academic settings. Women are also disproportionately tasked with caregiving responsibilities, both domestically and institutionally, limiting their time to dedicate to research (Aarons et al. 2023; Giakoumi et al. 2021; Kloot 2004; Maia et al. 2024; Staniscuaski et al. 2021).

A particularly significant barrier for women in Ocean science is related to participation in maritime expeditions. These missions are highly valued and play a crucial role in the development of scientific careers, offering opportunities for in-situ data collection, international collaboration, and academic visibility. However, during WWII and subsequent years, women in Western European and North American countries were often barred from participating in scientific cruises. In many cases, they were limited to field research in shallow or nearshore waters, while men had access to scientific ships and deep-sea missions (Bonatti and Crane 2012). Most bans in Western Europe and North America ended in the 1970s, but women's leadership in maritime research remains below 30% (Orcutt and Cetinic 2014). Even with the end of these bans, the legacy of these historical exclusions persists, reflected in a power and leadership structure that favors men, perpetuating disparities in access to these opportunities.

Overcoming these barriers requires coordinated efforts that involve implementing effective public policies, to ensure women to have equitable access to all opportunities in Ocean science, from education to leadership positions. These policies should include, in addition to measures to increase female participation, the creation of safe, inclusive, and welcoming work environments where women can develop their careers without facing discrimination or harassment.

Another essential aspect is recognizing and valuing the historical and contemporary contributions of women to Ocean science. By giving visibility to the trajectories and achievements of these scientists, we address historical injustices and create inspiring role models for future generations of women, which encourages their active participation and leadership in marine sciences. Celebrating these journeys is crucial to building a more diverse, innovative science capable of addressing global challenges more inclusively.

11.3 The Role and Potential of Ocean Literacy in Promoting Gender Equity

As discussed in the previous section, women face a range of structural and cultural barriers that hinder their full involvement in Ocean science. Historical restrictions on maritime expeditions, lower female representation in leadership positions, and the lack of recognition for their contributions exemplify a scenario that requires profound changes. In light of these realities, promoting gender equity should not only be viewed as a matter of social justice but also as a scientific and institutional strategy to enrich the field with diversity and innovation. In this context, Ocean Literacy emerges as a powerful pathway to transforming these disparities, offering a new lens through which societal attitudes towards gender roles in Ocean sciences can be reexamined and reshaped.

The Ocean Decade (2021–2030), a United Nations initiative aligned with the 2030 Agenda and SDG 5 (Gender Equality), aims to achieve gender equality and empower all women and girls. This global movement highlights the urgent need to address gender disparities within Ocean sciences. Although traditionally male-dominated, Ocean science offers immense opportunities to reduce gender gaps, provided that deliberate and sustained actions are taken to ensure equal access to education, professional development, and leadership roles for women. Fostering these opportunities is not only about equality but about building a stronger and more innovative scientific community capable of addressing the challenges faced by our Ocean.

However, these gender disparities are often exacerbated by unequal access to essential resources, such as training, mentorship, and career progression opportunities. Women often do not receive the same level of support as men, limiting their ability to fully participate in strategic decisions that shape the future of Ocean research and conservation. Without comprehensive public policies and institutional actions to address these barriers, the field of Ocean sciences will continue to underutilize women's potential, slowing the progress needed to protect our marine ecosystems.

The Barcelona Declaration (UNESCO 2024) proposes a comprehensive approach to addressing these inequalities, emphasizing the removal of gender barriers in Ocean sciences while also considering other axes of diversity, including geographic, cultural, and generational disparities. Ocean Literacy emerges here as a transformative tool, not only fostering a deeper societal understanding of the Ocean but also challenging traditional gender dynamics that have historically marginalized women in this field. By promoting a new relationship between society and the Ocean, Ocean Literacy helps to create an inclusive framework that encourages and empowers girls and women to fully engage with Ocean-related sciences.

The Ocean Decade has emphasized Ocean Literacy to drive societal change towards Ocean sustainability (McKinley et al. 2023). Ocean Literacy is not simply an educational concept; it is a strategic tool that can help shift societal perceptions, making the Ocean an integral part of our daily lives and future survival. By embedding gender equity within Ocean Literacy initiatives, a more inclusive environment is fostered, one where the active and meaningful participation of girls and women is promoted at every level of Ocean science.

For example, adapting school curricula to include comprehensive Ocean education and encouraging girls to engage with subjects such as science, technology, engineering, and mathematics (STEM) is crucial for sparking early interest and breaking down gender stereotypes. Furthermore, the implementation of mentorship programs that provide continuous support and guidance can empower women to navigate challenges throughout their academic and professional journeys, preparing them for leadership roles and key decision-making positions in Ocean governance.

Ensuring a sustainable future for both the Ocean and humanity requires that gender inequalities be addressed at all stages—from basic education, where gender biases often take root, to the highest echelons of governance and scientific leadership. Promoting the equitable participation of women in Ocean science is not merely a social justice issue. Research shows that diverse teams are better equipped to innovate and solve complex environmental challenges. Including more women in Ocean science brings broader perspectives and expertise to the management and preservation of marine resources, helping to design more effective and sustainable solutions for global environmental issues.

In this context, Ocean Literacy plays a pivotal role in dismantling gender stereotypes and fostering a more inclusive scientific community. By promoting diversity in the sciences, these initiatives help create spaces where women can express themselves freely, lead research efforts, and fully contribute to scientific advancements. This expands opportunities for girls and women and also enriches the entire field of Ocean science, bringing in new perspectives and approaches that drive innovation.

Finally, integrating gender equity into Ocean Literacy initiatives is essential for achieving several Sustainable Development Goals (SDGs), particularly SDG 5 (Gender Equality) and SDG 14 (Life Below Water). By promoting the inclusion of women in Ocean sciences, we contribute to a more just and equitable society and enhance the global capacity to respond effectively to the multifaceted challenges threatening Ocean sustainability.

To exemplify how these best practices can be applied in a practical way, the next section will discuss the challenges and opportunities of promoting gender equity in Ocean Literacy initiatives in Brazil, with a focus on the best practices implemented by the Maré de Ciência Program and its partners.

11.4 Good Practices of Ocean Literacy for the Ocean Decade in Brazil

In previous sections, we discussed the importance of Ocean Literacy as a powerful tool for promoting gender equity, especially in the field of Ocean sciences, historically been dominated by men. Throughout history, women have faced structural barriers that prevented many from occupying prominent positions, and we have seen how, even today, many of these barriers continue to hinder women's progress in areas such as leadership and Ocean governance. However, initiatives like those promoted by the Ocean Decade and the 2030 Agenda offer a promising path to transforming this scenario. Incorporating inclusive practices and creating targeted opportunities for girls and women are essential steps to counteract the 'scissors effect,' a phenomenon where women encounter progressively fewer opportunities as they advance in their careers (Giakoumi et al. 2021).

In this context, Brazil, through the Maré de Ciência Program and its partners, has developed a series of best practices aimed at Ocean Literacy, with the goal of reducing gender disparities and creating more inclusive and diverse environments. These practices not only contribute significantly to the goals set by the Ocean Decade but are also aligned with the achievement of SDG goals, such as SDG 4, SDG 5, and SDG 14.

The Maré de Ciência Program and its partners have been instrumental in creating inclusive spaces, providing concrete opportunities for girls and women to actively participate in a wide range of activities. Among the main initiatives are the Planning Workshops for the Ocean Decade in Brazil, the Escola Azul Brasil Program, the Ocean Olympics, the Ocean Youth Ambassadors Forums, and community projects developed in vulnerable areas of Santos/Brazil. All these actions have been carefully structured to engage local communities and to encourage female participation, consolidating the role of women and girls as agents of transformation in Ocean protection and conservation.

Through a methodology that combines quantitative and qualitative data, the analysis presented here aims to provide a comprehensive view of female representation in these actions and to discuss the importance of an inclusive and diverse approach to advancing the 2030 Agenda. The analysis of girls' and women's representation in Ocean Literacy initiatives was conducted based on participation data in events and forums related to the theme, as shown in Table 11.1. To assess female representation, the number of registered participants, speakers, and other roles performed were categorized by gender, including the options "female," "male," "non-binary," "other," "prefer not to say," and "not informed," as self-identified by individuals during event registration. In addition, the specific proportion of participation of girls and women in each event was calculated; this proportion was expressed as "female representation." Additionally, the data was segmented by age groups to identify the specific participation of girls (ages 11 to 18) and adult women (over 18 years old).

The data indicates a high participation of girls and women in these initiatives. It is noteworthy that these impressive numbers of female participants are observed even though the count includes

Table 11.1 Participation data in events and forums related to the theme of Ocean Literacy in Brazil

Event	Role	Female	Male	Non-binary	Other	Prefer not to inform	Not informed	Women and girls' representation	Age
Ocean challenge in education	Registered	49	18	0	0	0	0	73.1%	> 18
Young ambassador forum	III forum participant	17	8	0	0	0	0	68.0%	11 to 18
	III forum speaker	4	8	0	0	0	0	33.3%	> 18
	V forum participant	80	44	0	0	0	0	64.5%	11 to 18
	V forum speaker	4	9	0	0	0	0	30.8%	> 18
	VI forum participant	41	16	0	0	0	0	71.9%	11 to 18
Workshop AA-BSN	participant	13	6	0	0	0	0	68.4%	> 18
Alliance forum	I forum participant	37	21	0	0	0	0	63.8%	> 18
	II forum participant	21	12	0	0	0	0	63.6%	
Ocean Decade regional workshop	Participant	291	174	0	3	6	14	59.6%	> 14
Ocean Literacy dialogues	Participant	109	47	0	0	0	0	69.9%	> 18
	Speaker	149	109	0	0	0	9	55.8%	
I Ocean Literacy multiplier training	Registered	343	170	0	0	1	0	66.7%	> 18
	Participant	197	71	0	0	0	0	73.5%	
O2—knowledge modality	2021 participant	463	340	9	5	6	3	56.1%	9 to 18
	2021 medalist	318	232	7	3	4	0	56.4%	
	2022 participant	2178	2025	37	10	39	91	49.7%	
	2022 medalist	1506	1269	19	4	21	66	52.2%	
	2023 participant	10,805	9945	71	41	3006	1027	43.4%	
	2023 medalist	7027	6042	26	20	1783	586	45.4%	
	Jr. scientific initiation—2022 fellows	24	19	0	0	0	0	55.8%	
	Jr. scientific initiation—2023 fellows	60	37	0	0	0	0	61.9%	
Science Fair—Blue School 2023	Fellow teacher	10	5	0	0	0	0	66.7%	> 18
	Fellow student	46	32	0	0	0	0	59.0%	11 to 18
	Young ambassador student fellow	19	9	0	0	0	0	67.9%	
Arte no Dique workshop	2018 participant	24	1	0	0	0	0	96.0%	8 to 13
	2019 participant	13	7	0	0	0	0	65.0%	
	2021 participant	10	2	0	0	0	0	83.3%	
	2022 participant	18	11	0	0	0	0	62.1%	
	2023 participant	11	5	0	0	0	0	68.8%	
	2024 participant	8	3	0	0	0	0	72.7%	
Ocean in education—2020	Participant	76	53	0	0	0	0	58.9%	9 to 18

people who did not report the gender with which they identify. In other words, female participation is high, and it may be underreported, as we included people who did not disclose their gender.

Despite these significant numbers, one aspect that still deserves attention is the representation of women in leadership and visibility roles, such as speakers and moderators at events. For instance, in the Youth Ocean Ambassadors

Forums, female representation among speakers was below 40%, which highlights the need for additional efforts to ensure that women are also present in leadership spaces, where they can inspire future generations.

Participation by age group also reveals interesting data. Events aimed at young people show a high participation of girls, demonstrating that, from an early age, there is genuine interest and willingness to get involved in scientific initiatives. However, as the participants' ages increase, we observe a decrease in female representation in forums with adults. To delve deeper into this discussion, we can use the metaphor of the "scissors effect," which reflects the gender barriers women face at different stages of life. The first "scissor" may emerge in childhood and adolescence, when gender roles and stereotypes discourage girls from getting involved in scientific and leadership areas. This initial cut may reduce their interest and engagement in STEM (Science, Technology, Engineering, and Mathematics) fields from an early age. However, despite this discouragement, many girls, upon entering adulthood, manage to break through these barriers, motivated by a sense of "now things will be different." This breakthrough is reflected in the growing numbers of women in higher education and PhD programs, marking an attempt to break established paradigms. At this stage, personal determination and individual effort become more evident, allowing them to advance in their careers, despite well-known difficulties and remaining challenges.

In summary, the data analysis shows a positive scenario regarding female participation in Ocean Literacy initiatives in Brazil. However, there is still room for improvement, such as increasing female representation in roles as speakers and moderators and creating more inclusive environments for all gender identities. It is essential to continue promoting actions that encourage women's participation and leadership in Ocean-related discussions and actions.

One of the highlights of best practices in Ocean Literacy in Brazil is the Brazilian Ocean Olympics (O2). O2 stands out as a powerful catalyst in promoting practices that not only raise awareness and engage society but also actively involve girls and women in Ocean Literacy-related activities. By creating space for the entire community, O2 fosters an inclusive and inspiring environment where girls and women can discover their passion for science and the Ocean.

Since its first edition, O2 has introduced the approach of "cross-cutting themes," offered for the Social-Environmental Project and Artistic, Cultural, and/or Technological Production categories, allowing participants to discuss contemporary topics and integrate them into the context of Ocean Literacy. Among these themes, "Women in Science" stands out, aiming to promote gender equity, in line with institutional guidelines and the SDGs established by the 2030 Agenda. This initiative plays a fundamental role in deconstructing gender stereotypes, encouraging more and more girls to pursue careers in STEM fields.

Starting with the 2022 call for projects, junior scientific initiation scholarships were made available to public school students who won medals in the Knowledge category. These scholarships represent an important milestone in encouraging girls, as a significant portion of them is allocated to this group. O2, by valuing the potential of these young women, opens paths for them to begin their scientific research journeys, contributing to a fairer and more equitable future in science.

In addition to offering learning and development opportunities, O2 fosters the creation of collaborative networks among educators, researchers, and other women involved in Ocean causes. By sharing knowledge and experiences, these women inspire others to follow in their footsteps, becoming agents of change for a more sustainable future for the Ocean. Through initiatives like O2, we are contributing to the formation of a generation of women scientists who will have a significant impact on Ocean protection and conservation.

11.5 Contribution to the Ocean Decade and the Sustainable Development Goals

Ocean protection and conservation are central issues on the global stage, particularly at a time when the planet is under increasing pressure from climate change, pollution, and over-exploitation. To address these challenges, an integrated approach involving education, gender equity, and technological innovation is essential. These areas are directly related to the UN SDGs, especially SDGs 4, 5, and 14, which are interdependent and, when worked on together, have the potential to create significant changes.

Integrating Ocean Literacy initiatives with the promotion of gender equity makes a significant contribution to advancing several SDGs. By combining Ocean science education with policies that promote gender equality, we create a stronger foundation for sustainable marine conservation and the construction of a more just society. Scientists and educators recognize the growing importance of including Ocean education in school curricula as a key strategy for achieving global sustainability, and UNESCO (2021b) highlights the need to integrate Ocean Literacy into educational systems by 2025 as part of the Ocean Decade goals.

This integration increases scientific knowledge about the Ocean (SDG 14) and contributes to the development of more conscious and responsible global citizenship (SDG 4). Educating citizens with an interdisciplinary and diverse vision is essential to tackling global challenges, particularly regarding Ocean conservation. Moreover, quality education that incorporates Ocean sciences can equip future generations with essential skills such as critical thinking, problem-solving, and the ability to lead change. These competencies are fundamental to building innovative and sustainable solutions that ensure Ocean protection.

Actively promoting the participation of girls and women in Ocean sciences is an effective strategy not only to achieve gender equality (SDG 5) but also to enhance research and marine conservation. Including more women in STEM fields is essential to foster an environment of innovation and diversity, enabling different perspectives to collaborate on solving complex Ocean-related challenges. This diversity is particularly important in Ocean sciences, where the nature of the issues requires creative, interdisciplinary, and transdisciplinary solutions.

Additionally, by increasing the visibility of successful women in Ocean sciences, we provide role models who inspire the next generation of scientists, breaking down barriers and creating a cycle of empowerment and inclusion. Initiatives like those promoted by the Maré de Ciência Program and its partners, which encourage female participation in Ocean science projects and actions, have a direct impact on reducing gender disparities and creating opportunities for career development in the scientific field.

Encouraging female participation in this field helps reduce gender disparities and creates meaningful opportunities for girls and women to develop their careers. Moreover, increasing the visibility of successful women in Ocean-related fields can inspire others to follow similar paths, creating a positive cycle of empowerment and inclusion.

Quality Ocean education, combined with the promotion of gender equality, not only enriches individuals' knowledge and skills but also ensures more inclusive and equitable participation in global issues. By uniting these initiatives, we contribute to the protection and conservation of the Ocean while promoting the construction of a more just, equitable, and sustainable society.

11.6 Conclusion

This chapter sought to analyze the challenges and opportunities related to promoting gender equity in Ocean sciences, focusing on Ocean Literacy initiatives conducted in Brazil. As explored throughout the text, gender inequality remains a critical issue in many areas of science, with Ocean sciences being one of the fields where these disparities are particularly evident. The analysis revealed that, while there have been significant advances in recent decades, especially regarding access to education and women's participation in scientific programs, structural and cultural barriers persist, limiting the full development of women's careers in the marine field.

The concept of Ocean Literacy, by promoting greater public understanding of the importance of the Oceans and the relationship between humans and marine ecosystems, offers a powerful tool for fostering social change and, consequently, addressing these disparities. The initiatives analyzed, such as those promoted by the Maré de Ciência Program and its partners, demonstrate that it is possible to create more inclusive and diverse spaces where girls and women can fully engage and contribute meaningfully to the advancement of Ocean science.

However, for these initiatives to have a lasting impact, they must be combined with effective public policies aimed at ensuring equal access for women at all levels of science, from primary education to leadership positions in scientific and governmental institutions. This includes implementing policies that promote safe and inclusive work environments, free from harassment and discrimination, that encourage women's full participation in all stages of scientific development. It is also necessary to recognize and value the historical and contemporary contributions of women in Ocean science, giving visibility to their achievements and creating inspiring role models for future generations.

Moreover, these Ocean Literacy initiatives must align with the United Nations' Sustainable Development Goals (SDGs), particularly SDG 5 (Gender Equality) and SDG 14 (Life Below Water). Integrating gender equity into Ocean Literacy initiatives is a matter of social justice and a necessary strategy to ensure a more sustainable and resilient future for the Oceans and humanity. Ocean sciences, by their very interdisciplinary nature, require the collaboration of diverse perspectives to tackle the complex challenges threatening marine ecosystems. Including more women in science contributes to building a more diverse and innovative science and strengthens society's ability to respond more effectively and creatively to global problems related to Ocean sustainability.

Throughout this chapter, we discussed how good gender equity practices, promoted in initiatives like the Brazilian Ocean Olympics (O2) and the Ocean Youth Ambassadors Forums, have played a fundamental role in encouraging the active participation of girls and women in Ocean science projects. These practices not only broaden opportunities for women but also enrich the field of science with new perspectives and approaches. However, as shown by participation data, much work remains to ensure that women participate and occupy leadership and visibility roles, such as speakers and moderators at scientific events.

The analysis of participation data in Ocean Literacy initiatives in Brazil revealed a strong female presence at events aimed at younger audiences, indicating that girls are interested and motivated to engage with science from an early age. However, as the participants' age increases, a decline in female representation is observed in forums with adults, reflecting the phenomenon known as the "scissors effect." This effect, which describes the progressive reduction of opportunities for women as they advance in their careers, shows that continuous efforts are needed to support the development of women's professional trajectories, from childhood to the highest levels of leadership in Ocean sciences.

Therefore, promoting gender equity through Ocean Literacy initiatives should be seen as part of a broader social transformation strategy. By integrating girls and women equitably into science, we are not only correcting a historical injustice but also strengthening both science and society. A more inclusive and diverse Ocean science will be better equipped to address global challenges and propose innovative solutions for Ocean preservation and climate change mitigation.

Finally, building a fairer and more sustainable future for the Ocean depends on concrete actions that ensure gender equity in all spheres of Ocean science and marine governance. The initiatives discussed throughout this chapter are just the beginning of a path that needs to be continuously expanded and supported for us to see real and lasting changes. The continuation and expansion of these initiatives are essential to ensuring that girls and women can participate and lead and innovate in Ocean sciences, benefiting society as a whole and contributing to the preservation of the planet's health for current and future generations.

11.6.1 Summary

Key concepts explored in this chapter include:

- Gender Inequality in Ocean Sciences. Despite advances, women face barriers like limited leadership opportunities, lower publication rates, and exclusion from maritime expeditions. These challenges are rooted in historical and structural biases that persist globally, including in Brazil.

- Role of Ocean Literacy in Promoting Equity. Ocean Literacy serves as a transformative tool, educating society about Ocean sustainability while challenging gender stereotypes. Integrating gender equity into these initiatives enriches the field with diverse perspectives, fostering innovation and inclusion.
- Brazil's Initiatives and Best Practices. Programs like Maré de Ciência and the Brazilian Ocean Olympics (O2) have successfully promoted female participation in Ocean-related sciences through mentorships, education, and inclusive events. However, challenges remain in achieving equal representation in leadership and visibility roles.
- Intersection with Global Goals. Aligning Ocean Literacy initiatives with the UN's Sustainable Development Goals (SDGs), particularly SDG 5 (Gender Equality) and SDG 14 (Life Below Water), enhances efforts to create a more sustainable and equitable Ocean science community.
- Strategic Pathways for Progress. To advance gender equity in Ocean sciences, policies must ensure safe, inclusive environments, equal access to opportunities, and visibility for women's contributions. Ocean Literacy must empower girls and women at all stages, from education to leadership, driving long-term systemic change.

11.6.2 Recommended Resources

The following resources provide further insights and support on this topic:

- Maré de Ciência. (n.d.). *Maré de Ciência: Discover the program*. Retrieved November 10, 2024, from https://maredeciencia.eco.br/
- Science at Sea. (n.d.). *Science at Sea: Brazil initiatives*. Retrieved November 10, 2024, from https://ciencianomar-mctic-gov-br.translate.goog/?_x_tr_sl=auto&_x_tr_tl=en&_x_tr_hl=en
- Intergovernmental Oceanographic Commission. (2020). *The science we need for the Ocean we want: The United Nations Decade of Ocean Science for Sustainable Development (2021–2030)*. Retrieved November 11, 2024, from https://unesdoc.unesco.org/ark:/48223/pf0000265198
- Phillips, K. W. (2014). How diversity makes us smarter. *Scientific American*, 42. https://doi.org/10.1038/scientificamerican1014-42
- United Nations. (2015). *Transforming our world: The 2030 Agenda for Sustainable Development*. Retrieved from https://sdgs.un.org/2030agenda

Acknowledgements We thank the participants of the Maré de Ciência initiatives, as well as Federal University of Sao Paulo (UNIFESP), UNESCO Brazil, the Ministry of Science, Technology, and Innovation (MCTI), Fundação Grupo Boticário, CNPq, All- Atlantic Blue School, the Interministerial Commission for Marine Resources (CIRM), and Euc eano Chanel for their invaluable support in promoting Ocean Literacy and sustainability.

References

Aarons S et al (2023) Ad hoc task force on space allocation. Scripps Institution of Oceanography. https://diversity.ucsd.edu/_files/2023-01-18-SIO-space-allocation-report-edits.pdf

Bonatti E, Crane K (2012) Oceanography and women: Early challenges. Oceanography 25(4):32–39

Cavaleri Gerhardinger L et al (2024) A call for a cultural shift in Oceanography. Coast Manag 52(4–5):1–13

Giakoumi S et al (2021) Persistent gender bias in marine science and conservation calls for action to achieve equity. Biol Conserv 257:7

Kitada M, Rodríguez-Chaves M (2024) Advancing gender equality in contemporary Ocean affairs. Int J Mar Coast Law, [s. l.] 39(3):508–518

Kloot L (2004) Women and leadership in universities: A case study of women academic managers. Int J Public Sect Manag 17(6):470–485

Lewis S, Holloway A, Yarincik K (2023) Assessing diversity in U.S. Ocean science institutions: Insights from fifteen years (2007–2021) of OSER data. Oceanography 36(4):10–20. https://tos.org/Oceanography/article/assessing-diversity-in-us-Ocean-science-institutions-insights-from-fifteen-years-20072021-of-oser-data

Lippi A (2024) Women in marine sciences: A case study on the field and postgraduate programs in Brazil. *Master's Dissertation in Ocean Science and Technology*. UNIFESP, Santos, SP

Maia MC et al (2024) Harassment and bullying aboard: Impacts of gender inequality on Ocean professionals. Mar Policy 160:105946

McKinley E, Burdon D, Shellock RJ (2023) The evolution of Ocean Literacy: A new framework for the United Nations Ocean Decade and beyond. Mar Pollut Bull 186:114467. https://doi.org/10.1016/j.marpolbul.2022.114467

Orcutt B, Cetinic I (2014) Women in Oceanography: Continuing challenges. *Oceanography* 27(4):5–13

Oreskes N (2000) "Laissez-tomber": Military patronage and women's work in mid-20th-century Oceanography. Hist Stud Phys Biol Sci 30(2):373–392

Oreskes N (2021) Science on a mission: How military funding shaped what we do and don't know about the Ocean. University of Chicago Press, Chicago

Staniscuaski F et al (2021) Gender, race and parenthood impact academic productivity during the COVID-19 pandemic: From survey to action. Front Psychol 12:663252

UNESCO (2021a) Global Ocean science report 2020: Charting capacity for Ocean sustainability. United Nations. https://www.un-ilibrary.org/content/books/9789216040048

UNESCO (2021b) Ocean Literacy framework for the UN decade of Ocean science for sustainable development 2021–2030. UNESCO, Paris. (IOC Ocean Decade Series, 22)

UNESCO (2024) Barcelona statement identifies the priority areas of action for the Ocean Decade in coming years. *United Nations Decade of Ocean Science for Sustainable Development.* https://Oceandecade.org/news/barcelona-statement-identifies-the-priority-areas-of-action-for-the-Ocean-decade-in-coming-years/

Women Occupy 5% Of Maritime Leadership Roles (2021) In: SPINNAKER. 9 fev 2021. https://spinnaker-global.com/0_09-02-2021-women-in-maritime-leadership-salaries. Accessed 13 Nov. 2024

Author Contributions and Biographical Profiles

[1]*Tatiana Martelli Mazzo: Co-conceptualization; Writing—Original Draft, Review, and Editing.
[2]Adriana Lippi: Writing—Review and Editing; Methodology—Data Collection.
[3]William Remo Pedroso Conti: Writing—Review and Editing; Methodology—Data Analysis.
[4]Andrezza Justino Gozzo Andreotti: Co-conceptualization; Writing—Review.
[5]Camila Keiko Takahashi: Writing—Review and Editing.
[6]Barbara Lage Ignacio: Co-conceptualization; Writing—Review; Methodology—Data Collection.
[7]Marilia Bueno: Writing—Review and Editing; Methodology—Data Collection and Analysis.
[8]Ronaldo Adriano Christofoletti: Co-conceptualization.
[9]Leandra Regina Gonçalves: Writing—Review and Editing.
*Corresponding Author at tatiana.mazzo@unifesp.br

Tatiana Martelli Mazzo is a professor at the Institute of Marine Sciences at the Federal University of São Paulo (UNIFESP). She coordinates interdisciplinary research projects spanning the fields of chemistry, marine sciences, materials engineering, nanotechnology, biotechnology, Ocean Literacy, active citizenship, and sustainability. She engages in scientific literacy in Ocean Literacy for children and young people in formal and informal educational settings. She coordinates actions for the dissemination and popularization of science. She is involved in training and capacity-building activities in Ocean Literacy, active citizenship, and sustainability. She leads the Women in Science Project.

Adriana Lippi is an oceanographer (USP), has a master's (UNIFESP) in Marine Science and Technology, researching women in marine science. She is co-founder of the Women's League for the Ocean and works as a communicator on marine science and the climate emergency.

William Remo Pedroso Conti has a degree in Physics from the Physics Institute of the University of São Paulo, a master's degree, and a doctorate from the same institution. He has been a professor at the Federal University of São Paulo since 2015. He has experience in Mathematical Physics, with an emphasis on Statistical Physics.

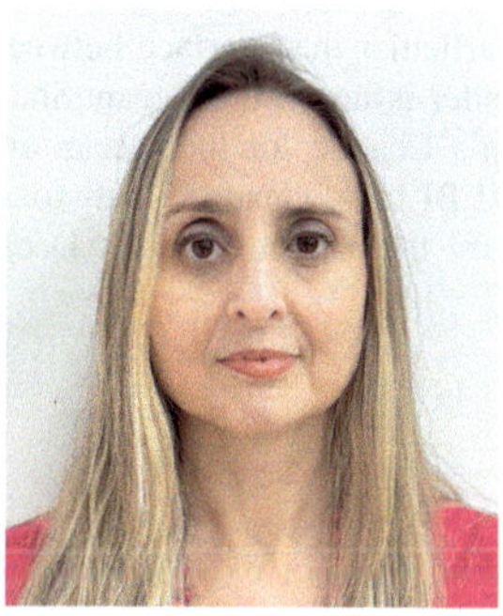

Andrezza Justino Gozzo Andreotti is a scientist and associate professor at the Institute of Marine Sciences at the Federal University of São Paulo (UNIFESP). She develops projects in the areas of biochemistry, biotechnology, sustainability, and Ocean Literacy, always thinking about how to turn scientific findings into benefits for society. She has been passionate about the sea and teaching for as long as she can remember.

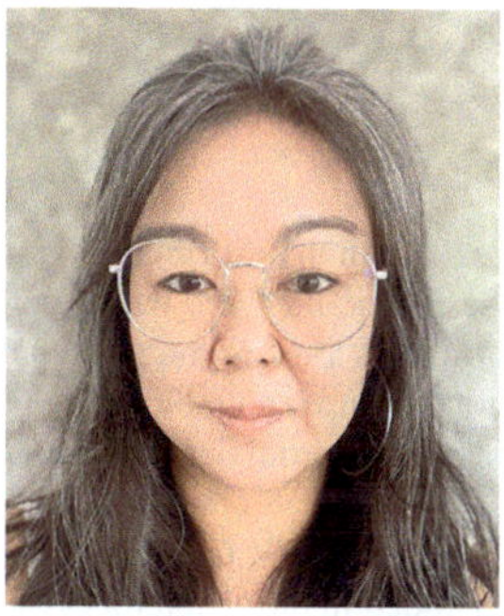

Camila Keiko Takahashi is the Program Manager at Maré de Ciência, mother of Mel and Noah, biologist and MSc in Water and Coastal Management with a background in marine science and marine conservation, working on strengthening the interface between science, public policies, and society. She is co-founder of the Women's League for the Ocean and co-secretariat of the Brazilian Ocean Literacy Alliance.

Barbara Lage Ignacio is a scientist and professor at the Federal University of São Paulo, actively engaged in transdisciplinary projects and initiatives. Her work centers on the human impact on coastal ecosystems, science literacy, citizen science, active citizenship to drive positive socio-environmental change, and the promotion of collaborative learning and action. The Maré de Ciência program reflects her long-standing commitment to social-action projects, her expertise in Marine Sciences, her efforts in the democratization of science and the concept of science everywhere, science with all people, and science for everyone's benefit.

Marilia Bueno is a postdoctoral researcher at the Institute of Marine Sciences at the Federal University of São Paulo (UNIFESP) with a background in marine ecology and Ocean Literacy. She is part of the Maré de Ciência team and is currently engaged in the Ocean Olympics.

Ronaldo Adriano Christofoletti is a researcher with a background in marine sciences and transdisciplinary approaches. In addition to his contribution to marine ecology and climate change, Ronaldo has been working with strategic communications and Ocean Literacy, with special attention to the inclusion of Global South contributions in these discussions. As chair of the Group of Experts on Ocean Literacy of the UNESCO-IOC and co-chair of the Strategic Communications Group of the UN Ocean Decade, he has been advocating for the integration of Ocean Literacy and communications to strengthen their societal impact.

Leandra R. Gonçalves is an Assistant Professor in the Institute of Marine Sciences at the Federal University of São Paulo. Biologist has a PhD from the Institute of International Relations and a postdoctoral degree from the Oceanographic Institute of University of São Paulo. She has been researching the different dimensions of coastal and marine management and governance for over 10 years, in particular the interface between science and policy and gender issues in the Ocean. She is co-founder of the Women's League for the Ocean and one of the members of WEBLUE—Gender Equity for Ocean, a project endorsed by the United Nations Decade of Ocean Science.

12 Cultivating Ocean Guardians: Evaluating a Novel Pedagogical Model for Enhanced Ocean Citizenship in Basic Education

Caroline Schio and Pedro Reis

Abstract

The implementation of the pedagogical model developed by Schio and Reis (2024) offers a transformative approach to fostering Ocean citizenship among students in basic education. This model integrates systemic educational approaches, Ocean Literacy, and citizenship theories into a hands-on citizen science framework. Through activities such as coastal monitoring, data collection, advocacy actions, and interdisciplinary projects, the model has succeeded in raising awareness and promoting sustainable Ocean practices among students, teachers, and their communities. Over two school years, the model engaged over 1500 students and 76 teachers across 36 Portuguese schools, demonstrating its scalability and adaptability to different educational contexts, including schools without direct access to the coast. The model's contribution to the Ocean Decade challenges is evident through its diverse educational outcomes, which build knowledge, develop skills, and foster a personal connection to the marine environment, while instilling a strong sense of Ocean citizenship in students. Piloting the pedagogical model provided a robust framework for integrating Ocean Literacy into basic education, encouraging a generation of Ocean stewards committed to preserving marine ecosystems. These efforts represent a vital step towards the Ocean Decade's broader objectives and the development of a blue curriculum in basic education.

Keywords

Pedagogical model · Design-based research · Ocean citizenship · Ocean Literacy (OL) · Citizen science · Students · Teachers · Basic education · Blue curriculum · Ocean Decade

12.1 Introduction

According to the International Oceanographic Commission of UNESCO (IOC-UNESCO 2021), a key challenge in fostering societal engagement in Ocean sustainability, in order to achieve the goals of the Ocean Decade, is developing holistic educational strategies that deepen personal connections with the Ocean and inspire behavioral changes, ultimately leading to the emergence of Ocean citizenship. Ocean citizenship overall involves recognizing the link between our daily actions and marine health, taking responsibility for informed actions that support

C. Schio (✉) · P. Reis
Instituto de Educação, Universidade de Lisboa, Lisbon, Portugal
e-mail: caroline.schio@edu.ulisboa.pt; preis@ie.ulisboa.pt

T. J. Kennedy (ed.), *Ocean Literacy: The Foundation for the Success of the Ocean Decade, Volume I*, Sustainable Development Goals Series, https://doi.org/10.1007/978-3-031-98145-6_12

marine sustainability, and exercising the right to advocate for improved marine governance (Buchan et al. 2023; Fletcher and Potts 2007; McKinley and Fletcher 2012).

But how can we tackle this global challenge and foster a more committed and informed generation when Ocean-related issues are scarcely represented in formal education worldwide, and youth lack opportunities to engage with the Ocean from an early age? Squarcina and Pecorelli (2017) suggest that we need to adopt an idealistic vision, thinking in terms of a cultural revolution that fosters both individual and collective responsibility towards this blue vastness. This shift demands a systemic educational approach that nurtures a multifaceted understanding of the interdependencies within the Human-Ocean System (Brennan et al. 2019; Tran et al. 2010).

Experiential learning activities hold immense potential for strengthening bonds with the Ocean and deepening our understanding of marine ecosystems, offering numerous avenues to engage new generations in informed decision-making about the Ocean and fostering interest in marine-related careers (Chen and Tsai 2016; NOAA 2021; IOC-UNESCO 2022). The pedagogical model designed by Schio and Reis (2024) (Fig. 12.1) aligns with these goals, providing a structured and practical approach to fostering Ocean citizenship in basic education. Grounded in systems thinking principles and informed by literature on Ocean Literacy and citizenship, it integrates conceptual and practical elements through citizen science and youth activism, involving the school community in monitoring and conservation activities along their coastlines.

The model's conceptual framework consists of four dimensions—training, experiences, involvement, and activism—each tied to particular actions and competencies development, accompanied by a toolkit to encourage these initiatives. Meanwhile, the practical component of the model includes six educational activities connected to a coastal monitoring project, carried out by students under the guidance of their teachers. These activities include: (1) contextualizing the issues related to the Ocean; (2) conducting three field trips to monitor the beach; (3) evaluating the

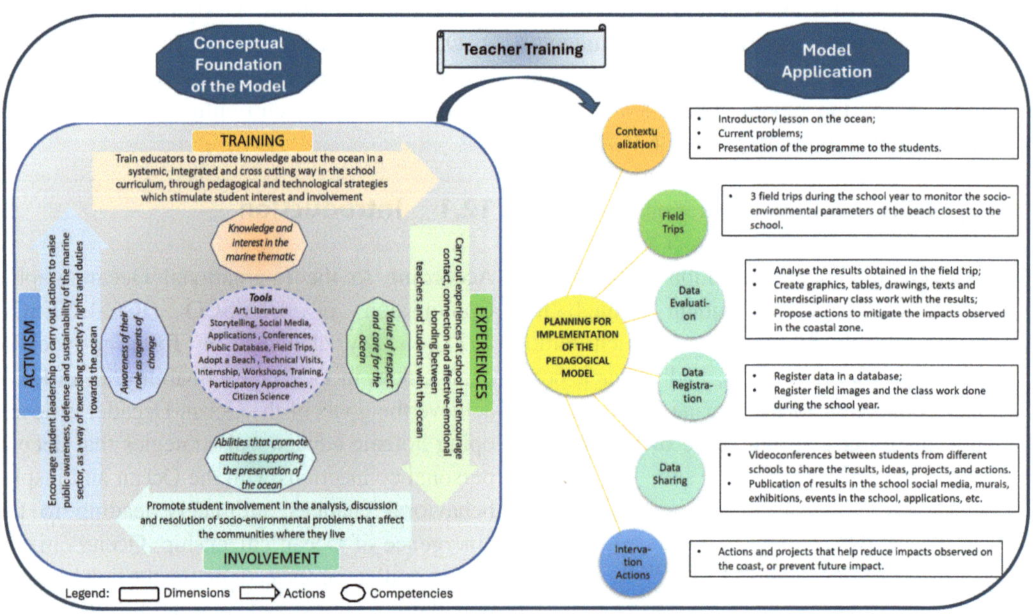

Fig. 12.1 Pedagogical model for promoting Ocean citizenship in basic education as proposed by Schio and Reis (2024, p. 16)

data obtained in the field; (4) recording the results; (5) sharing the results; and (6) implementing intervention actions within the school or community. Acting as an essential connector, teacher training links these components, facilitating the implementation of a pedagogical model in formal education (Fig. 12.1).

By evaluating the educational outcomes of this model's implementation over two school years (2021/2022 and 2023/2024) in Portuguese schools, this study aims to provide valuable insights and pedagogical strategies to enhance Ocean Literacy and Citizenship in formal education. The findings are intended to contribute to global goals such as SDG 14—"Protect the Oceans and marine resources through Ocean Literacy and action," while also addressing key Ocean Decade Challenges, particularly Challenge 9—"Skills, Knowledge, and Technology for All," and Challenge 10—"Change Humanity's Relationship with the Ocean."

12.2 Model's Application and Evaluation

The Design-Based Research (DBR) methodology was employed throughout both the development and implementation phases of the pedagogical model. A complete design cycle is composed of four main phases: (1) diagnosis, (2) planning, (3) action and (4) evaluation (Coghlan and Brannick 2001). The conceptual framework of the model was designed by applying phases 1 and 2, as described in Schio and Reis (2024), while the implementation and evaluation of the model's outcomes corresponded to phases 3 and 4. This chapter will focus on this last 2 phases, assessing the applicability and educational contributions of the model to foster Ocean citizenship in basic education.

The model was piloted over two school years (2021/2022 and 2023/2024). In the first year, applications were open exclusively to Portuguese Blue Schools to test the model's activities, while in the second year, the application process was opened to all Portuguese schools. Teachers enrolled in the model's training through a public application process using a Google Forms questionnaire at the beginning of each school year. The training consisted of one in-person workshop and three online workshops, during which teachers learned about marine topics and coastal monitoring practices that will enable them to implement the six activities of the model with their students, as outlined in the programmatic content shown in Fig. 12.2.

At the end of the training, teachers received a set of 8 video classes featuring various practical experiments. Using simple ingredients such as salt, oil, coloring food, vinegar, bicarbonate, among others, the experiments teach children about different oceanic processes through hands-on scientific activities (Fig. 12.3). These set of video classes were created to support teachers in delivering Ocean-related content throughout the school year in an interactive and engaging manner.

For the coastal monitoring activities, the Monitoramento Mirim Costeiro (MMC) project (https://www.monitoramentomirimcosteiro.com.br/) served as a reference. MMC is a pioneering citizen science program that has been monitoring beaches with Brazilian basic education students since 2012, and in 2023, it became an officially endorsed UN Ocean Decade project, integrated into the Ocean Literacy With All—OLWA Programme. The environmental parameters recorded by students—such as weather and Ocean conditions, water quality, pollution levels, and socioeconomic activities in the coastal zone—are detailed on the four pages of MMC field sheets used by the schools (Fig. 12.4).

All of this data is recorded by students in an Excel sheet, which is later integrated into the MMC dashboard (https://tinyurl.com/mmc-dashboard). The data is also shared on citizen science platforms such as *Biocollect* and *CIVIS*. The MMC's approach not only enhances students' scientific literacy but also actively involves them

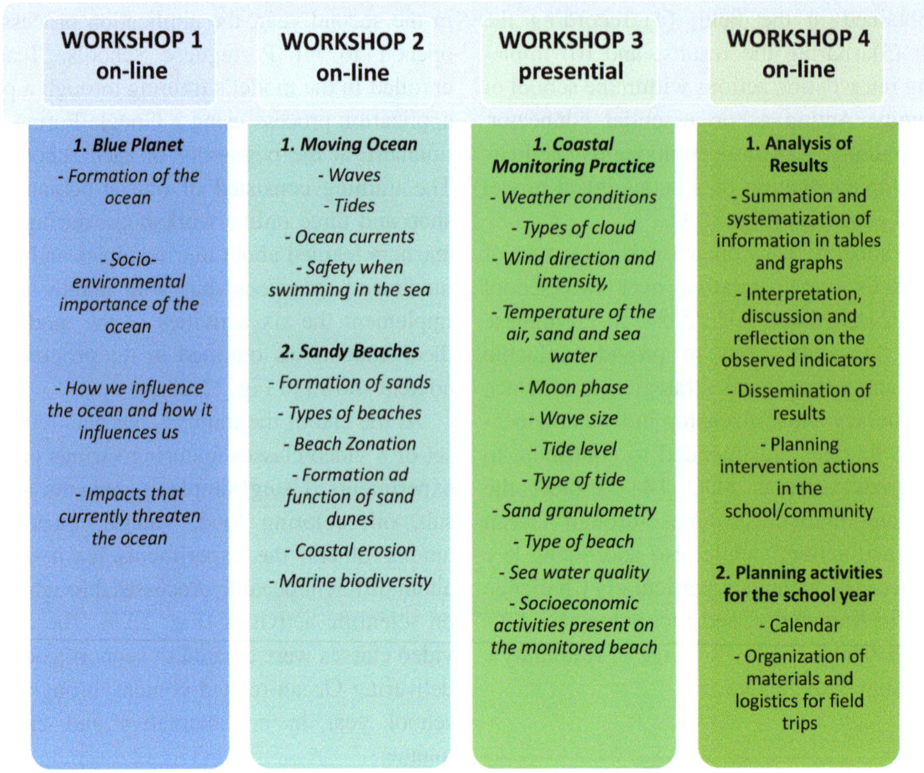

Fig. 12.2 Diagram of the programmatic content of phase 1 of the training course for teachers

Fig. 12.3 Experiments conducted by children addressing oceanic-related content in a practical way

Fig. 12.4 Data recording by students during coastal monitoring excursions, based on the Brazilian Monitoramento Mirim Costeiro Programme

in Ocean stewardship through hands-on data collection and environmental monitoring.

To annually assess the applicability and educational contributions of the model in fostering Ocean citizenship, a combination of online questionnaires, focus groups, and direct observation were utilized. Teachers were engaged through online questionnaires and virtual focus group meetings, providing feedback on their experiences running the model's activities. Concurrently, students participated in in-person focus group meetings for each class, allowing for a deeper understanding of their learning outcomes and engagement levels. This comprehensive evaluation approach ensured that multiple perspectives were considered, enhancing the overall assessment of the model's impact on promoting Ocean Literacy and citizenship.

A bonus workshop, named *Exploring the Marine Mini Museum as a Scientist*, was run at the end of each student focus group. In this workshop, children explored several biological and mineralogical samples using various research tools, further enhancing their hands-on learning experience, scientific skills, and Ocean knowledge (Fig. 12.5).

12.3 Educational Outcomes of the Model's Implementation

12.3.1 Overall Metrics

Over the course of two school years, the model's implementation involved 1565 students from 83 first to eighth-grade classes (aged 6 to 14), along with 76 science and math teachers across 36 public schools located in 28 Portuguese localities (Table 12.1). A total of 80 sites were monitored by the schools, including Ocean beaches (80%), estuary beaches (8.75%) beaches, and fluvial beaches, rivers, and lakes (11.25%). By accommodating a wide range of water-based settings, the model demonstrates its adaptability in monitoring diverse aquatic environments and its potential for implementation in schools with limited access to the sea.

This ensures that students from different regions can engage in citizen science initiatives, fostering a broader understanding, appreciation, and commitment to the preservation of their local aquatic ecosystems. Such an inclusive approach not only enhances the environmental educational

Fig. 12.5 Students participating in the workshop *Exploring the Marine Mini Museum as a Scientist*

Table 12.1 Pedagogical model application metrics per school year

School year	Students	Teachers	Schools	Classes	Grades	Monitored spots	Localities
2021/2022	543	23	10	26	5th–6th	15	10
2023/2024	1022	53	26	57	1th–8th	65	18
Total	**1565**	**76**	**36**	**83**	**8**	**80**	**28**

experience but also contributes to a greater sense of citizenship, stewardship, and responsibility towards these environments, which is crucial for achieving the Ocean Decade goals (IOC-UNESCO 2021).

It is important to highlight that the main implementation metrics of the model more than doubled in the second year of its application in Portuguese schools, as shown in Table 12.1. This increase resulted from several factors, such as:

1. Teachers shared their experiences and motivation with colleagues, sparking interest among others, even those from different grades and subjects.
2. The compatibility of the model's activities with school curriculum subjects allowed teachers to work in interdisciplinary groups throughout the school year.
3. Students' acceptance, interest, motivation, and engagement with all proposed activities, particularly with field trips and hands-on activities.
4. The training course officially offered 50 credits towards teachers' professional career progression, which served as a strong motivational factor and further incentivized participation.
5. Other schools became aware of the coastal monitoring project through the Blue Schools Program evaluation meetings, where participating schools shared their motivations and experiences in conducting the model's activities.
6. Schools noted that implementing the model's activities could help them obtain or maintain the Blue School certification, as the model application meets all the necessary requirements for a school to become a Blue School.
7. In the second year, the application process was opened to the entire school community, including the Azores Island region, and extended to additional primary grades.

This significant increase demonstrates the potential for the model's expansion across Portuguese school community. The natural interest shown by teachers in participating, combined with the project's self-promotion among schools, highlights its capacity to spread organically. This suggests that as more educators and schools recognize the educational benefits and positive outcomes of implementing the model's activities, its adoption will continue to grow, further enhancing Ocean Literacy and citizenship among students nationwide.

On the other hand, it also emphasizes the importance of continuously offering teacher training, facilitating experience-sharing meetings, and developing new pedagogical activities and materials. These efforts provide innovative tools that enhance both teaching practices and the model's educational impact on promoting Ocean citizenship. Fostering practical professional development courses is crucial, as it not only strengthens teachers' knowledge and pedagogical skills in marine science but also empowers them to create more dynamic and engaging learning environments, leading to improved educational outcomes and student performance (Lestari et al. 2023; Wulff and Johannesson 2019). This is further supported by teachers' feedback, which highlights how the training provided new skills that contributed to their professional development (Fig. 12.6).

12.3.2 Nurturing Interdisciplinarity

Teachers emphasized the model's interdisciplinary approach, which enhanced collaboration across subjects and introduced active, project-based learning focused on Ocean Literacy and citizenship. Hands-on coastal monitoring activities, including fieldwork, data collection, and

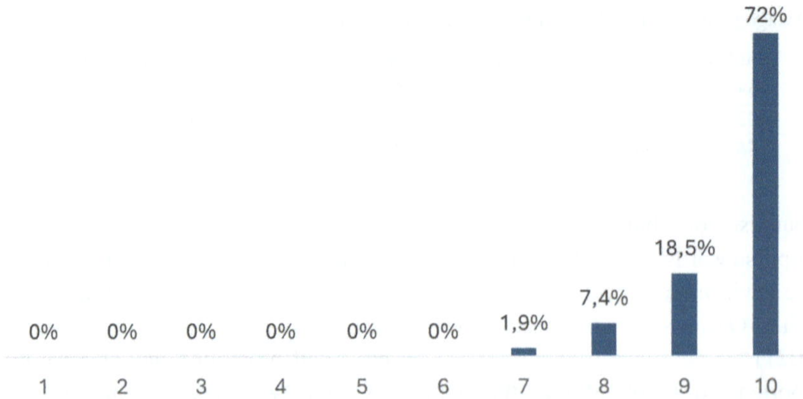

Fig. 12.6 How much teachers believe the training contributed new skills to their teaching career on a scale from 0 to 10

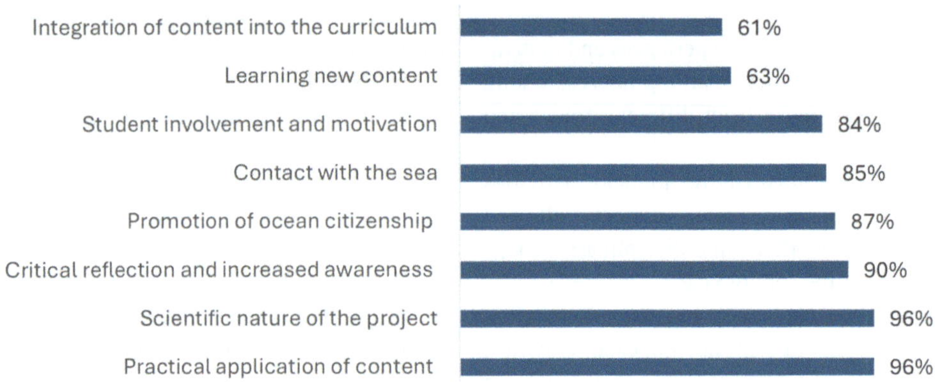

Fig. 12.7 Key aspects of the pedagogical model's implementation highlighted by teachers

critical reflection, were also highlighted by teachers as key aspects that provided them with practical knowledge and tools to effectively integrate Ocean conservation into the educational curriculum (Fig. 12.7).

The diverse set of activities offered by the pedagogical model demonstrated how its application fostered interdisciplinarity across different grades. Teachers were able to integrate the model's activities with various subjects, working collaboratively in pedagogical teams and engaging educators from different disciplines (Figs. 12.8 and 12.9). This collaborative approach not only enriched the learning experience but also promoted a more cohesive implementation of Ocean citizenship throughout the curriculum.

12.3.3 Learning by Exploring

The teachers´ feedback indicates that students were more motivated and enthusiastic about Ocean conservation when they could physically observe and interact with their coastal environment, making the learning more memorable and meaningful. Field trips consistently emerged as the most engaging hands-on activity, providing opportunities for students to directly interact with the coastal environment, applying their learning in a practical and tangible way. Other activities, such as the mini marine museum workshop and interdisciplinary school projects, were also crucial to student engagement (Fig. 12.10).

The scientific and practical nature of the model's activities naturally sparks students' curiosity,

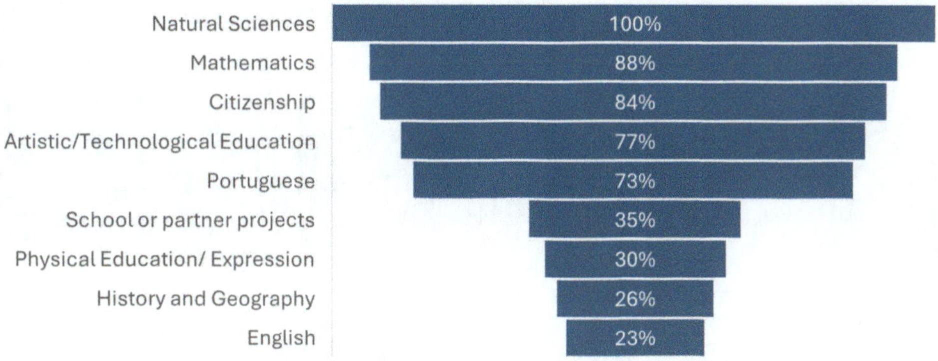

Fig. 12.8 Subjects in which the model's activities could be integrated according to teachers´ evaluation

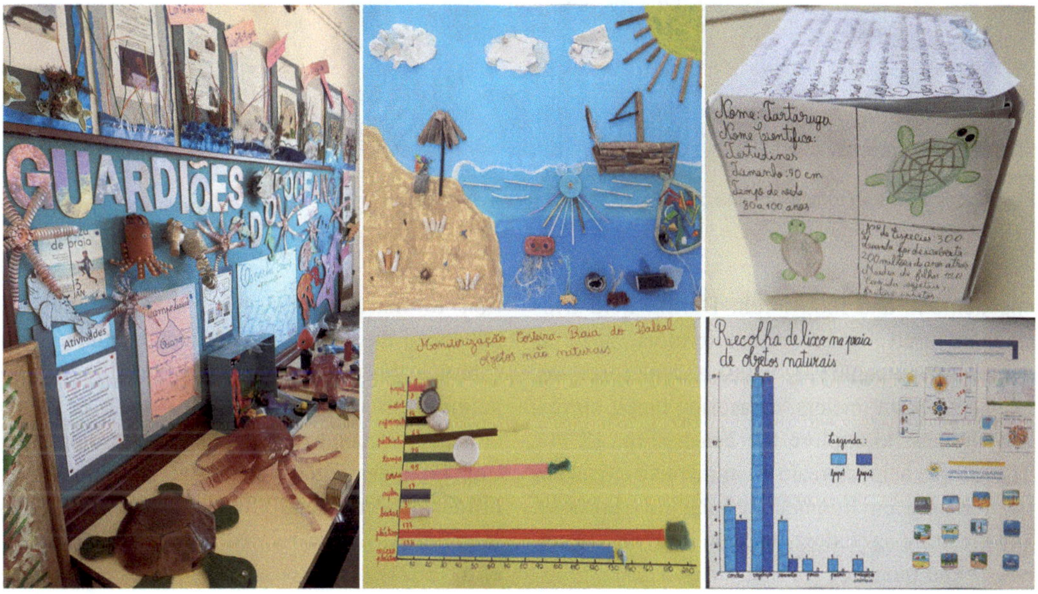

Fig. 12.9 Examples of interdisciplinary work classes conducted by students across different subjects during the school year

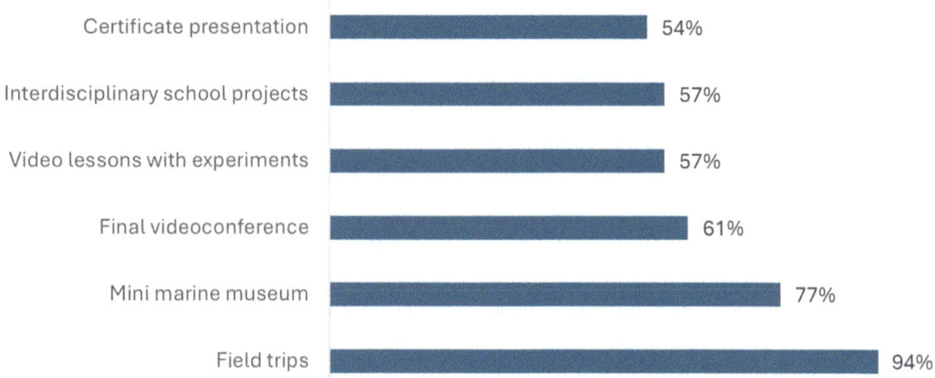

Fig. 12.10 Most significant moments for the students according to the teachers' opinions

Fig. 12.11 Children monitoring the coast using several research tools

engagement, and motivation. During field trips students used various research tools, such as compasses, magnifying glasses, microscopes, magnets, thermometers, measuring tapes, sieves, and shovels, to explore the beach (Fig. 12.11). These tools helped create an immersive experiential learning environment, giving joy, purpose, and meaning as students take on the role of "scientists and monitors" of their coastal zone.

Feedback from students about what they enjoyed most during the coastal monitoring project was systematized in the word cloud (Fig. 12.12), and the main aspects highlighted by students are exemplified with quotes in Fig. 12.13, underscoring the value of hands-on experiences. Activities like cleaning the beach, digging, sifting, finding things in the sand, conducting experiments from the video classes, and participating in interdisciplinary work were the most frequently mentioned. These activities not only instilled a sense of purpose in the actions they were undertaking but also highlighted the role of children as active participants in their learning process and as agents of change within their community.

This hands-on involvement in real-world scientific research, combined with the experience of being outdoors, deepened students' connection to marine ecosystems and significantly enhanced the learning experience. Studies corroborate this impact, showing that direct interaction with the marine environment not only enhances Ocean Literacy and deepens appreciation for marine ecosystems but also positively influences eco-friendly behaviors and increases interest in marine-related careers (Boaventura et al. 2021; Guest et al. 2015; Garcia-Vazquez et al. 2022).

More detailed information about the students' experience in participating in the activities of the pedagogical model and how this process contributed to the development of competencies in Ocean citizenship can be found in Schio and Reis (2024b).

12.3.4 Passing of the Torch

Furthermore, these activities empowered students as agents of change and knowledge multipliers within their communities. For instance, in the second year of the model's implementation, some schools that participated in the first year facilitated a "passing of the torch" from the previous year's students to the new participants.

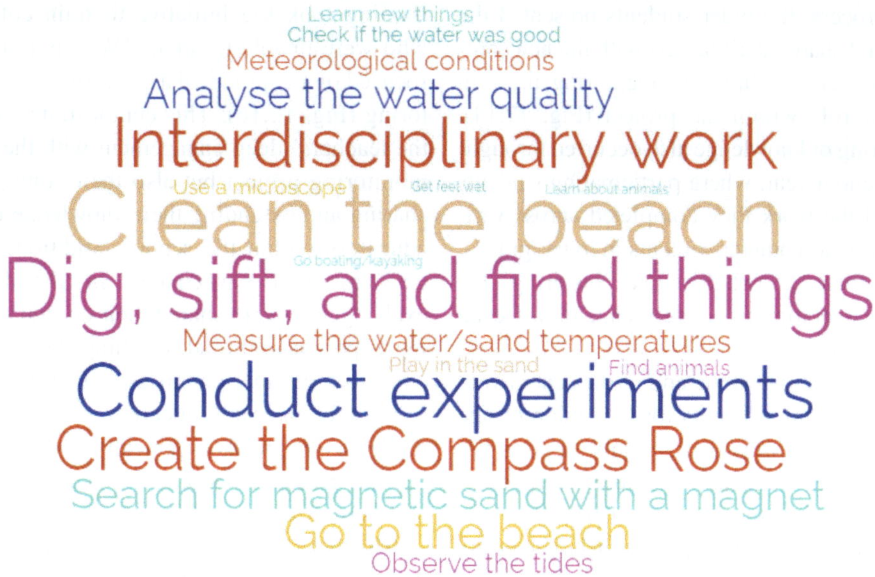

Fig. 12.12 Activities that students enjoyed most during the coastal monitoring project

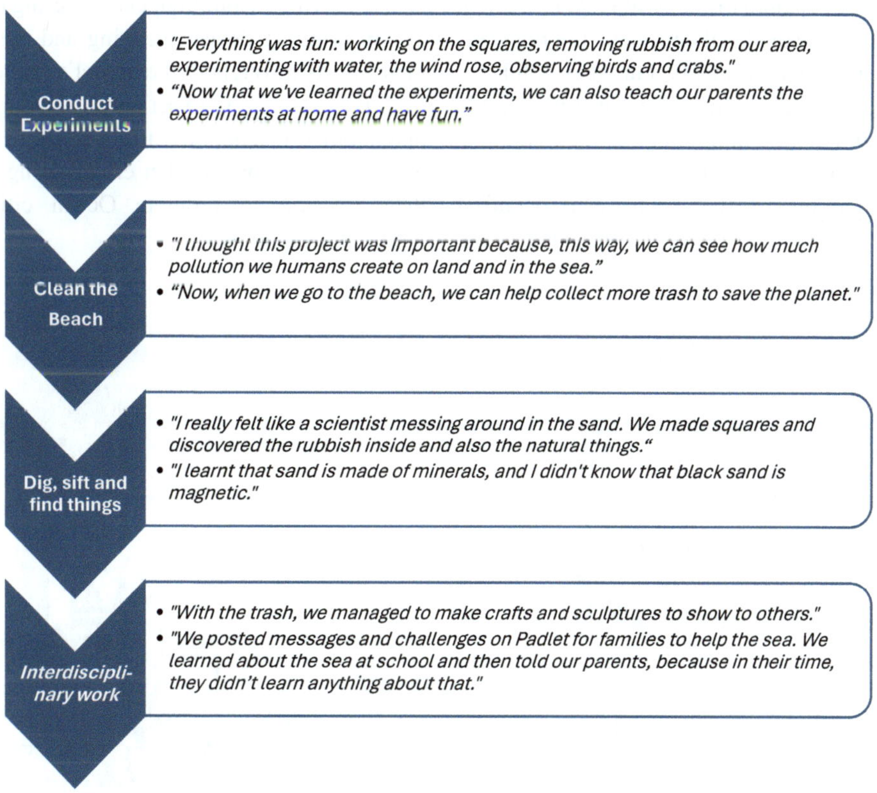

Fig. 12.13 Quotes from students highlighting key elements of their experience

In this process, the older students presented the project and shared their results with the newcomers, effectively guiding and mentoring them in their new role within the project (Fig. 12.14). This sharing of knowledge also occurred throughout the school year, where participating students presented the work they completed across various subjects to younger students in lower grades, sharing their knowledge and inspiring the younger generation to actively engage in Ocean conservation.

Videoconference at the end of the school year were also pivotal moments for the children. The video conference provided a unique opportunity for students to connect with peers from other schools, allowing them to see their work in a broader context and feel part of a larger group dedicated to Ocean preservation. This interaction fostered a sense of unity and collective purpose, reinforcing the importance of their contributions. The certificate presentation, meanwhile, served as a powerful symbol of recognition (Fig. 12.15). Receiving a personalized certificate not only celebrated their achievements but also solidified their roles as Ocean Guardians, instilling pride and encouraging them to continue their commitment to marine conservation.

The "passing of the torch" also occurred among teachers. Those who participated in the training took the initiative to train colleagues who were unable to attend Workshop 4, which focused on the practical aspects of coastal monitoring (Fig. 12.16). This demonstrates not only the teachers' deep engagement with the coastal monitoring project but also their confidence in sharing and extending their knowledge to other educators within the school. Additionally, two teachers who participated in the Erasmus exchange program shared their coastal monitoring experience abroad, teaching the method to educators in other countries. This further highlights the natural replication potential of the model, as it spreads organically both within and beyond national borders, expanding the reach and impact of Ocean Literacy initiatives.

The "passing of the torch" was also observed through the continued communication in the WhatsApp group, created during the second year, which helped strengthen the unity among teachers. This platform became a space for sharing ideas, offering inspiration, and fostering collaboration, bringing deeper meaning and purpose to their efforts throughout the model's implementation. By exchanging experiences and supporting one another, teachers felt more connected to the model's aspirations, further empowering them in their mission to promote Ocean citizenship within their schools.

Fig. 12.14 Students sharing their experiences and inspiring the younger generation

Fig. 12.15 Students celebrating their certification as Ocean Guardians at the end of the school year

Fig. 12.16 Teachers instructing each other in coastal monitoring practices

12.3.5 Integration with Other Projects

Another important factor observed was the compatibility of integrating the pedagogical model's activities with other projects and events in which the school participates, such as Kids-Dive, Eco-Escolas, Science Clubs, eTwinning, World Oceans Day, International Coastal Cleanup, Blue Schools Olympics, among others. This compatibility further facilitates the integration of the model into the school curriculum, allowing its activities to connect smoothly with existing projects and initiatives. Additionally, the model's flexibility enables its adaptation to different grade levels, expanding its impact across various age groups and educational contexts.

This integration is crucial, especially when considering the importance of fostering Ocean Literacy from an early age and structuring this learning progressively throughout the school years (Halversen et al. 2021). By introducing concepts of Ocean citizenship and sustainability early on, it is possible to build a solid foundation of knowledge that deepens as students advance in their education, providing innovative tools that both motivate and engage them with marine issues.

This continuous approach not only enriches the curriculum but also contributes to the development of a blue generation prepared to tackle the global environmental challenges currently threatening Ocean health (IOC-UNESCO 2021). These efforts also align closely with Ocean Decade Challenge 9—"Skills, Knowledge, and Technology for All," by fostering in students the development of essential competencies in environmental monitoring and conservation advocacy.

12.3.6 Activism Actions

Although some teachers mentioned that, due to lack of time, they were unable to carry out intervention actions in their communities (the sixth activity of the model), activism improved significantly in the second year of the model's implementation, reflecting greater student engagement in conservation initiatives. Multiple examples of activism actions are detailed in Table 12.2 and illustrated in Fig. 12.17, further highlighting the schools' commitment to fostering active student involvement in real-world advocacy. Fostering advocacy programmes to train young ambassadors as agents of change is crucial for developing future scientists and responsible decision makers, cultivating the next generation of professionals committed to inclusive and ethical Ocean governance (Cappelletto et al. 2021).

Table 12.2 Examples of activism actions undertaken by schools

Activism actions	Examples
Theatrical and artistic presentations	Projects included theatrical dances, public performances, and artistic installations. Posters and recycled marine creatures with awareness messages were placed in local shops as part of a larger campaign to draw attention to Ocean issues.
Beach clean-ups	Multiple schools engaged in beach clean-ups, some involving families and the wider community.
Public awareness campaigns & community engagement	Some schools organized marches advocating for Ocean preservation. Students created posters, held exhibitions, and participated in mobilizations with performances and demonstrations, as well as events like the "Night of Scientists" and "Olympic Blue Schools". Schools also collaborated on government projects such as the "Reduna Project" (dune preservation).
Advocacy letters	Letters and emails were sent to local government officials advocating for Ocean-friendly policies, such as reducing cigarette butts on beaches and requesting waste disposal infrastructure near beaches.
Assembly and public presentations	Presentations were made at assemblies, including Lisbon's children's assembly. A marine-focused book created by students was recognized with an award at the municipal level.

12.3.7 Families´ Engagement

Schools positively mobilized students´ families, and the wider community around these efforts, creating a ripple effect that extended awareness beyond the classroom. According to teachers' feedback, many families participated directly in school activities, such as beach clean-ups, artistic projects, and even helping to bake cakes to raise money for students' transportation to the beach, among other initiatives. Teachers noted that parents showed enthusiasm in supporting their children's involvement. Families were also engaged through various channels, including informal conversations, email updates, school meetings,

Fig. 12.17 Illustrations of activism actions undertaken by schools

and social media platforms like Facebook, WhatsApp, and Padlet, where they followed the progress of the project.

Teachers reported that students often shared their learning experiences at home, sparking discussions on marine conservation, and in some cases, even observing behavioral changes, such as increased environmental awareness and improved waste management practices, with students encouraging their parents to adopt environmentally friendly behaviors. By engaging in these discussions and behavior shifts, students demonstrate the powerful role they can play as agents of change, influencing the behaviors of those around them and contributing to broader community shifts toward more sustainable practices and a deeper commitment to environmental stewardship. However, there is still a need to further reinforce these efforts to ensure even broader participation and long-term impact.

12.3.8 Fostering Blue Schools

An interesting observation that is important to highlight is that even though the training program in the second year was open to all schools, only 5 out of 36 participating schools were not Blue Schools—this included all 3 participating schools from the Azores, as well as 2 others from Northern Portugal. This demonstrates the interest, and commitment of these certified schools to nurturing Ocean Literacy within their curricula. As the model continues to be replicated nationally, it is expected that more non-Blue Schools will become aware of it and express interest in participating. This growing interest could potentially lead to more schools pursuing recognition through the national Blue School certification, as demonstrated by a school in the Azores.

For instance, São Miguel Island now has its first Blue School, thanks to a participating school that implemented the coastal monitoring project and subsequently applied for Blue School certification (Fig. 12.18). Similarly, a school in northern Portugal followed the same process. These examples highlight how the model can effectively expand the network of schools committed to Ocean Literacy on a national scale. By providing a structured and practical approach to Ocean education, the model not only supports existing Blue Schools but also encourages other schools to join the network.

Fig. 12.18 First São Miguel Blue School certification ceremony with students receiving their Ocean Guardian Certificates

12.3.9 Meeting the Ocean

An important observation during the 2 years of implementing the model was that 35 Portuguese students had never been to the beach before participating in the coastal monitoring project. Remarkably, even on the island of São Miguel, where the Ocean is an ever-present part of life, 3 students had never experienced the beach firsthand. The project provided these students with their first opportunity to visit the Ocean, underscoring the significant potential of citizen science projects in fostering and strengthening both students' and teachers' bonds with the sea. Moreover, it aligns with Ocean Decade Challenge 10—"Change Humanity's Relationship with the Ocean," by nurturing a sense of personal connection and responsibility toward marine environment.

12.4 Challenges, Improvements and Next Steps

The most common challenges faced during both years of project implementation were related to transportation (71%) and meteorological conditions (72%), which often affected the ability to conduct field activities. Additionally, issues with field materials (26%), time constraints (35%), student behavior (6%), the practical application of content in both field and classroom settings (4%), and the need for more support from school and colleagues (4%) were also mentioned as challenges by teachers. However, despite these challenges, teachers were able to overcome these constraints, finding alternative solutions with the help of their colleagues, partners and the students' parents. This collaborative effort ensured that the project could continue successfully, demonstrating the resilience and commitment of both the school community and the families involved.

The majority of teachers (72%) expressed a strong interest in continuing the coastal monitoring project in the next school year. Additionally, 27% indicated that their participation would depend on the school in which they will be teaching, while 1% expressed no interest. Many teachers highlighted specific areas for improvement, such as planning field trips with greater anticipation to address logistical issues, especially in relation to transportation, and starting the project earlier in the school year to allow for more time to explore the project's different phases.

Some teachers also expressed the desire to expand the project to include more classes and involve more disciplines, creating a more integrated, interdisciplinary approach. Others mentioned the importance of involving local businesses and external partners, including transportation companies, to enhance the project's reach and visibility. Additionally, several teachers suggested introducing new civic activities and expanding the project to other beaches and even freshwater environments.

When asked if they believed it was possible to develop a community of practice among schools that could constitute an active Ocean Guardians Network within Portuguese education, 100% of the teachers in both years of the model's implementation responded affirmatively. These reflections highlight the potential to develop a robust Ocean citizenship network, where teachers collaborate, share best practices, and support one another in overcoming challenges. To maintain a dynamic and active network between schools, teachers proposed several strategies aimed at fostering collaboration, communication, and engagement. Their suggestions included:

- Foster the ongoing WhatsApp group for sharing activities and ensuring continuous communication among educators.
- Raise the visibility of the project through social media, community engagement, and partnerships with local environmental associations.
- Hold semi-annual online meetings where students from different schools can share experiences and resources.
- Seek support from local authorities, such as municipal governments, to help organize events and provide resources.
- Provide teacher training and promote the project in schools that have not yet joined.
- Create a transdisciplinary portfolio to integrate activities across different subjects.
- Establish a collaborative platform for sharing resources, results, and best practices.
- Organize end-of-year trips to schools within the same region, where students prepare activities and exhibitions, with the location changing each year to promote diversity and collaboration.

A common thread across these suggestions was the importance of creating a shared *sense of purpose* through ongoing collaboration, mutual support, and sharing best practices, all of which could be supported by events, digital tools, and local partnerships. All these strategies can help outline a concise strategic plan to foster student, teacher, and school participation in the coming years, enhancing the growth and dissemination of the pedagogical model within the Portuguese school community.

The development of an educational digital platform that congregates online training classes, teaching tutorials, pedagogical materials, games, challenges, and a chat for interaction would be crucial to nurturing the Ocean Guardian Network and maintaining the engagement, motivation, and commitment of schools in fostering Ocean citizenship. This digital platform could not only help disseminate the project at a national level, but also has the potential to catapult it onto the international stage by adapting its content for different languages and educational contexts. Such a platform would further contribute to the Ocean Decade goals by expanding its reach and impact globally, encouraging schools worldwide to integrate Ocean citizenship in a blue curriculum.

12.5 Main Contributions of the Pedagogical Model in Supporting Ocean Decade Goals

The pedagogical model promoting Ocean citizenship in schools aligns strongly with the goals and challenges of the UN Ocean Decade and SDG 14: Life Below Water. Through citizen science, Ocean Literacy workshops, beach cleanups, and activist actions, the model fosters active student participation in reducing marine pollution, protecting ecosystems, and promoting the sustainable use of marine resources. By emphasizing active engagement, the model not only raises awareness and enhances scientific knowledge, but also instills a deep sense of responsibility, transforming students into Ocean stewards who drive bottom-up public engagement for Ocean sustainability.

The model's ability to foster personal connections with the Ocean highlights its potential to address the challenge of changing humanity's relationship with the Ocean, making a significant contribution to its long-term preservation. Table 12.3 illustrates the detailed correlations between the model's activities, Ocean Decade goals, challenges, and SDG 14 targets, highlighting the key contributions of the model to the Ocean Decade goals.

12.6 Conclusion

The implementation of the pedagogical model across two academic years has demonstrated significant contributions to fostering Ocean citizenship and supporting the goals and challenges of the UN Ocean Decade. The model's integration of citizen science, interdisciplinary learning, and real-world activism has effectively engaged students and teachers in meaningful marine conservation activities. These initiatives have not only deepened participants' understanding of Ocean ecosystems but also instilled a strong sense of responsibility and commitment to sustainable practices.

The model's adaptability and success in various educational contexts, including its expansion across different grades and regions, highlight its potential for broader application within the Portuguese school system and beyond. By addressing key Ocean Decade Challenges, particularly in changing humanity's relationship with the Ocean and enhancing Ocean Literacy, the model serves as a powerful tool in cultivating a generation of informed, engaged, and active Ocean stewards. As schools and educators continue to integrate this model into their curricula, the initiative represents a vital step towards achieving global sustainability goals.

12.6.1 Summary

Key concepts explored in this chapter include:

- Fostering Ocean Citizenship through Experiential Learning. Engaging students in Ocean Literacy, citizen science, and activism through Monitoramento Mirim Costeiro (MMC) program helps develop a sense of responsibility for marine conservation.
- A Structured Framework for Ocean Education. The pedagogical model based on Monitoramento Mirim Costeiro (MMC) incorporates training, experiences, involvement, and activism to enhance student engagement in Ocean Literacy.
- Hands-on Scientific Inquiry Strengthens Learning. Activities such as beach monitoring, data collection, and environmental intervention provide students with practical experience in marine science.
- Scalability and Impact Across Schools. Implemented in 36 schools, the pedagogical model successfully engaged over 1500 students and 76 teachers, demonstrating its adaptability to different educational settings.
- Empowering Educators through Teacher Training. Professional development workshops equipped teachers with effective strategies and resources to implement Ocean-focused lessons, enhancing student engagement and strengthening their connection to marine environments through hands-on activities.
- Interdisciplinary Learning Enhances Engagement and Expands Educational Opportunities. Collaboration across subjects and grade levels fosters a holistic approach to Ocean education, integrating science, social studies, and environmental studies to deepen student understanding and engagement.
- Community Involvement is Key to Success. Beach clean-ups and awareness campaigns organized by schools strengthen both student and community participation in marine conservation efforts.

Table 12.3 Contributions of the pedagogical model to Ocean Decade goals through correlation of activities, challenges, and SDG 14 targets

Ocean Decade Challenges	Ocean Decade outcomes	SDG targets	Model's activity	Model's contributions to Ocean Decade goals
Challenge 1: Understand and beat marine pollution	A clean Ocean	SDG 14.1: Reduce marine pollution	Contextualization, field trips and intervention actions	Awareness about marine pollution is raised through hands-on activities, such as coastal monitoring and beach clean-ups, prompting critical reflection on consumption habits, waste disposal, and the need for behavior change and community engagement campaigns.
Challenge 2: Protect and restore ecosystems and biodiversity	A healthy and resilient Ocean	SDG 14.2: Protect and restore marine ecosystems	Intervention actions	Through targeted intervention actions, such as beach clean-ups, environmental marches, participation in dune restoration projects, and the creation of posters, exhibitions, and marine animals with awareness messages, students are empowered to actively participate in preserving marine ecosystems and biodiversity.
Challenge 5: Unlock Ocean-based solutions to climate change	A healthy and resilient Ocean	SDG 14.2: Protect and restore marine ecosystems	Contextualization and intervention actions)	Students are educated about the role of Oceans in climate regulation and participate in activities that highlight Ocean-based solutions to climate change.
Challenge 7: Expand the Global Ocean Observing System	A predicted Ocean	SDG 14.8: Increase scientific knowledge, research and technology for Ocean health	Field trips and data evaluation	Students actively participate in the collection and analysis of environmental data, contributing to the broader understanding of coastal and marine conditions. These activities align with the goals of expanding the Global Ocean Observing System by engaging students in citizen science, which helps to increase scientific knowledge and supports the ongoing monitoring and prediction of Ocean health.
Challenge 8: Create a digital representation of Ocean	An accessible Ocean	SDG 14.8: Increase scientific knowledge, research and technology for Ocean health	Data evaluation and data sharing	Students analyzed data collected during field trips using Excel sheets. The data from all schools is then shared on the Monitoramento Mirim Costeiro dashboard, which is integrated with citizen science platforms, enhancing the accessibility of coastal data and contributing to the development of digital representations of the Ocean.

(*continued*)

Table 12.3 (continued)

Ocean Decade Challenges	Ocean Decade outcomes	SDG targets	Model's activity	Model's contributions to Ocean Decade goals
Challenge 9: Skills, knowledge, and technology for all	An accessible Ocean	SDG 14.8: Increase scientific knowledge, research and technology for Ocean health	Field trips, data evaluation, and data sharing	Students, even those with limited access to the sea, develop essential skills in environmental monitoring, data analysis, evaluation, and communication. By sharing their findings through various methods such as murals, posters, exhibitions, letters, work classes, video conferences, and citizen science platforms, students contribute to increasing scientific knowledge and making Ocean data and other aquatic environments more accessible.
Challenge 10: Change humanity's relationship with the Ocean	An inspiring and engaging Ocean	SDG 14.2: Protect and restore marine ecosystems	All the six model's activities	By engaging students in hands-on, real world experiences, the model fosters a deep, personal connection with the Ocean, inspiring responsibility, stewardship, and a long-term commitment to marine conservation.

- Addressing Challenges for Wider Implementation. Overcoming logistical hurdles, such as transportation, availability of field materials, and scheduling, is essential for expanding the pedagogical model to additional schools.
- **Alignment with the Ocean Decade.** The pedagogical model supports global goals for sustainable marine education and conservation by fostering Ocean Literacy, community engagement, and hands-on environmental stewardship.

12.6.2 Recommended Resources

The following resources provide further insights and support on this topic:

- Monitoramento Mirim Costeiro: https://www.monitoramentomirimcosteiro.com.br/
- Monitoramento Mirim Costeiro Ocean Decade Action: https://Oceandecade.org/actions/monitorament-o-mirim-costeiro/
- REDE Guardiões Mirins do Oceano: https://tinyurl.com/mmc-dashboard
- Monitoramento Mirim Costeiro Instagram: https://www.instagram.com/monitoramento_mirim_costeiro/

Acknowledgements We deeply thank the schools, teachers, students, and supporting institutions for their invaluable contributions and dedication, which were crucial for the success and enrichment of this research on Ocean citizenship. This research was funded by a Ph.D. fellowship from FCT—Portuguese Foundation for Science and Technology, grant number 2022.10641.BD, and by National Funds through FCT— Portuguese Foundation for Science and Technology, I.P., under the scope of UIDEF—Unidade de Investigação e Desenvolvimento em Educação e Formação, UIDB/04107/2020, https://doi.org/10.54499/UIDB/04107/2020.

References

Boaventura D, Neves AT, Santos J, Pereira PC, Luís C, Monteiro AT, Cartaxana A, Hawkins SJ, Caldeira MF, Carvalho AP (2021) Promoting Ocean Literacy in elementary school students through investigation activities and citizen science. Front Mar Sci 8:675278. https://doi.org/10.3389/fmars.2021.675278

Brennan C, Ashley M, Molloy O (2019) A system dynamics approach to increasing Ocean Literacy. Front Mar Sci 6:360. https://doi.org/10.3389/fmars.2019.00360

Buchan PM, Evans LS, Pieraccini M, Barr S (2023) Marine citizenship: the right to participate in the transformation of the human-ocean relationship for sustainability. PLoS One 18(3):e0280518. https://doi.org/10.1371/journal.pone.0280518

Cappelletto M, Giuffredi R, Kastanidi E, Vassilopoulou V, L'Astorina A (2021) Grounding Ocean ethics while sharing knowledge and promoting environmental responsibility: Empowering young ambassadors as agents of change. Front Mar Sci 8:717789. https://doi.org/10.3389/fmars.2021.717789

Chen C-L, Tsai C-H (2016) Marine environmental awareness among university students in Taiwan: A potential signal for sustainability of the Oceans. Environ Educ Res 22(7):958–977. https://doi.org/10.1080/13504622.2015.1054266

Coghlan D, Brannick T (2001) Doing action research in your own organization. Sage

Fletcher S, Potts J (2007) Ocean citizenship: An emergent geographical concept. Coast Manag 35(4):511–524

Garcia-Vazquez E, Garcia-Ael C, Mesa MLC, Dopico E, Rodriguez N (2022) Enhancing marine citizenship as a strategy to promote the reduction of single-use plastics consumption in different cultures. Front Mar Sci 9:941694. https://doi.org/10.3389/fmars.2022.941694

Guest H, Lotze HK, Wallace D (2015) Youth and the sea: Ocean Literacy in Nova Scotia, Canada. Mar Policy 58:98–107

Halversen C, Schoedinger S, Payne D (2021) A handbook for increasing Ocean Literacy: Tools for educators and Ocean Literacy advocates. National Marine Educators Association

IOC-UNESCO (2021) *Ocean Literacy framework for the UN decade of Ocean science for sustainable development 2021–2030*. UNESCO. (IOC Ocean Decade Series, 22)

IOC-UNESCO (2022) *A new blue curriculum—A toolkit for policy-makers*. UNESCO. (IOC Manuals and Guides, 90)

Lestari IW, Surtukanti HK, Rahman T (2023) Science teacher's professionalism in planning practical learning activities. Syntax Literate Indones Sci J 12(2):1392–1402. https://doi.org/10.36418/syntax-literate.v8i9.11443

McKinley E, Fletcher S (2012) Improving marine environmental health through marine citizenship: A call for debate. Mar Policy 36(3):839–843

NOAA (2021) *The essential principles and fundamental concepts for learners of all ages*. https://static1.squarespace.com/static/5b4cecfde2ccd188cfed8026/t/6101cb7536e2ed6426ba15b6/1627507591681/OceanLiteracyGuide_V3_2020.pdf

Schio C, Reis P (2024) Design of a pedagogical model to foster Ocean citizenship in basic education. Sustainability 16:967. https://doi.org/10.3390/su16030967

Schio C, Reis P (2024b) Implementation of a pedagogical model of Ocean citizenship: Students' perspective. Aust J Environ Educ 40(5):863–885. https://doi.org/10.1017/aee.2024.57

Squarcina E, Pecorelli V (2017) Ocean citizenship: The time to adopt a useful concept for environmental teaching and citizenship education is now. J Res Didact Geogr 2:45–53

Tran LU, Payne DL, Whitley L (2010) Research on learning and teaching Ocean and aquatic sciences. NMEA Special Rep 3:22–26

Wulff A, Johannesson H (2019) Teaching marine sciences through practical experiences: Impact on students' learning and attitudes. J Environ Educ 50(2):98–111. https://doi.org/10.1080/00958964.2019.1564210

Author Contributions and Biographical Profiles

[1]*Caroline Schio: Co-conceptualization; Writing—Original Draft, Review, and Editing; Methodology—Data Collection and Analysis.

[2]Pedro Reis: Co-conceptualization; Writing—Review; Methodology—Review; Supervision.

*Corresponding Author at caroline.schio@edu.ulisboa.pt.

Caroline Schio is a PhD student in Science Education at the University of Lisbon, Portugal. She holds a bachelor's degree in oceanography from the University of Vale do Itajaí, Brazil, a postgraduate degree in Fisheries Economics and Management from the University of Barcelona, and a master's degree in Agroecosystems from the Federal University of Santa Catarina, Brazil. With over a decade of experience in Ocean Literacy and environmental education, she founded and has coordinated the Monitoramento Mirim Costeiro Programsince 2012, engaging elementary school students in coastal monitoring. She has also served as the president of the Monitoramento Mirim Costeiro Institute for five years. From 2021 to 2023, she coordinated the educational activities of the Maçarico Ocean Literacy Program at CoLab + ATLANTIC in Portugal. Her work focuses on socio-environmental education, citizen science, and hands-on learning experiences that promote Ocean Literacy and youth environmental activism.

Pedro Reis, PhD, is a professor at the Instituto de Educação, Universidade de Lisboa, Portugal. His research focuses on science education across diverse contexts, the nature of science, socioscientific issues, teacher professional development, and ICT integration in science education. He supervises Ph.D. and master's students in these fields and has led research, teacher training, and curriculum development projects across Europe, Africa, and Brazil. He earned his Ph.D. in Science Education (2004), M.Ed. (1997), and B.Sc. in Biology (1988) from the Universidade de Lisboa. Before becoming a professor, he taught science at the elementary and secondary levels. He currently coordinates the Research Group on Didactics at the Instituto de Educação, as well as the Ph.D. program in Science Education and the master's program in Innovation in Education.

13 The Ocean We Have: A Case Study at the Ocean Research College Academy

Josh Searle and Ardi Kveven

Abstract

To imagine the Ocean we want, it is helpful to better understand the Ocean we have. This is one of many lessons learned in two decades at the innovative, Ocean-centered educational program called the Ocean Research College Academy (ORCA) at Everett Community College in the United States. Another lesson learned is that empowering young people to ask and investigate their own informed questions enhances their Ocean Literacy. Inspired by the vision of the United Nations' Ocean Decade Challenge 10, this case study sought to determine the extent to which graduates who participated in a National Science Foundation-funded experience within the ORCA program significantly changed their current understanding of the Ocean. A qualitative analysis of responses revealed five themes: the power of questions; importance of community; experience in place; interdependence of parameters in the ecosystem; and the role of data in transforming understanding. Quantitative and qualitative results confirm participants' understanding of the Ocean changed significantly, including new appreciation for the Ocean as a dynamic place full of unanswered questions. These results illustrate powerful components that are adaptable to formal and informal education efforts to inspire change in the way people understand the Ocean.

Keywords

Ocean Literacy (OL) · Ocean Decade · ORCA · Community · GEOPATHS · Undergraduate education · Estuary

13.1 Introduction

During the first week of students' experience at the Ocean Research College Academy (ORCA) in Everett, Washington, they are asked for their definitions of *nature*. Instructors at ORCA have asked this of students for the past 19 years, and those same instructors work hard to compartmentalize their own definitions—it is more important to acknowledge the value of each student's definition as a snapshot of what they know at that moment. Instructors have an additional purpose, too: connecting with students' prior knowledge is a crucial first step to engaging in genuine reflection and to meeting students where they are as opposed to where instructors think they should be.

J. Searle (✉) · A. Kveven
Everett Community College, Everett, WA, USA
e-mail: jsearle@everettcc.edu; akveven@everettcc.edu

T. J. Kennedy (ed.), *Ocean Literacy: The Foundation for the Success of the Ocean Decade, Volume I*, Sustainable Development Goals Series, https://doi.org/10.1007/978-3-031-98145-6_13

Nine weeks after initial discussions about various definitions of nature, our students have read multiple essays written by nature writers, from Ralph Waldo Emerson through Rachel Carson, Robin Wall Kimmerer, and Carolyn Finney. They have conducted basic water chemistry sampling off a research boat in the nearby estuary. They have been introduced to a few essential oceanographic sampling tools. They have begun the process of analyzing their own data and comparing it to data collected by previous students. They have also written essays exploring their own relationship with nature. So, when they are asked again to define nature, 9 weeks after starting their two-year program, nearly all of them—even if some of them have not significantly changed their definitions — recognize the enormous complexity of *nature* as a construct influenced by time, space, culture, and personal beliefs. To us, this recognition is success because the goal is never to change their definitions. Instead, the goal is to begin the process of creating a community in which genuine conversation, direct experience, and exploration of various forms of evidence become core values. That these goals are grounded in a study of the marine environment outside our classroom windows enriches the experience, though the authors of this chapter posit that this approach to education is applicable in any formal and informal education setting, especially for programs located near the Ocean.

This first quarter exercise at ORCA serves as a microcosm of our students' two-year experience with the Ocean. Just as instructors do not set out to change student definitions of nature, the goal of the program is not to overtly change students' relationship with the Ocean. Our program has committed itself to helping students realize their own goals as they develop into active, responsible, and inquisitive learners. In the process of realizing the program's outcomes through a transdisciplinary experience, however, students do in fact change their views of the Ocean—from a static, surface, and shore-based understanding to an understanding of the Ocean as a dynamic, multi-dimensional and essential component of our entire ecosystem. The transformation happens on their terms and is a product of discussion, field work, background reading, repeated exposure, and genuine analysis of data. One question that emerges, however, is whether this significant shift in students' understanding of the Ocean remains after they graduate. The case study that is the focus of this chapter seeks to understand how graduated ORCA students feel about their ORCA experience from 1 to 5 years after graduation.

The Ocean Research College Academy

To better understand the context of students' survey responses, it is critical to know about the program itself. The Ocean Research College Academy, an innovative early college at Everett Community College in Everett, Washington (NW United States), was developed over two decades ago with the support of a Bill and Melinda Gates Foundation grant and founders, including the authors, who embraced the power of getting students in the field to conduct research while empowering young people to ask, investigate, and share their own questions and results. Multiple National Science Foundation (NSF) grants, including a grant to build our own research vessel to conduct environmental monitoring in our nearby salt wedge estuary, have magnified the impact of students' field research on their own learning. Today, sixty 15 and 16-year-old students are admitted each year, and over 85% of those incoming students complete the program in 2 years, earning an associate degree (a two-year college degree) at the same time they earn their high school diploma. They are full-time college students, taking classes primarily with ORCA's college faculty across the disciplines of science, math, English composition and literature, and social sciences. Because we have focused on engaging historically underrepresented students in science, technology, engineering, and math (STEM), we typically see 70% of our students identify as female while students of color, first-time college attenders, and low-income students make up roughly 25% of the student population.

Students engage in research across disciplines, and all students will have engaged in significant individual research in a discipline of their choice by the end of year 2. That said, roughly 2/3 of students now conduct research directly related to the local estuary. For the past 14 years, the program has been located at the Port of Everett Marina, which is situated at the confluence of the Snohomish River and Possession Sound, an arm of the larger Salish Sea that extends from Washington State to British Columbia, Canada to the north (Fig. 13.1).

Independent student research has been a primary goal at ORCA from the beginning. It has been central to guided instruction in year one, but over time, we realized too few of our students were engaging in original undergraduate research during their second year. Our efforts to increase the number of second year researchers accelerated when we were selected to join the NSF-funded Community College Undergraduate Research Initiative, coordinated by faculty from Finger Lakes Community College in New York. The grant allowed us to extend some students' individual research time over two quarters (20 weeks) rather than asking them to conduct an original research project (and write about it) in 10 weeks (see Searle 2018). This effort proved successful, with ORCA students presenting their research at undergraduate research symposia and one student publishing results focused on her analysis of local river otter diet.

Fig. 13.1 Faculty and student researchers on the R/V Phocoena in Possession Sound, Washington, USA (*Photo by H. Clayton*)

We hypothesized we could create still greater student research outcomes, and an NSF Pathways into Geoscience through extra-curricular experiences and training activities (GEOPATHS-EXTRA) grant provided resources to test our hypothesis. The stated goal of this grant funding was to increase knowledge of and interest in geosciences, especially among historically underserved young people. The grant also allowed a subset of our overall student population to begin an independent research experience in the summer before their second and final year in our program. In practical terms, students' independent research experience went from 10 weeks in our original model to 40 weeks with the support of National Science Foundation funding. In our own quarterly surveys, students in the NSF-funded program told us overwhelmingly that the experience was powerful (Kveven 2022b, October). We also observed students' deepening level of research, measured by increased sophistication of language in their final research papers, clarity and thoughtfulness of figures, and confidence in their evolving understanding of issues such as currents, tides, heavy metals, bathymetry, and water chemistry. Still, we wondered whether those learning gains transferred to experiences after graduation, and we wondered whether those experiences led to longer-term changes in students' views of the Ocean. These are the questions motivating our case study.

13.2 The GEOPATHS Alumni Survey

The authors approached the question of whether our graduates maintained a changed view of the Ocean through both quantitative and qualitative survey prompts. We developed a framework of both Likert scale and free response questions with support from an external evaluator who has worked with longitudinal alumni surveys of other programs. Additionally, we received feedback on the survey instrument from a collaborating qualitative data scientist and our own statistics instructor. (The survey tool is available on request.) The survey was sent to 53 ORCA GEOPATHS alumni, providing a month for response time and 3 reminders. We received responses from 36.

We followed procedures to maintain respondent confidentiality, and the authors read all the free response comments multiple times. Developing codes after modest training with MAXQDA and consulting primary literature, the authors looked for overlap with the UN Ocean Decade #10, educational learning theory, and the words of our former students. We asked them to reflect on their experience in our program as well as to determine how much their experience at ORCA continues to affect them. As we discuss below, our first major effort at collecting these data suggests the experiences of students in our program positively affects their lives and their understanding of the Ocean after graduation.

While the survey was developed to measure programmatic impact, a few of the questions also reflected the key attributes of Ocean Literacy (McKinley et al. 2023). Of the twelve Likert scale questions, which used a numeric scale of 1–5 and ranged from "not at all" to "great extent," one in particular asked, "To what extent did your experience in GEOPATHS in particular change the way you view the Ocean?" Descriptive statistics for this Likert scale question had a 95% confidence interval of 3.9–4.6 around an average of 4.2 out of 5.

In addition to quantitative data, the survey was designed to collect more nuanced responses in 9 free response questions, each of which was associated with the Likert scale questions. Following guidelines recommended for assessing the impact of research experiences (Reed et al. 2021), as well as arguments for the value of qualitative approaches to get at actual beliefs and perceptions (Bercht 2021), the authors sought to familiarize themselves with the survey responses, considering context and paying close attention to what respondents wrote. What emerged from this analysis are five core themes: the power of questions, importance of community, experience in place, interdependence of parameters in the ecosystem, and role of data in transforming understanding.

13.2.1 Theme #1: The Question Is the Answer

One of the pervasive messages sent to students by their previous schools and society at large is that success is measured by knowing all the answers. While knowledge acquisition is essential, we choose to emphasize the power of the question to unlock explorations of the known and unknown. The scientific endeavor serves as a marvelous reinforcement of the idea that good questions are essential components of any knowledge acquisition. This emphasis has been a staple of planning and teaching at ORCA, but we have long wondered whether this focus on the questioning nature of science transferred well after graduation, as it has for this graduate: "*The greatest thing ORCA did for me was to push and enable my curiosity—to this day I still get told I ask a lot of questions, something I can thank ORCA for.*"

Not every respondent singled out the centrality of questions, but for 11 of the 36 respondents, some version of the importance of questions emerged as a critical part of their ORCA experience. The quote above connects questioning with curiosity and suggests the ORCA approach "enable[d]" that trait. Another respondent made a similar connection: "*[T]he emphasis on asking questions and pushing curiosity has always pushed me towards collaborative spaces. The ability to ask questions towards individuals some may see as intimidating is perhaps the greatest life skill and has allowed me to accelerate my career in unbelievable ways.*" Both of these respondents demonstrate a degree of empowerment that continues to positively influence their lives.

A second sub-theme that emerges in responses focused on questions is the realization that so many questions about the Ocean environment remain. Participant comments ranged from: "*I discovered how many research questions are still unanswered and how beautiful the ecosystems are*" to "*Before ORCA, I assumed that scientists had already figured out most aspects of the Ocean – how all living things functioned and interacted. However, upon arriving at ORCA, I discovered so many research opportunities and unanswered questions within the marine ecosystem.*"

In similar responses, participants pointed to the excitement of learning how much was unknown about the Ocean, and while the search for meaning and the quest for knowledge continues to motivate, the complexities and size of the marine environment helped students find inspiration in the unknown. Really, these two subthemes are closely connected because they are both related to an intrinsic interest in learning more about the environment around us.

Recognizing the power of questions is no surprise, as it is foundational to the ORCA approach (e.g., Kveven 2009). From the first day of students' experience in the program, they are asked to explore the world around them and pose questions—especially if the answer isn't readily available. The point is to establish a cultural value for questions over knowing all the answers at the outset. When a student wonders why a beetle has an iridescent green hue, the rest of us say: "*Interesting question! Where might you look for information on that? What else do you wonder?*"

This approach to honoring the question is critical in all disciplines, and it sets the stage for investigations during the rest of their ORCA experience.

The benefit of students who are empowered to ask their own questions manifests in the way they approach complicated challenges. Those who are inquisitive make the time to explore their questions fully. They are willing to make multiple graphs while seeking to best understand their data. They look further for resources that not only help them explain their own ideas but explore different ideas, too. That is the mark of an inquisitive person, and the comments from our survey suggest this trait is important in their lives after ORCA.

13.2.2 Theme #2: Lasting Impact of Community

Asking students to take risks in and out of the classroom, which they do when they ask questions and try out new ideas, doesn't happen in a vacuum. An environment that is more collaborative than competitive and a place that values the search for understanding over a desire to separate and rank will make any educational environment safer for learning and growth (see Ooi and Cortina 2023). This, at least, has been a guiding philosophy at ORCA from the beginning (Kveven and Searle 2008). It also mirrors the collaborative ideal of the scientific endeavor.

The role of community in graduates' experiences at ORCA appears over and over again. Among the 36 separate survey responses, 50 strands emerged about some form of community. Of those examples, 21 focused on the degree to which graduates recalled the benefits of working together in groups at the start of their experience, including learning from each other. This illustrates the degree to which their experiences in creating a community of learners and researchers continues to influence their desire to engage in research communities in their new environments. One response represents this category well: "*The research of other students in my cohort, as we worked together through research problems, also expanded my perspective on Possession Sound's human interaction...*"

This response, like others, points specifically to peers' research as a key vehicle for expanding their own knowledge. Their memory of the collaborative experience was likely aided by the way they were introduced to community building from the first 10 weeks of their academic work: group science work projects, construction of group norms in their social science and humanities courses, and deliberate reflection about their group work and all-class discussions. The point is that their grant-funded community-building in GEOPATHS was an extension of the culture they had already been introduced to throughout their ORCA experience.

Students' sense of community with each other was broadened by our focus on mentorship and networking with scientists in academia and environmental consulting. As students met and then worked with different researchers and environmental consultants, they enlarged their circle of contacts. Respondents remembered: "*presenting to professionals in the field,*" and they realized "*that most people are super open to giving help[;] just being polite and asking the right questions is important.*"

Respondents commented over and over about the value of networking to create ever larger communities. One respondent wrote: "*The concept of networking really remains with me from my time at ORCA ... By talking to those who presented to us at GEOPATHS, I have continued to make connections into my collegiate career. It always amazes me how all the people I meet at different times know each other or someone I want to contact.*"

These views about the importance of connecting to others were true even for those participating mostly online during the global pandemic.

What is revealed in these responses is the degree to which these students embraced the strength of a community of researchers and then expanded that community beyond their ORCA experience. For those incorporating these networks in their academic career and beyond, these comments point to advantages in experiencing early networking and community-building.

We hypothesize, as well, that community is essential to encourage exploration and to model science as an inherently collaborative experience. Research builds on the work of previous researchers, and the more students experience a welcoming environment, the more they will want to be a part of that research. In addition, the more collaborative the environment, the more likely it is that students will take individual risks and seek out feedback and critique. Student comments didn't exactly state the above benefit to building community and larger networks, but it is a central tenet of the ORCA approach.

13.2.3 Theme #3: Experience in Place

For decades, recommendations for environmental education reform have suggested engaging students in place-based, experiential, and real-world problems (Glithero et al. 2024). Further, the UN goals outlined in the 2024 white paper suggest, "The Ocean ... must become inextricably linked with our sense of self, our sense of place, and our collective well-being" (Glithero et al. 2024). These best practices have been at the core of ORCA since inception and are reflected in survey responses. The survey tool asked students to recall their research experiences and key takeaways, and 28 out of 36 (75%) survey respondents pointed to their localized experience as memorable due in part to the place-based field experiences: "I remember growing to love the process of research through both the hands-on experience gained during cruises, along with the skills that I gained analyzing data that [led] to a better understanding of Possession Sound." Similarly, another student wrote: "*I loved being able to go out and collect research and do a lot of hands-on experience[s].*"

These experiences help instill a sense of self and self-confidence, articulated frequently by responses such as (Fig. 13.2): "*Having the experiences of ORCA has definitely helped me feel more confident in my abilities as a scientist.*"

Centering research in place provides agency for students, as a large percentage will stay local, and these experiences may catalyze a desire to effect change where they live. When they begin our program, most of our students know little about the local estuary. This is why we embed

Fig. 13.2 Student researchers observing a surfacing gray whale in the distance (*Photo by A. Kveven*)

our location into multiple aspects of our program, including a field experience on a human-made island near campus on the first day of each academic year. Within a month, first year students are conducting boat-based water sampling profiles of a tidally-driven estuary at the mouth of a river while also learning to identify seabirds and local pinnipeds. Experiences in the local estuary likely led to survey results such as this student response: "*Before starting at ORCA, I was way more interested in what was in (emphasis added) the Ocean, rather than just the Ocean itself. I came to ORCA because I wanted to study sharks and I came out with a love for learning about phytoplankton, nutrients, upwelling, water stratification, and so much more*."

We harness student interest in the "seen," the biology that they can see (like the student quote above), and connect it to the "unseen," where the physical and chemical processes are often the drivers in an estuarine system. Students uncover layers of understanding about water density and tides through the collection of data, at the same time reading the works of nature writers, looking at the historical context, and developing skills by applying introductory statistics to data they collected. Essential to these activities is that they occur in real time and in the place where we are located (Fig. 13.3):

Fig. 13.3 A group of ORCA students collecting and entering data for a number of oceanographic parameters (*Photo by A. Kveven*)

> "*[B]eing able to experience hands on research collection made the process of analyzing and reporting data feel more meaningful for me and made it so I better understood the ways in which the data worked, since I was there for the collection.*"

This sense of "meaningful" connection increases the potential for a "well balanced human-Ocean relationship" (Glithero et al. 2024) and creates an opportunity to inspire students to apply their localized learning globally. While our focal experience starts in the local estuary, instruction emphasizes connections, and applications are made to the global Ocean system. The connection of local systems and global systems is evident in this student response: "*ORCA gave me an in depth look at more aspects of the Ocean, especially the Pacific Northwest ecosystem. I had a better understanding of weather patterns like the ENSO and how Ocean acidification happens and why it has such an impact on different ecosystems.*"

Mirroring the complexity of the ecosystem while building student knowledge and connection to a place is quite an undertaking. We endeavor to make local connections with students while they develop transferable learning, which has been a key tenet of environmental education for decades. The Senegalese forestry engineer, Baba Dioum said in 1968, "In the end we will conserve only what we love, we will love only what we understand, and we will understand only what we are taught" (as cited in Lotze 2020). We see evidence of this transference through the power of the experience in place when our students responded to the prompt, "Share two or more ways your understanding of Possession Sound changed (if at all)" yielded impressive observations. The associated Likert scale question had the highest mean (4.7 out of 5 with a 95% confidence interval of 4.5–4.9). One student wrote: "*I learned about how ever-changing even such a relatively small part of the Salish Sea can be, and how we can measure these changes. I also learned how human involvement along the shoreline and within Possession Sound can change the way [its] ecosystem works, particularly in regards to eelgrass beds.*"

The expanding knowledge base for students is cultivated through the ORCA research experiences grounded in place with ripple effects when they start to apply their learning in new contexts.

Respondents frequently commented on increased understanding and connection to our place (Possession Sound) with a nod toward how they fit in the system: "*I had no idea how complex it was. ORCA also changed my understanding of how I fit into Possession Sound and where I belonged in marine science.*"

Ocean Literacy supports the notion that learning and connecting to a place is as relevant as the fact that all water is connected (Glithero et al. 2024).

13.2.4 Theme #4: Interdependence of Ecosystem Parameters

Just as the learning is scaffolded over 2 years, we coordinate instruction and field experiences with a goal of facilitating student understanding of the complex interdependence of the local ecosystem. This layering effect is evident in the responses from students who are well on their way "to support the health of the Ocean by recognizing it as a living ecosystem" (Glithero et al. 2024). Though the survey did not specifically ask about their ecosystem understanding and the way we lay the foundation for it, many students pointed it out unprompted. For example, one graduate wrote: "*I also enjoyed how the experiences ramped up in coordination with what we were learning in class - first, a relatively simple test project from the river trip, then a more complex question based on physics/chemistry, and finally linking biology into the story.*" And, "*Overall, ORCA exposed me to the immense complexity of the Ocean and how interdisciplinary the field of oceanography is.*"

The underpinnings of the complexity are revealed like the layers in the estuary as students experience repeated exposure to broadening their understanding, as highlighted here: "*I gained a complex understanding of halocline dynamics at the mouth of the Snohomish, a river that I continued to learn about and appreciate during all my time at ORCA, not just during GEOPATHS.*"

The connection the following student is making to the impact of water chemistry on the biology represents the importance of repeated experiences connected to a widening view of the Ocean and the interconnectedness of everything in it: "*GEOPATHS gave me a new way to look at the Ocean through profiles of the data we collected. I began to look at the water and think about how the temperature changes with depth in different areas and how that may affect what life lives in that water.*"

Students' deepening understanding as they build connections from the local system to the broader global system is represented in frequent quotes similar to this response: "*In learning about the Ocean during my time at ORCA, I further have a deeper appreciation of the ways in which it supports the global climate and local ecosystems.*" Another graduate wrote, "*My view of the Ocean became deeper with the understanding of how changes to its ecosystem invariably lead to a chain reaction throughout.*"

The more students experience the function of the local system, the more they look for parallels in other systems, like this response: "*Learned that even smaller scale bodies of water have complex dynamics and are affected by local trends on land.*"

While these quotes strongly support students' understanding of ecosystem interdependence beyond graduation, there is also powerful evidence of their learning from the Possession Sound Student Showcase and Talks. This annual event, which brings together the local scientific community, our collaborators, and funders, is a primary platform for students to present their research. Having to present to a wider audience is a crucible moment where students hone the depth of their understanding and assess their ability to explain the connections that exist in their research, which showed up in student takeaways on the survey: "*I also remember spending a lot of time working with many people to figure out how to analyze and present data in a way that didn't oversimplify it but also was easy to understand.*"

Student research titles reinforce the themes we've established, with an emphasis on their understanding of ecosystem interdependence with these showcase titles: "Dissolved oxygen, water temperature, and chlorophyll profiles (horizontal, vertical, and temporal) in Possession Sound," "Seasonal delay of chlorophyll, sunlight and dissolved oxygen over the water column," "The impacts of combined sewer outlet overflows on turbidity and dissolved oxygen," and "Seasonal relationships between light penetration and water chemistry variables from 2016–2023."

The showcase provides the embodiment of science communication through the strategic ambition of Challenge 10 where the student research is on display in a conference-type format with posters, lightning talks, and 250 local scientists and community members in attendance. As the organization with in-the-water sensors monitoring water quality hourly coupled with biweekly research cruises, ORCA students are the knowledge holders of the data of this place, and they proudly share it with the community (Fig. 13.4).

When students have the opportunity to prepare and explain their work to an audience beyond ORCA, deeper learning occurs and broader connections to global systems are often the result (Kveven 2022a, April).

13.2.5 Theme #5: Role of Data in Transforming Understanding

In 64 coded segments from 28 of the 36 respondents, the word *data* was featured. In part this is due to the survey asking students to reflect on their use of oceanographic tools as well as tools for analyzing increasingly large data sets. But, many responses pointing to data emerged in their responses to questions about takeaways in general. In the early days of ORCA and GEOPATHS, ORCA faculty focused on Excel and ArcGIS as the primary vehicles for data analysis. The two tools were free to our community college students, and faculty were experienced in problem-solving data analysis with students using these tools. Over time, however, GEOPATHS students in particular had access to the kinds of data that easily overwhelmed Excel. We introduced students to moored CTDs collecting data all day long, every day of the week, and we deployed, with the help of one of our partners, an acoustic doppler current profiler (ADCP). This tool enlarged our exploration of the physical features of the estuary, and it created a significant amount of data. We also introduced students to hydrophone data, which required the addition of new data analysis tools (RavenPro and

Fig. 13.4 An ORCA student presents second-year research on vertical stratification to an audience of researchers and community members (*Photo by A. Kveven*)

Pamguard). So, by the final years of the first GEOPATHS grant, we introduced students to R as a tool much better equipped to analyze large data. For students in the current GEOPATHS program, R is an expectation rather than a special tool, and students are responding to the R challenge in amazing ways.

The students in our current alumni survey had less exposure to R, yet 10 of the 36 respondents specifically mentioned it. Those students who did work with R illustrate how it deepened their understanding of data—and of the estuary being sampled. One student wrote: "*I remember learning how to use R to create graphs, being creative with figures, and finding explanations for puzzling and unexpected trends in data.*" Another wrote, "*I really enjoyed learning, working and problem solving in R to create figures and analyze data.*"

Though R is a powerful tool, the process of learning to code for specific features in any one figure and the degree to which students have to ask the program to run specific statistical tests changes students' relationship with data. Though we are not measuring the degree to which these positive comments about R illustrate a particularly positive effect on typically underserved students, the authors continue to observe, anecdotally, that young women and people of color begin to see themselves as capable of coding and statistical analysis when they previously did not see themselves as capable. Given that 26% of respondents identify as non-white while 69% of respondents self-identify as women and another 12% identify as non-binary, the GEOPATHS experience played an important role in advancing more historically underrepresented people into geoscience fields.

These comments about R fit with the rest of the theme in that students see joy in exploring data from the Ocean environment. When asked to reflect on their time using oceanographic tools to collect data, one representative student responded:

> "*Firstly, that learning to properly use such equipment to collect valid, useful data is attainable and, in my opinion, rather exciting … secondly, that you have to take care that your data is accurate, useful, and organized, because not all data is.*"

Many other respondents touched on this experience of collecting, managing, and then analyzing data. The process contributes to greater understanding of the Ocean environment as well as the tools available to monitor parameters that are otherwise hard to see. One student wrote:

> "*After my GEOPATHS research, I felt like my understanding of how much information you could get from big datasets was improved since I got to practice first-hand how to use these datasets.*" Another wrote, "*[The GEOPATHS experience] involved a continuous cycle of data collection, analysis, and discussions with [one of my mentors] to derive meaningful conclusions.*"

The theme that emerges from many respondents is the centrality of data in their changing understanding of the Ocean.

13.3 Conclusion: The Estuary Is Us

We began our analysis fueled by the vision of Ocean Decade's Challenge 10: Driver #3. It resonates with the work the authors have engaged in with nearly 1000 students over the past 20 years. Challenge 10 acknowledges the call to make "ecological learning … place-based, relational, and experiential." It recognizes that "[e]xisting relationships with the Ocean are varied, shaped by our unique experiences…." It proposes Ocean Literacy as "an outcome—a society that understands, values, and cares for the Ocean." Perhaps equally important, "Challenge 10 emphasizes the need for all other challenges to place people at the centre of their approaches." Each of these components of Challenge 10 is inherently a part of the approach developed at ORCA, and this approach continues to evolve when new tools become accessible, when new students and questions arrive, and when faculty change and grow.

What students tell us is that the experience at ORCA changes them in positive ways. We are encouraged by analysis conducted by our external evaluator at the Social and Economic Sciences Research Center–Puget Sound Division, who noted about our GEOPATHS-EXTRA program: "*Overall, it appears that this funded program has had a profound impact on the students' knowledge and skills; influenced their educational and career plans; helped them begin to create a rich network of scientists/researchers; provided exposure to the scientific community through site visits, research cruises, conferences, and presentations; and contributed to them beginning to take steps towards attaining their educational dreams through receiving scholarships, internships, and awards.*"

We therefore wondered whether students perceived their ORCA experience to fundamentally change the way they view our local estuary and the larger Ocean environment. For nearly all of the respondents to our most recent survey, the answer is yes.

The free response version of the survey gave respondents an opportunity to explain how, what, and why change happened. The preceding analysis explored five of the themes that emerged, but we were also intrigued by their more direct responses to the question about changing their understanding of the Ocean. One respondent wrote: "*[I]nstead of seeing it as this flat water you can look at for miles – I now see it as the zooplankton, the phytoplankton, the fish and invertebrates that live throughout it and linger beneath the surface.*" Another wrote, "*I also learned more about the characteristics of the water chemistry that I can apply to thinking about boats or how nutrients cycle through the ecosystem. I think about these things every time I visit the Everett marina....*"

Both of these quotes represent a category of responses that focuses on the ways their own research and the research of their peers enlarged their sense of the dynamic processes at work in the estuary and beyond. These students have changed the way they view the Ocean.

For a number of respondents, their changes were more closely related to action as they became increasingly aware of the complex relationships among the natural world and human activities. We know some students begin our program with the idea that humans affect the environment, but for many of the respondents in our survey, the relationship between humans and the Ocean became more real (and more complex) through their own research. One respondent wrote: "*I became increasingly aware of environmental issues affecting marine life, including lesser-discussed factors such as water currents, river discharge in the salt wedge estuary, and noise pollution. These aspects can significantly impact the health and stability of the Puget Sound ecosystem, revealing complexities that were new to me before arriving at ORCA.*" Another wrote, "*I also learned how human involvement along the shoreline and within Possession Sound can change the way its ecosystem works, particularly in regards to eelgrass beds.*"

What we notice in these responses is both a growing awareness of human impacts and an awareness of how interrelated physical, chemical, and biological factors are when considering environmental impacts. This is the kind of complexity of thinking that is a goal for students in the program, and it is the kind of thinking necessary to develop Ocean Literacy.

As a formal academic program, we continue to evolve in response to opportunities, student feedback, and our environment. We have also made significant changes recently to capitalize on the power of data analysis to inspire growth in understanding. Students now begin the program with an introduction to statistical tools that are more closely connected to their data analysis, and this is reinforced in their second year when they are introduced to R to help them explore larger datasets. Like everyone else in formal and informal education, we face limitations in terms of cost. Oceanographic tools, lab tests, research vessels—all are helpful in catalyzing student transformation, but they carry a cost. Using creativity and building networks of like-minded researchers in the community are helpful in overcoming those limitations. We are fortunate to have a supportive college, and we are fortunate that a number of partners continue to support the program, too.

Our program's infographic highlights the many components that make up the local estuary as well as the *estuary* that is our educational program. The food web in the center illustrates the biological aspect of the environment, the part that typically draws our students to the study of the Ocean. Behind it, however, exist the tidal signature connecting all that we do and a color gradient representing the pycnoclines that define so much of the dynamic environment. Our program name, and the descriptions that correspond to each word in that name, are purposely integrated in the rest of the environment. We humans are also part of the estuary (Fig. 13.5).

Fig. 13.5 The ORCA infographic

For formal and informal academic settings around the world, this case study suggests a number of keys to successfully transforming students' views of the Ocean: putting student questions at the center of any practice, anchoring those questions in the dynamic environment in which the program is located, investigating those questions in a community of researchers, considering those questions through interdisciplinary contexts, and engaging those questions through the collection and analysis of real-time, large datasets. We believe these keys are applicable in any program, regardless of the age of the student or the location of the program. Making the full spectrum of Ocean parameters visible to students is the opportunity we see in educational approaches like the one at ORCA, but our experience tells us Ocean Literacy starts by creating a culture in which others are invited to ask and investigate complex questions about the Ocean we have.

13.3.1 Summary

Key concepts explored in this chapter include:

- Experiential, place-based learning fosters deeper Ocean Literacy. Engaging students in field research, ecosystem monitoring, and data-driven inquiry strengthens their understanding of Ocean processes and their role in sustainability.
- Qualitative analysis revealed five key themes in student learning. The power of questions, the impact of community, the importance of researching in place, recognition of the interdependence of ecosystem parameters, and data's role in transforming understanding all contribute to long-term student engagement with Ocean Literacy.
- Ocean Literacy extends beyond education to societal transformation. Programs like ORCA not only prepare students for STEM careers but also equip them with the skills and awareness to make informed decisions that impact marine conservation and policy.
- Equity and inclusion are essential for strengthening Ocean Literacy. Providing hands-on Ocean experiences to historically underrepresented groups in STEM broadens participation and fosters diverse perspectives in marine science.
- Localized research connects to global Ocean systems. Studying regional ecosystems, such as Possession Sound, enhances student understanding of broader Ocean-climate interactions, reinforcing the interconnectedness of global marine processes.

13.3.2 Recommended Resources

The following resources provide further insights and support on this topic:

- Cejda, B. D. & Hensel, N. Eds. (2009). *Undergraduate Research at Community Colleges.* Council on Undergraduate Research. https://www.cur.org/wp-content/uploads/2023/08/Undergraduate_Research_at_Community_Colleges_-_Full_Text_-_Final.pdf
- The Council on Undergraduate Research: https://www.cur.org/
- *Current: The Journal of Marine Education Current*. https://current-journal.com/
- Handelsman, J., Miller, S., & Pfund, C. (2007). *Scientific teaching*. Macmillan.

- Hensel, N. Ed. (2018). *Course-Based Undergraduate Research: Educational Equity and High-Impact Practice*, Stylus Publishing.
- Karp-Boss, L., E. Boss, H. Weller, J. Loftin, and J. Albright. (2009). Teaching Physical Concepts in Oceanography: An Inquiry Based Approach. *Oceanography 22*(3), supplement, pp. 48. https://doi.org/10.5670/Oceanog.2009.supplement.01
- Mapes, L. V. (2021, April 26). Ghost shrimp, humpbacks, tiny plankton: See Puget Sound surge with springtime life. *The Seattle Times* https://www.seattletimes.com/seattle-news/environment/ghost-shrimp-humpbacks-tiny-plankton-see-puget-sound-surge-with-springtime-life/
- Parker, S. & Holladay Van Damme, L. (2022, December 8). Charismatic megaeducator Ardi Kveven Revolutionizes Marine Education, *National Marine Educators Association.* https://www.marine-ed.org/news/charismatic-megaeducator-ardi-kveven-revolutionizes-marine-education
- Yager, R. E. Ed. (2009). *Inquiry: The Key to Exemplary Science*, NSTA Press

Acknowledgements We would like to thank and acknowledge the entire ORCA team for their work in the program and particularly to Marla Hyder for her editing prowess and Madelyn Voelker for statistics assistance with the survey instrument. The National Science Foundation has been instrumental in executing our student research initiatives. Our external evaluators for the NSF grants at Washington State University Social & Economic Sciences Research Center assisted with the survey tool, as did foundry10's Travis W. Windleharth. Thanks to the current and former GEOPATHS students who provided such insightful feedback (even though the survey was long). Everett Community College's unwavering support of the ORCA program over two decades has been instrumental in this work.

References

Bercht AL (2021) How qualitative approaches matter in climate and Ocean change research: Uncovering contradictions about climate concern. Glob Environ Chang 70:102326. https://doi.org/10.1016/j.gloenvcha.2021.102326

Glithero LD, Bridge N, Hart N, Mann-Lang J, McPhie R, Paul K, Peebler A, Wiener C, Yen C, Kelly R, McRuer J, Hodgins D, Curtin F (2024) Ocean Decade vision 2030 white papers – Challenge 10: Restoring society's relationship with the Ocean. UNESCO-IOC, Paris. (The Ocean Decade Series, 51.10.). https://unesdoc.unesco.org/ark:/48223/pf0000390126

Kveven A (2009) Inquiry at the Ocean research college academy. In: Yager RE (ed) Inquiry: The key to exemplary science. National Science Teachers Association Press, pp 1–14

Kveven A (2022a, April) Enhanced student engagement in research through presenting at the Salish Sea Ecosystem Conference. https://cedar.wwu.edu/ssec/2022ssec/allsessions/516/

Kveven A (2022b, October) Project outcomes report. *Award Abstract #1801658, GP:EXTRA—Engaging Community College Students in Undergraduate Research*, U.S. National Science Foundation. https://www.nsf.gov/awardsearch/showAward?AWD_ID=1801658&HistoricalAwards=false

Kveven A, Searle J (2008) Ocean Research College Academy. Current J Mar Educ 24(3):24–28. https://current-journal.com/issues/8/files/8-37-PB.pdf

Lotze HK (2020) Combining love and knowledge to heal the Ocean. Ethics Sci Environ Polit 20:33–39. https://doi.org/10.3354/esep00193

McKinley E, Burdon D, Shellock RJ (2023) The evolution of Ocean Literacy: A new framework for the United Nations Ocean Decade and beyond. Mar Pollut Bull 186:114467. https://doi.org/10.1016/j.marpolbul.2022.114467

Ooi SX, Cortina KS (2023) Cooperative and competitive school climate: Their impact on sense of belonging across cultures. Front Educ 8:1113996. https://doi.org/10.3389/feduc.2023.1113996

Reed MS, Ferre M, Martin-Ortega J, Blanche R, Lawford-Rolfe R, Dallimer M, Holden J (2021) Evaluating impact from research: A methodological framework. Res Policy 50(4):104147. https://doi.org/10.1016/j.respol.2020.104147

Searle J (2018) Experience is education. In: Hensel NH (ed) Course-based undergraduate research: Educational equity and high-impact practice. Stylus, pp 99–110

Author Contributions and Biographical Profiles

[1]*Josh Searle: Co-conceptualization; Writing—Introduction, Themes 1, 2, 5, Conclusion, Review and Editing; Project Administration; Methodology—Lead qualitative data analyst.

[2]Ardi Kveven: Co-conceptualization; Writing—Themes 3 and 4, Review and Editing; Methodology—Lead survey creation and implementation; Supervision.

*Corresponding Author at jsearle@everettcc.edu

Josh Searle is a founding faculty member at ORCA. He engages students across the disciplines in humanities, social sciences, and in the natural sciences as a research mentor. Josh earned English and Political Science degrees at the University of Washington and his Master's in Teaching at Seattle University. He also earned National Board Certification in English/Language Arts. For twenty-nine years, Josh has worked with students of all ages, engaging them in the active process of critical reading, thinking, and writing. At ORCA, Josh has developed innovative pathways for students to conduct their own research over multiple quarters.

Ardi Kveven is the founder and executive director of the Ocean Research College Academy (ORCA), an innovative early-college program at Everett Community College. Ardi earned her bachelor's degree in biology with a marine emphasis, along with her teaching certificate from the University of Washington. She holds a master's degree in science education from Western Washington University and a USA Coast Guard 100-ton Master's License. Innovative work engaging students in experimental research in the Snohomish River estuary led to four National Science Foundation grants totaling nearly 1.5 million dollars, including one for a custom designed 34-foot research vessel.

14 Whale Tale: Big Challenges Facing the Ocean's Biggest Species

Tammy D. Lee, Bonnie Glass, Gail Lemiec, and Fernando Martín Velazco

Abstract

A team of educators and researchers designed a course centered on Ocean Literacy and climate science, aiming to foster a deeper understanding of marine systems and environmental challenges. The course, including its 7-day field experience in Baja California Sur, Mexico, focused on Ocean Decade Challenge #10, restore society's relationship with the Ocean. The purpose of the course was to examine impacts of climate change from a local to a global perspective by studying whales and their migratory patterns. Between the months of November and April, right whales visit the North Carolina coast on their migratory journey to Georgia and Florida where they mate, calve, and spend the winter. This same phenomenon occurs with gray whales as they migrate the Pacific Ocean to the protected lagoons of Baja California Sur, Mexico. Using whales as a focus, students met for weekly seminars exploring contentious environmental issues facing the marine mammals in both areas. Students heard from various stakeholders who provided multiple perspectives on the complex issue of human impacts to both marine ecosystems. This chapter discusses the rationale behind the course creation, its framework, student learning objectives, and course assignments. It examines the course's impact on students' understanding of whale migratory patterns, their engagement in advocacy projects for species protection, and their knowledge of Ocean Literacy. Additionally, it highlights students' sense of connectedness to the Ocean, their agency in marine conservation, and includes reflections and future plans for course revisions.

Keywords

Ocean Literacy (OL) · Climate change · Whales · Systems thinking · Immersive experience · Study abroad · Ocean challenge · Conversation · Migratory patterns · Human interactions

T. D. Lee (✉) · B. Glass
East Carolina University, Greenville, NC, USA
e-mail: leeta@ecu.edu; glassb@ecu.edu

G. Lemiec
North Carolina Aquarium at Fort Fisher, Kure Beach, NC, USA
e-mail: Gail.Lemiec@ncaquariums.com

F. M. Velazco
Stultifera Navis Institutom, Baja California Sur, Mexico
e-mail: c@stultiferanavis.institute

14.1 Introduction

In the context of global Ocean challenges, providing opportunities for humans to interact and connect with Ocean species like the gray whale has the potential to promote Ocean Literacy and

T. J. Kennedy (ed.), *Ocean Literacy: The Foundation for the Success of the Ocean Decade, Volume I*, Sustainable Development Goals Series, https://doi.org/10.1007/978-3-031-98145-6_14

increase Ocean advocacy efforts. The course described in this chapter was developed to address these challenges and to promote Ocean Literacy and the agency of our students.

14.1.1 Why Whales? History of Human Interactions with Right and Gray Whales

Similar migratory behaviors and historical events along the east and west coasts of North America have resulted in gray (*Eschrichtius robustus*) and right whales (*Eubalaena glacialis*) having a parallel relationship with humans throughout history. The proximity of their migratory paths to the coastline and their specific behaviors have resulted in these being among the first whale species to be observed, described, hunted, and represented (Laist 2017; Jones et al. 1984). Unlike other species of migratory whales, whose migration follows food availability and can change dramatically from 1 year to the next, gray and right whales have a consistent pattern of migration throughout their life cycles (Corkeron and Connor 1999). Both species travel north to feed in cooler arctic waters each summer then return to mate and calve in subtropical regions each winter. The stability of their migratory patterns correlates with their survival and reproductive success (Hildebrand et al. 2021). These consistent migration paths have helped humans observe and predict the behaviors of right and gray whales since ancient times and to develop connections with these species.

Compared to species of larger whales, which are more human-reluctant and generally found in deeper waters, right and gray whale behaviors increase the likelihood of their human interaction: (1) their tendency to swim close to the shore, making them easy to spot; (2) their proclivity to visit closed bodies of water such as bays, inlets, coves, and lagoons; and (3) their specific behaviors including, but not limited to hunting; for example, the very slow escape efforts of right whales, and, in the case of gray whales, the complete immobilization of the speared mothers to save their calves (Scammon 2007; Velazco 2022a).

According to historical record, encounters between sailors and these species in antiquity were common. Tales of these encounters may have given shape to European folklore and literature related to sea creatures, sometimes portraying them as infernal beasts and other times depicting them as playful and worthy of kindness (Coleman 1983; Papadopoulos and Ruscillo 2002; Parsons 2004). These conflicting attributes are also present in ethnographies of North American indigenous people (Kuhnlein and Humphries 2017). These ethnographies confirm the predisposition of these whale species to close encounters with humans and note observable variety in whale responses to human interactions (Van Ginkel 2004; Brewster 2004). Stories of whales and their interactions with humans have been further confirmed by detailed descriptions collected by sailors during the modern industrial whaling era (Velazco 2022b). Multispecies ethnography recognizes the interconnectedness and inseparability between humans and other species like plants and animals. This perspective would frame differently, for example, the fact that the once-considered hostile gray whale or 'devil fish' as it was called in the whaling era, became known as seasonally docile and 'friendly' in the calving lagoons in Baja California Sur, Mexico (Swartz 2016).

Although they share a common history of interacting with humans and of being hunted to near extinction, the current conservation status for both species is in contrast. While gray whales of the East Pacific population, once considered extinct, recovered spectacularly during the twentieth Century and are now officially declared in 'least concern;' right whales still struggle as 'threatened with extinction,' with fewer than 500 living specimens, according to NOAA's recent estimations (2023, 2024a, 2024b). Despite differences in how the two species have rebounded, both face increasing threats from entanglements, vessel strikes, Ocean noise, and the effects of climate change.

Even today, with both species under special protection by international law, the diversity of preservation efforts among different coastal communities and nations is remarkable. While in the USA and Canada, strict regulations govern whale-watching activities, imposing sanctions on any attempt to approach cetaceans too closely (Marine Mammal Commission 2019; Canada 1993), whale-watching tour operators in Baja, Mexico, have developed non-invasive strategies to take advantage of the gray whales' 'friendly' attitude with regards to approaching humans. The regulated and limited encounters provided in the protected lagoons of El Vizcaino in Baja California Sur, Mexico, allow local licensed guides to provide tourists uniquely personal experiences initiated by the gray whales. Through this model of sustainable tourism, locals are invested in the preservation of the species and protecting the winter habitats of the gray whales (OECD 2016; Hoare 2023).

14.2 Framework of the Course and Learning Objectives

This section outlines our Honors College seminar course focused on a systems science issue, the migratory patterns of right and gray whales, using climate science and Ocean Literacy principles, as students compare the challenges facing our North Carolina marine ecosystems and the marine ecosystems of Baja California Sur, Mexico.

East Carolina University has an Honors College which provides scholarships and unique experiences for high-achieving students. Honors students participate in a specialized curriculum that aims to enhance student engagement and their college experience. Honors students are required to take at least one seminar. Honors College seminars offer unique course experiences that allow students to pursue a variety of interesting topics outside their major. Seminars are writing-intensive and satisfy general education and global or domestic diversity requirements. The Honors College annually invites faculty from across the university to propose seminar courses.

14.2.1 Framework for Course Development

Our Honors College Seminar course used the frameworks of system thinking, climate science and Ocean Literacy. Systems thinking provided the teaching perspective for how to approach the study of climate science, and the Ocean Literacy principles provided the conceptual ideas to address.

Systems Thinking

Complex systems, in this case our Ocean, impact our everyday lives, particularly the ecosystems we inhabit and share with other living organisms (Lee et al. 2019). Our Ocean encompasses many different natural systems with variations, causes, and effects that necessitate a systems thinking approach. A systems thinking approach requires skills such as examining, evaluating, and inventing solutions to complex problems (Frank 2000). A systems thinking approach provided students with a more holistic view of the issues facing right and gray whales as they migrate through our Ocean, including how humans have and continue to impact these yearly journeys.

Climate Change Science Framework

Our course addressed the national call to educate students on the causes, effects, and methods of mitigating human impacts on our climate. Teaching climate change with a systems thinking approach emphasizes the interactions and relationships between the Earth's spheres (geosphere, biosphere, atmosphere, and hydrosphere), focusing on the impacts on the ecological and biological systems. It is essential that citizens understand global climate change to make informed decisions about policies that impact societies and natural ecologies, both locally and globally.

Ocean Literacy Framework

The Ocean Literacy principles emphasize the interconnectedness of systems within the Ocean and their impacts on terrestrial systems, and thus represent content that supports a systems science approach (Ocean Literacy 2005). Research supports incorporating Ocean science concepts though focusing on megafauna such as whales to help students form personal connections to the Ocean (NOAA 2013). Promoting Ocean Literacy is a key component of resolving complex issues, as when people are more knowledgeable, they are more inclined to support healthy Ocean policies (NOAA 2013). Ocean Literate individuals understand the fundamental functions of the Ocean, communicate their understanding in meaningful ways, and can make informed decisions about the Ocean and its resources (Cavas et al. 2023).

14.3 Whales as Our Focus

In planning this course and field experience, we reached out to a local aquarium educator with whom we have collaborated. She connected us with an ecotourism group, Born to Baja, from Baja California Sur, Mexico. Through our discussions, we learned that if we visited Baja during February and March, we could travel with local permitted guides into the lagoons near Guerrero Negro to observe gray whales calving. They shared videos in which whales came close to their boats, and participants had very intimate and personal interactions with them. This was very exciting for us. Born to Baja developed an itinerary for our March dates. Due to the capacity of the van and boats in the lagoons, we agreed to bring ten students, two instructors and the aquarium educator. Born to Baja provided an all-inclusive per person cost which included in-country lodging, transportation, excursions, guides, and meals. With our Office of Global Affairs, we proposed our course and set our budget including a group flight to Loreto, Mexico. Once approved, we recruited for the course and enrolled students whose cost was subsidized by the Honors College.

To develop our course and field experience with gray whales, we engaged in discussions with our aquarium colleague about whale migration on the East Coast of North America, where we are located. We discovered that the North Atlantic right whale experiences a similar migratory pattern as the Northern Pacific gray whale; therefore, our course was designed to compare and draw parallels between these two whale species. Thus, through our conversations and research, we decided that to promote Ocean Literacy and advocacy in our course, we would focus on how humans and climate change are impacting the yearly migration of right and gray whales. We invited local marine biologists and Ocean scientists to share their expertise on right whales, and we worked with Born to Baja to design an immersive field experience to learn and interact with the gray whales of Baja California Sur, Mexico.

14.3.1 Course Catalog Description

Focusing on systems science and Ocean Literacy principles, students will investigate climate-related challenges facing the marine ecosystems of NC and Baja California Sur, Mexico. Students will meet once a week to explore climate impacts on marine mammal migration as they prepare for an immersive 6-night/7-day spring break field experience. In Baja, students will travel by boat to observe whales in the Gulf of California and in the Ojo de Liebre lagoon. They will experience this region's unique landscape and biodiversity, exploring the shores of Loreto observing dolphins, sea lions, blue-footed boobies, blue whales and orcas. Students will travel across the Baja peninsula through the desert and the Sierra de la Giganta Mountains to the salt lagoons near the west coast town of Guerrero Negro. The lagoons are recognized as the world's most important place for the reproduction of the once endangered Eastern subpopulation of the North Pacific gray whale. In the Ojo de Liebre lagoon they will observe gray whale mothers and their calves. They will discuss human impacts as they make cross-curricular connections and informed decisions about the complex issues facing the coastal communities of Baja California Sur, Mexico, and NC.

14.3.2 Student Learning Objectives (SLO)

This course was open to all Honors College students of any major; thus, it was designed to have wide appeal. Additionally, to meet Honors College seminar requirements, it had to include multiple competencies. See Appendix A for the complete list of student learning objectives, the various university goals met, and OL principles addressed.

14.4 Course Schedule and Assignments

Within this section, we provide a detailed description of our course assignments. The assignment descriptions address systems science, climate science, and Ocean Literacy principles.

14.4.1 Class Meeting Schedule

The class met weekly for 1 h 15 min for each of the 7 weeks before our spring break field experience. We used class meeting time to build community and to provide foundational content knowledge on climate change and human impacts on the North Atlantic right whale population and migration patterns on the East Coast of the United States. Following the field experience, we met weekly for 5 sessions.

Scheduled Weekly Meetings

- **Week 1:** Intro to the Course—Introductions, course requirements, overview of field experience, discussion of prior knowledge regarding the Ocean and whales and expectations of what we might learn. *Note: Since we were researching the impacts of this course on students' Ocean Literacy, we asked them to complete a pre-assessment following our first-class meeting.*
- **Week 2:** Guest, Gail Lemiec, Unique Experiences Coordinator, NC Aquarium at Fort Fisher, introduced the Ocean Literacy Principles and why they were created. She guided students in a jigsaw activity in which she assigned principles to pairs of students. Groups were rearranged so these "experts" could share what they had learned about their assigned principles with their peers. Ocean Literacy was a foundational component of this course so it was important to have students explore these principles early on.
- **Week 3:** *No face-to-face meeting.* Students met with their Climate Change Discussion partner to prepare.
- **Week 4:** Climate Change Jigsaw Discussion Assignment. Note: Prior to this class meeting, students were assigned a partner and different climate-related topics to research. During this class meeting, students led their peers in a discussion in which they shared talking points and visuals related to their assigned topic. Next, they engaged peers in a discussion related to the topic they shared. See assignment description below.
- **Week 5:** Guest, Keith Rittmaster, Director of the Bonehenge Whale Center in Beaufort, NC, shared his knowledge of whale species we find off our NC coast, current advocacy efforts (e.g. local efforts to ban balloon releases), as well as the efforts of the Bonehenge Whale Center to educate the public.
- **Week 6:** Guest, Tiffany F. Keenan, PhD, Marine Mammal Stranding Program Coordinator, University of North Carolina Wilmington, shared her knowledge of right whales and national advocacy efforts like slow zones to protect migrating whale populations.
- **Week 7:** Guest, Andy Reid, PhD, Director of the Duke University Marine Laboratory, Beaufort, NC, shared current research projects regarding right whale species off NC coast.
- **Week 8:** 6-night/7-day Field Experience in BCS, Mexico
- **Week 9:** Deconstructed field experience in a group discussion
- **Week 10:** No face-to-face meeting. Students worked on partner presentations
- **Week 11:** Whale Migration Advocacy Presentations—Students shared their partner presentations on a selected advocacy effort. See assignment description below.
- **Week 12:** Peer Review of Papers—Students met in pairs to share final course reflections then read their papers aloud to their partners while partners evaluated their writing using the rubric.

14.4.2 Assignment Descriptions

Students completed journals, engaged in jigsaw activities, created presentations, and wrote advocacy papers.

Journals

Students completed weekly journals prior to spring break and daily during our field experience documenting observations and reflecting on assigned readings and experiences. Part I of each journal entry was a reflection on our previous class meeting. Students were to explain any specific "take-aways" from our class meeting/guest speakers. Part II involved a reading from our texts (*Lagoon Time* and *Whales: Their biology and behavior*) and using evidence from the reading to support their responses to specific prompts (e.g. Create a timeline of events impacting the Pacific gray whale and explain the significance of each of the events and create a map of the migration pattern of gray or right whales). In some journal entries, they were also required to do some "pre-work" to prepare for guest speakers, for example (e.g. Explore the Bonehenge Whale Center in Beaufort, NC, prior to the director's visit and to write two questions for him).

Climate Change Jigsaw

Climate change is an issue students tend to think they understand, but often they over-generalize (e.g. the hole in the ozone layer causes climate change) or cannot articulate specific scientific evidence to support their views. This assignment's purpose was to identify misconceptions and gaps in knowledge about climate change and to engage their peers in meaningful discussions about local and global climate change issues. To complete this assignment, students were first to research their assigned topic.

Topics included: (1) the scientific argument of the general causes and evidence of climate change, (2) the impacts of climate change on Eastern NC and BCS, Mexico, and (3) local and global (BCS, Mexico and beyond) mitigation strategies to combat climate change. In partner groups they created talking points, printed visuals to use with their groups, and developed questions for a group discussion.

To begin the class meeting, students completed the Page Keeley Global Warming probe, and we briefly discussed initial ideas. Next, students divided into two groups, and one partner from each topic pair led the group as an expert on his/her assigned topic. They used their talking points, visuals, and questions to facilitate a discussion on their topic. After the jigsaw portion of the discussion, we held a whole group discussion to synthesize each groups' ideas. This assignment aligned clearly with the climate science framework. Through the group discussions, the interconnectedness of components of our Ocean and climate and the complexity of related issues was highlighted.

Whale Migration Advocacy Presentation

With a partner, students created a presentation on one existing advocacy effort designed to help their assigned whale (right or gray). They selected a medium (e.g., digital movie or digital poster) to creatively share their research. In their presentation, they were required to include a map, explain the project, identify its successes and failures, share their rationale for choosing it and offer suggestions as to how to promote the project.

Ocean Advocacy Paper

Our final paper represented an assimilation of learned Ocean Literacy principles and whale migration advocacy efforts. The paper was divided into three parts, (1) Students selected an existing environmental initiative supporting the migratory journey of the right and/or gray whale, advocated for the initiative, and explained related complexities, (2) Students reflected on how the course impacted their ideas of Ocean Literacy and discussed which OL principles connected to their selected environmental initiative, and (3) Students reflected on the field experience and its impact on their feelings of connectedness to the Ocean. (Appendix B).

14.5 Field Experience in Baja Mexico

This section outlines the immersive field experience with the daily itinerary, including in-country experiences and journal prompts (Appendix C). We close with highlights of the trip.

- **Day 1:** Arrive Loreto, BCS—Following transport to hotel, we enjoyed a group lunch and free time to explore the town.
- **Day 2:** Loreto, BCS—7 am-2 pm Whale-watching excursion in the Gulf of California where we followed blue whales, a pod of orcas, and fin whales. Lunch in Loreto followed by a walking tour with our guide, Martín.
- **Day 3:** Loreto, BCS—Boated to island while sighting blue whales, dolphin, sea lions, blue-footed boobies, etc. then picnic lunch. Afternoon drive and dinner at San Javier Mission.
- **Day 4:** Travel Day—Spent day driving from east coast of Baja peninsula through the Sierra de Gigantica Mountain range to Guerrero Negro on the west coast, viewing desert ecosystems along the way.
- **Day 5:** Guerrero Negro—Morning whale watching excursion in Ojo de Liebre lagoon to view gray whales and their calves. Spent afternoon exploring nearby sand dunes and salt flats.
- **Day 6:** Guerrero Negro, BCS—Morning whale watching in Ojo de Liebre lagoon. Depart for Loreto.
- **Day 7:** Loreto, BCS—Breakfast then return flight to NC

14.5.1 Highlights of Field Experience

The Born to Baja guides arranged for a local marine biologist and historian to spend the week with our group. Having their expertise on excursions was so impactful on our learning. We also shared all our meals with them and with our guides. These informal conversations built bonds between us and grew our understanding of the culture of BCS, Mexico.

As mentioned previously, the Baja trip was our sixth study abroad experience, yet for many reasons, it stands out as the most memorable. Most impactful were our days of whale-watching. Our experience on the water in Loreto was surprising, even to our Born to Baja guides. Our boat captains followed blow holes to position us to see blue whales, whose majestic tail flips signified they were diving deep, not to return for an hour or more. We also followed a pod of orcas who swam very near our boats. The marine biologist with us had spent her adult life researching whales and had been on many boating excursions yet had never observed orcas in the wild. She cried as she shared what this experience meant to her. We also experienced dolphins playfully swimming and jumping alongside our boats while we watched sea lions lazily sunning on the beach.

As magical as these days in Loreto were, our experience gray whale watching in the west coast lagoons was completely different. This experience is regulated, with limited boats allowed on the lagoons. Our guides explained they had to reserve their spots for this unique experience a year in advance. On small wooden pangas, our captains followed blow holes and spy-hopping whales, careful not to get too close, as the whales had to initiate contact with us. At times, the whales would come up to the boats and nudge them, raising their bodies so we could pet and even kiss them. The local captains would sing to them and showed us how to rub them with our knuckles. It was an intimate experience with nature and our Ocean that we will not forget.

14.6 Research Methodology

A mixed methods methodology was used to answer three research questions about our course's impacts on students' knowledge of OL and their understanding of whale migration and advocacy efforts. The first research question addressed students' understanding of

Ocean Literacy. To answer this question, we adapted the Survey of Ocean Literacy and Experience (SOLE) Instrument (Greely and Lodge 2009). We (two science educators) and one Ocean specialist reviewed questions. Based on this review, we selected 22 items to use. The second part of the assessment adapted questions from the Blue Survey (Paredes-Coral et al. 2022). We used sections related to attitudes on Ocean stewardship (7 items), Ocean as economic resource (3 items), Ocean-friendly behavior (4 items), willingness to act (4 items), and Ocean and environmental literacy (8 items). The Revised SOLE and Blue Survey was administered after the first day of class through a link on Canvas, and again on the last day of class. Limitations of generalization of data are due to sample size of eight.

Research question 2 addressed students' understanding of whale migration and human advocacy efforts for these migration journeys. To answer this question, we used the *"Whale Migration Advocacy Presentation"* assignment (see description in Sect. IV). This assignment was completed by partner groups following our field experience. Our team of two science educators and a graduate student read and viewed the students' artifacts and created a coding scheme based on themes that emerged from an open coding process.

For our final question, we focused on the overall impact of the course on students' personal connectedness to nature and to the Ocean. To answer this question, we used students' personal reflections from their final papers, identifying common themes. The final papers were submitted on week 13. We used the same analysis process discussed above.

14.7 Research Results and Findings

This section presents our findings on the content assessment, survey data, as well as the qualitative results.

14.7.1 Research Question 1

Research question 1 addressed two aspects; first, how our course impacted students' understanding of Ocean science and OL principles, and second, how our course impacted students' personal attitudes and understanding of human impact on the Ocean and willingness to act upon these ideas.

Results of Revised Survey of Ocean Literacy and Experience (Revised SOLE) Instrument

The Revised SOLE was used for determining the impact of the course on students' knowledge of OL. Table 14.1 Revised SOLE questions alignment with OL principles.

We calculated the averages of the pre and post scores, which were pre 88% and post 89%, indicating no significant growth in students' knowledge of Ocean science concepts. The Revised SOLE results indicated that students had prior knowledge of Ocean science ideas and that our course continued to contribute to their overall knowledge of Ocean science and OL. Since there was no significant difference in the pre and post content knowledge of OL principles, we reviewed items in which correct responses went down or

Table 14.1 Groups of SOLE questions in alignment with the Seven Essential Principles of Ocean Sciences

Essential principle	No. of questions	Instrument
1. Size of Ocean	6	Revised SOLE (1–5, 7)
2. Ocean & its life shape earth	1	Revised SOLE (6)
3. Weather & climate	5	Revised SOLE (8–12)
4. Habitability	0	Revised SOLE
5. Biodiversity	5	Revised SOLE (13–17)
6. Human connections	4	Revised SOLE (19–22)
7. Oceans largely unexplored	1	Revised SOLE (18)
Total	22	

stayed the same and determined that these items related to topics that we had not focused on. We selected the items prior to having taught this course or having participated in this field experience. Noteworthy, however, as we will discuss in part 3, despite students having the content knowledge demonstrated on the pre-assessment, this knowledge had not impacted their advocacy of the Ocean or participation in sustainable projects or initiatives. The feedback from the reflections showed that, despite having knowledge of OL, without the field experience, their connection to the Ocean was not personal, and their advocacy for the Ocean was limited or non-existent.

Following the overall analysis of the pre and post assessment, we analyzed responses to items corresponding to the OL principles to determine if there was an OL principle that needed more emphasis within our course redevelopment. Included below are some noteworthy findings.

Figure 14.1 displays questions (#1–5, 7) that pertain to OL principle #1—Size of the Ocean.

These six questions addressed the size of the Ocean with questions 2, 3, and 7 discussing the processes of the water cycle, climate science, and geological processes that impact creation of landforms and landforms impacts on the Ocean. The back-and-forth feedback loops of these processes include how the Ocean shapes the land and how, in turn, the land impacts the Ocean. The concepts within these questions reflect systems thinking. For example, question 2 asks students to think about how the river systems impact the salinity of the Ocean. The answer is through the process of erosion of the landforms along the riverbanks that lead to the Ocean. For this question, our students scored 63% on the pre-assessment and 75% on the post-assessment. Question 3 was also a systems thinking question, as it focused on the processes that cause sea level change. On the pre-assessment, students scored 75% then on the post-assessment students scored 100%.

The next OL principle, #3, addressed students' knowledge of weather and climate and its' association with climate change. Figure 14.2 showed the results of questions 8–12.

Question 12 was a concern since students decreased in their understanding from the pre to post. This question required students to identify the percentage of the carbon cycle that occurs in the sunlit areas of the Ocean. This was not an idea that we discussed specifically in our course.

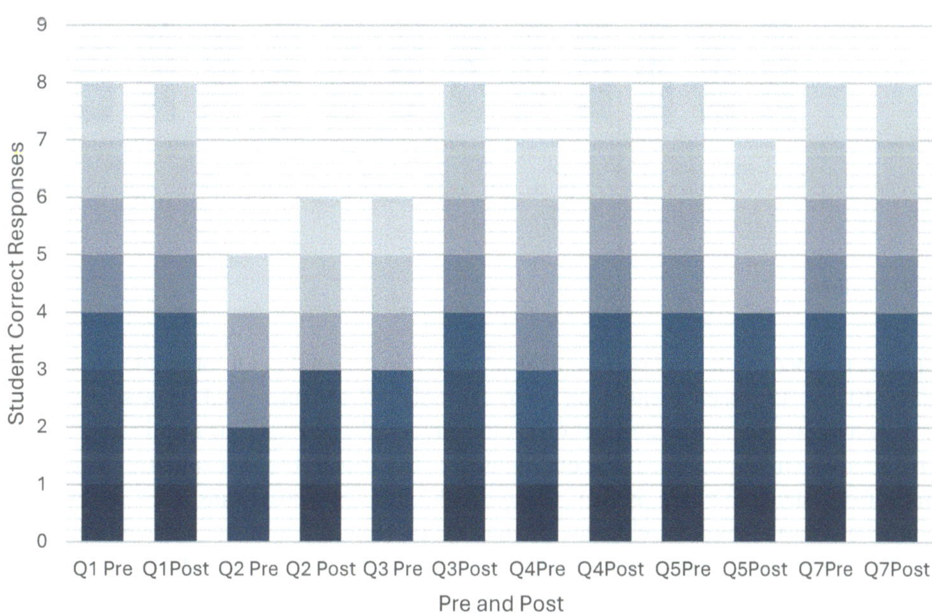

Fig. 14.1 Size of Ocean

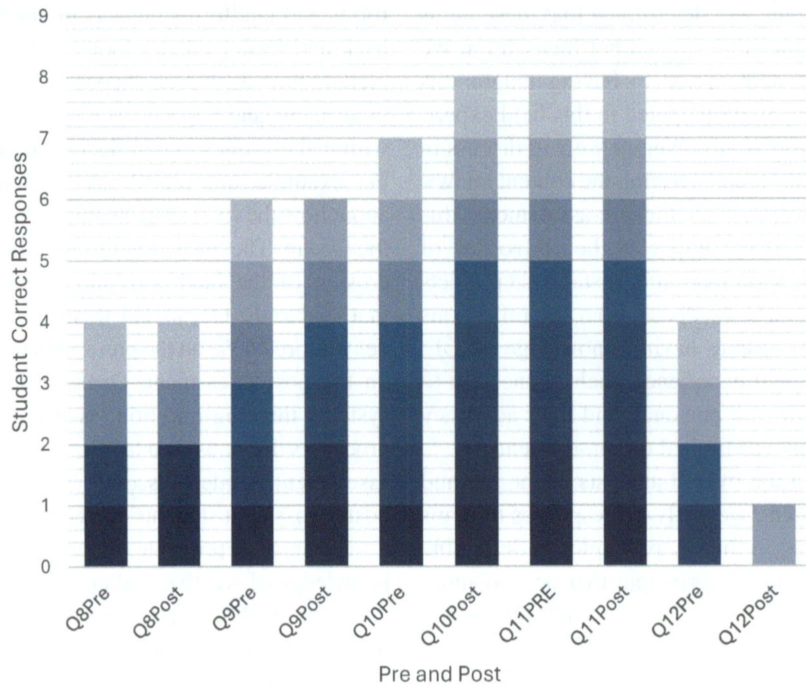

Fig. 14.2 Weather and climate

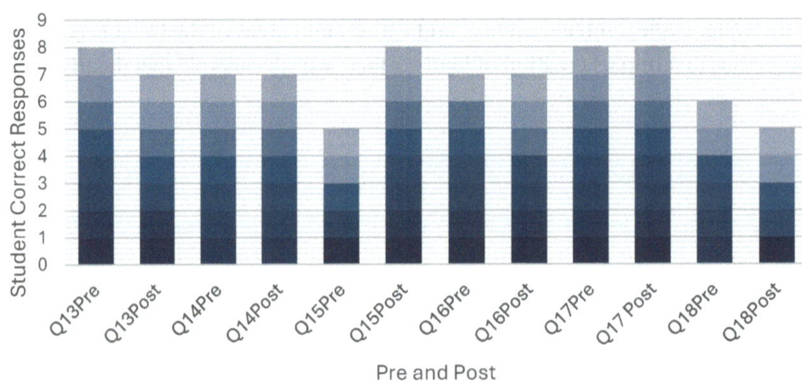

Fig. 14.3 Biodiversity

Students performed well on pre- and post-questions 10 and 11 which focused on natural systems such as the water cycle and carbon cycle and their impacts on the climate. These questions also reflected systems thinking as they focused on how the Ocean impacts climate change through the processes of storing, absorbing, and transporting water and carbon.

The next group of questions (#13–17) and question (#18) addressed the Ocean's biodiversity as reflected in OL principle #5 and OL principle #7. Figure 14.3 shows the results of students' understanding of this concept.

Questions 14–18 explicitly focused on the diversity of ecosystems such as estuaries and coral reefs as well as the size and number of species in the Ocean. Overall students scored either equal from pre to post or improved in their knowledge on these questions. Question 18 was of concern as students' understanding decreased

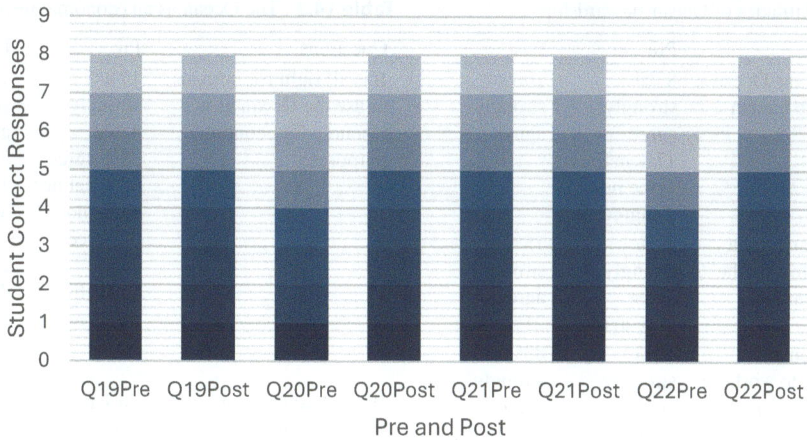

Fig. 14.4 Human connections

on the post-survey on this question focused on the biodiversity of coral reef ecosystems compared to other marine ecosystems such as seagrass meadows and mangrove forests.

The last set of questions (#19–22) focused on human connections to the Ocean (see Fig. 14.4 below).

The questions in this section discussed various natural resources we obtain from the Ocean, jobs and transportation provided by it, and its importance to natural security. These questions also focused on how humans need the Ocean for freshwater, oxygen, regulation of global temperatures and medicine. One question focused on the negative impacts of humans including pollution, physical changes to beaches, and the killing of large vertebrate marine animals. The last question referred to people living in coastal areas that are in the most danger due to climate change. According to the above scores, students seemed to have a good understanding of the connections of humans to the Ocean before the course and post-survey.

Research Question 1 Part 2

How did the course impact students' personal attitudes, understanding of human connection and impact on the Ocean and students' willingness to act on a personal level?

We used 26 questions from the Blue Survey (Paredes-Coral et al. 2022) to evaluate students' attitudes toward Ocean stewardship, Ocean as an economic resource, Ocean-friendly behavior, willingness to act, and Ocean and environmental literacy.

Results of the Blue Survey

The pre-survey was distributed to students on a link through Canvas after the first day of class. Six out of 8 students took the pre-survey. The post-survey was again distributed through a link on the last day of class. Eight out of the eight students took the post-survey. The results are displayed with the various sections of the survey as described above in Table 14.2.

Overall, the post survey results indicate a shift in students' attitudes toward Ocean stewardship. For example, 100% of students indicated on the post-survey that the health of the Ocean is important to human survival. 100% also agreed they have a personal responsibility to work towards the health of the Ocean and coastal areas. This sense of personal agency goes beyond the understanding that the Ocean impacts human survival and is an important realization for students. In contrast to this shift, it is interesting to compare the pre- and post-survey responses to the question of whether Ocean sustainability is more important than economic growth. On the post-survey, half of the students neither agreed nor disagreed with this statement while the other half agreed. Our team finds this result interesting but

Table 14.2 Attitudes of Ocean stewardship

Questions	Pre	Post
The health of the Ocean is important to human survival	83% strongly agree 17% neither agree nor disagree	100% strongly agree
My actions can have a significant effect on the health of Oceans and coastal areas	50% strongly agree 50% agree	75% strongly agree 25% agree
I have a personal responsibility to work for the health of Oceans and coastal areas	17% strongly agree 67% agree 17% neither agree nor disagree	88% strongly agree 13% agree
Business and industry should be responsible for Ocean sustainability	50% strongly agree 50% agree	75% strongly agree 25% agree
I understand the issues facing the global Ocean	17% strongly agree 67% agree 17% neither agree nor disagree	38% strongly agree 63% agree
Individual citizens should be responsible for Ocean sustainability	83% agree 17% disagree	25% strongly agree 63% agree 13% neutral
Ocean sustainability is more important than economic growth	50% strongly agree 33% agree 17% neither agree nor disagree	25% strongly agree 25% agree 50% neither agree nor disagree

not surprising since we stressed in our course the tension between the economy and Ocean sustainability. Our guest speakers (e.g., scientists, citizens, and naturalists) articulated the complexity of these issues and how sustainability efforts (e.g. working with fishing industry to adopt new fishing techniques or implementing "slow zones" in shipping routes during whale migration seasons) are often not easily implemented. See Table 14.3 for all results described.

Table 14.3 The Ocean as an economic resource

Questions	Pre	Post
It is all right for humans to use the Ocean as a resource for economic purposes	17% strongly agree 33% agree 33% neither agree nor disagree 17% strongly disagree	13% strongly agree 38% agree 13% neither agree nor disagree 25% disagree 13% strongly disagree
We should no longer use the Ocean as a resource for economic purposes	50% neither agree nor disagree 33% disagree 17% strongly disagree	25% neither agree nor disagree 63% disagree 13% strongly disagree
Maritime economic activities are compatible with Ocean sustainability	33% strongly agree 17% agree 33% neither agree nor disagree 17% disagree	13% strongly agree 38% agree 25% neither agree nor disagree 13% disagree 13% strongly disagree

The Ocean has been designated as the new economic frontier (OECD 2016); yet, as discussed previously, students report mixed feelings about the Ocean as an economic resource. The students' post-surveys indicate some movement in their views on the use of the Ocean as an economic resource. For example, students on the post-survey showed more diverse ideas on how they felt about the right of humans to use the Ocean for economic purposes; moving from more neutral to disagree/strongly disagree while 50% of students still agreed/strongly agreed with this idea. The same mixed result is seen for the last statement which proposed that maritime economic activities are compatible with Ocean sustainability. There were more students on the post-survey who indicated that they disagreed/

strongly disagreed with this statement where half of the students still agreed/strongly agreed with this statement. Again, our team recognizes that these results align with our discussion of the issues facing our Ocean as complex and multifaceted. Their varied and more diverse responses on the post-survey reflect that there is no right or wrong answer to these questions. Having a willingness to view each problem from various perspectives, as these students are doing, will be required to find solutions. Results are displayed in Table 14.4 below.

The post-survey results for the section on Ocean-friendly behaviors suggest that students may consider making some personal changes to their behaviors such as opting to use plastic-free alternatives and taking shorter showers. There were slight indicators that they may avoid using products that may be toxic to marine environments and may reduce energy consumption at home, as well (Table 14.5).

The willingness to act post-survey results displayed similar results as discussed above. Students seem willing to opt for using plastic-free alternatives, to take short showers, and to avoid toxic products for marine environments. The reduction of energy consumption does not seem to be an area where there is much change from pre to post (Table 14.6).

On post-surveys, students showed increase of understanding of issues facing the global Ocean and that the Ocean and coastal regions overall are "not" so vast and healthy to absorb pollution and man-made stressors. Following the course, more students were able to identify specific things they could do to help the Ocean, and students had more confidence in their ability to write a letter to a congressional representative about an issue impacting the Ocean.

Table 14.4 Ocean-friendly behavior

Questions	Pre	Post
I opt for plastic-free alternatives	67% agree 33% neither agree not disagree	25% strongly agree 63% agree 13% neither agree nor disagree
I avoid products with ingredients that are toxic for the marine environment or that are derived from endangered marine organisms	50% agree 33% neither agree nor disagree 17% disagree	50% strongly agree 13% agree 36% neither agree nor disagree
I reduce my energy consumption at home	17% strongly agree 33% agree 33% neither agree nor disagree 17% disagree	13% strongly agree 63% agree 13% neither agree nor disagree 13% disagree
I take short showers	17% strongly agree 33% agree 50% disagree	13% strongly agree 75% agree 13% disagree

Table 14.5 Willingness to act

Questions	Pre	Post
I would be willing to reduce my energy consumption at home	33% strongly agree 50% agree 17% neither agree nor disagree	38% strongly agree 50% agree 13% neither agree nor disagree
I would be willing to avoid products with ingredients that are toxic for the marine environment or that are derived from endangered marine organisms	50% strongly agree 33% agree 17% neither agree nor disagree	75% strongly agree 25% agree
I would be willing to opt for plastic-free alternatives	67% strongly agree 33% agree	50% strongly agree 50% agree
I would be willing to take short showers	50% strongly agree 33% agree 17% disagree	50% strongly agree 50% agree

Table 14.6 Ocean and environmental literacy

Questions	Pre	Post
I know some specific things I could do to help the Ocean.	83% agree 17% disagree	63% strongly agree 38% agree
I am familiar with the issues facing the global Ocean.	67% agree 33% disagree	38% strongly agree 63% agree
I have enough background knowledge to write a substantive letter to my congressional representative about an issue affecting the Ocean.	50% neither agree nor disagree 33% disagree 17% strongly disagree	25% strongly agree 50% agree 25% neither agree nor disagree
The Ocean and coastal regions overall are so vast and healthy that they can continue to absorb pollution and other kinds of man-made stresses for the foreseeable future.	17% agree 33% disagree 50% strongly disagree	50% disagree 50% strongly disagree
Human-made stresses are endangering coastal regions and the Ocean's ability to sustain itself and may well be leading to long-term damage and serious problems.	50% strongly agree 33% agree 17% strongly disagree	63% strongly agree 25% agree 13% neither agree nor disagree
We do not need to worry about the health of the Oceans, because we will develop new technologies to keep them clean.	50% disagree 50% strongly disagree	13% disagree 88% strongly disagree
What I do in my life doesn't impact the Ocean at all.	67% disagree 33% strongly disagree	63% disagree 38% strongly disagree
Government should be responsible for protecting marine environments.	67% strongly agree 33% agree	63% strongly agree 38% agree

14.7.2 Research Question 2

How did our course impact students' understanding of whale migration and advocacy efforts of humans?

To answer this question, our team first reviewed students' "Whale Migration Advocacy Presentation" assignments. Students were tasked with presenting an existing advocacy initiative focused on the migration of their assigned whale and sharing suggestions as to how to promote or alter this initiative to improve its effectiveness. Our team of two science educators and a graduate student read and viewed the students' artifacts and created a coding scheme based on themes that emerged from an open coding process.

Results

Each partner group demonstrated knowledge of their whale's migratory journey with a map illustrating the feeding and calving areas of their assigned whale (e.g., right or gray) (Figs. 14.5 and 14.6).

Table 14.7 below displays the results of the Whale Migration Advocacy assignment presentations. Two groups selected advocacy initiatives we categorized as direct action, meaning that the project is working directly to change human practices. For example, Group 1 selected the direct action of East Coast fishermen to implement ropeless fishing to help eliminate entanglement of right whales. Ropeless fishing allows a remotely activated trap to buoy up to the surface rather than pulling fishing lines up with a metallic line whales cannot tear through and can get entangled. This group demonstrated a systems thinking framing in their argument for ropeless fishing by stating the complexity of the issue for fishermen such as the price of changing equipment as well as the time it takes to educate themselves in using this new technology. Group 2 also advocated a direct action. This group advocated that NOAA should require all vessels to use an

Fig. 14.5 Example of a gray whale migration map included in student presentation

Fig. 14.6 Examples of right whale migration maps included in student presentation

automatic identifier system that would track and locate vessels to help monitor and enforce the speed rule in "slow zones."

As opposed to direct actions, the other two groups discussed organizations that worked on advocacy projects such as the Whale Sanctuary of El Vizcaino and the Gulf of Cataline Whale Foundation. Both organizations advocate for the gray whale on the West Coast. The Whale Sanctuary of El Vizcaino works with the government of Mexico to preserve the lands and waters of the lagoons where gray whales calve each winter. The other organization works with local schools in California to take students out on boats to have direct interactions with gray whales to impact these students' ideas of Ocean and whale conservation.

14.7.3 Research Question 3

What was the impact of the course on students' personal connectedness to nature and to the Ocean? To answer this question, our team reviewed part 3 of the final reflection which asked students to personally reflect on the course and the field experience (Appendix B).

Results

Eight students wrote their own personal reflection. Four themes emerged based on our analysis. The four themes were (1) increased connection to the natural world, (2) advocacy about the Ocean and understanding human impacts, (3) advocacy for personal growth about the Ocean, and finally, (4) increased connection to the Ocean. The first and second themes were mentioned by 7 students in their reflections. The third and fourth themes were mentioned by 6 students in their reflections. Student excerpts of themes in Table 14.8 signify the importance of the field experience on students' connectedness to nature, the Ocean, their willingness to advocate for a healthy and sustainable environment, and finally their own desire to continue to educate themselves on the environment.

Table 14.7 Results of Whale Migration Advocacy Presentation Assignment

Student groups	Whale	Advocacy project selected	Type of advocacy	Suggestion for promotion or change
1	Right	Ropeless fishing	Direct Action- Implementing a different fishing technique to prevent entanglements.	Promotion of educational projects to citizens with ropeless fishing (e.g., school students). Target areas along the eastern North Atlantic coastline.
2	Gray	Whale Sanctuary of El Vizcaino	Organization—working with the government to provide permanent protection of lagoons for calving allows ecotourism for people.	Promotion to increase international funding, advertise through various social media platforms, and creating digital posters.
3	Gray	Gulf of Cataline Whale Foundation	Organization—Preservation and education foundation providing students in Southern CA with field trips.	Promotion to try and connect other sources to expand project to other regions and improve website.
4	Right	Right Whale Slow Zones	Direct Action—National Oceanic and Atmospheric Administration (NOAA) promotion of slow zones along areas of the foraging and calving of the right whale.	Promotion for NOAA to require all vessels to use an automatic identifier system to track and locate vessels to monitor and enforce the speed rule.

Table 14.8 Personal connectedness to nature

Themes	# of students that mention themes in reflections	Excerpts of themes
Increased Connection to the natural world	7	"This course served as an indicator to me that there is so much more of the natural world for me to explore, and that I have only scratched the surface." "I have noticed that I have become more mindful and open-minded due to this field experience. I also find myself wanting to disconnect from my phone and spending more time outdoors. My screen time has significantly dropped since spring break and I have started taking any chance I can get to study outside." "This class, and the field experience has allowed me to reignite my passion for the environment."
Advocacy about the Ocean and Understanding Human Impacts	7	"Learning about whale migration patterns, marine ecosystems, and conservation challenges has made me more conscious of my impact on the environment and motivated me to advocate for responsible stewardship of marine resources." "Interacting with marine scientists, participating in research activities, and observing marine life in their natural habitats heightened my sense of awe and responsibility toward the Ocean."
Advocacy for Personal Growth about the Ocean	6	"After this trip, I see the Ocean as a beautiful habitat that is an important part of our planet. I am interested and eager to learn more about the Ocean and its inhabitants." "I am determined to continue learning for the rest of my life, even just to reduce my personal impact on harmful effects towards the Ocean." "I want to get outside more and take more time to appreciate nature around me, not just at the beach, while also wanting to continue finding ways to be involved with marine and coastal studies."
Increased Connection to the Ocean	6	"The field experience has dramatically increased my interest in the Oceans. It is impossible to see such beautiful wildlife and not have the deepest profound appreciation for what you are seeing. Experiencing the size of whales and bodies of water and just how massive both of those things are was a special moment." "Getting to experience these animals in their natural habitat increases my wonder on the behaviors of other Ocean animals in their natural habitat and not just on the surface but in the deep Ocean as well where we don't have as much knowledge."

14.8 Discussion of Results

The Revised SOLE assessment results suggest students maintained their conceptual understanding of Ocean science throughout our course. A review of OL principles #1–7 (#4 not addressed) from the assessment demonstrated students' understanding of these represented principles. The Blue Survey results showed that post-course, students had a good understanding of how humans are connected to the Ocean, positively and negatively, and increased in their understanding of how critical global issues impact the Ocean and coastal regions, while also displaying mixed ideas on using the Ocean as an economic resource. These results suggest that our coursework was successful in demonstrating that these constructs are complex and multifaceted and should be viewed through multiple perspectives.

Based on course artifacts, students expressed understanding of whale migration routes and human advocacy efforts. On reflections, students acknowledged their increased connectedness to the Ocean and began to see themselves as advocates for the Ocean, desiring to further educate themselves on Ocean conservation. While the small sample size of eight participants limits the generalizability of these findings, our qualitative data revealed patterns of understanding and increased awareness of connectedness to the Ocean suggesting potential for further investigation with a larger sample to confirm these observations.

14.9 Conclusion and Next Steps

This course examined the impacts of climate change from a local to global perspective by studying whales and their migratory journeys. Our course used a systems science approach focusing on climate science and incorporating the principles of OL. Our assignments prior to the field experience focused on climate change and presented the challenges of Atlantic right whales by introducing students to local marine scientists and activists who work with state and national organizations on conversation endeavors. The comparison of the right and gray whales' migration along the Atlantic and Pacific Oceans provided students with both a local and a global view of climate change challenges. Our immersive field experience in Mexico with gray whales transformed students' ideas about environmental conservation.

Due to its impact on students, the Honors College is supportive of us offering this course again. After reviewing our assignments, data, and conversations with our partners, we plan to make some adjustments. Even though the final reflections provided data on students' success with knowledge of OL principles, connectedness to nature, and advocacy for projects, we would like to build on the advocacy work and change this assignment to involve writing a letter to a government representative supporting or suggesting changes to an advocacy effort.

One of the co-authors and in-country partners has an ongoing research project on how poetry, music, and song is being used to communicate with the gray whales. This work stems from collaborations and relationships with indigenous people. We would like to incorporate readings about this work, the historical interactions of the indigenous people with the whales and have students create their own music, poem, or song to use on the boats, record the timing of the whale interactions, and the way the whale interacts.

In our world faced with global climate change and Ocean challenges, it is essential to provide opportunities for humans to interact and connect with the Ocean. A species like the gray whale offers the potential to create a personal relationship with humans that is transformative, leads to a sense of advocacy and a desire to learn more and do more for the species and the place that it inhabits.

14.9.1 Summary

Key concepts explored in this chapter include:

- Students learn about climate change and the impacts on right and gray whales' migratory journeys along their respective coastlines.
- Students learn about the importance of our Ocean through the exploration of Ocean Literacy principles.

- Students participate in an immersive field experience observing and interacting with marine mammals Baja California, Mexico.
- Students propose advocacy projects for right and gray whales based on course assignments and field experiences.

14.9.2 Recommended Resources

The following resources provide further insights and support on this topic:

Required Texts of Course

- Swartz, S. (2014). *Lagoon Time: Our Life and Times Among the Gray Whales of Laguna San Ignacio*. ISBN 9780916251420
- Hammond, P., Heinrich, S., Hooker, S., & Tyack, P. (2017). *Whales: Their Biology and Behavior.* Comstock Publishing Associates. ISBN: 1501716565

North Carolina Organizations Associated with Guest Speakers

- NC Aquarium Fort Fisher: https://www.ncaquariums.com/fort-fisher
- Bonehenge Whale Center Beaufort NC: https://bonehenge.org/
- University of Wilmington Marine Mammal Stranding Programhttps://uncw.edu/research/major-programs/marine-mammal-stranding/
- Duke Nicholas School of the Environment Duke University Marine Labhttps://nicholas.duke.edu/marinelab

Baja Organizations

- Born to Baja (ecotourism partner): https://www.borntobaja.com/
- Co-Author Fernando Martín Velazo's previous work: https://www.academia.edu/110744777/Northwest_Traveler_A_story_of_gray_whales_and_humans

Acknowledgements We would like to thank the Honors College, College of Education, and the Office of Global Affairs at East Carolina University for their support of our work and for their cooperation and sponsorship with the study abroad field experience of this course. We would like to thank Born to Baja for their partnership and collaboration in creating this magical and transformational experience for our students.

Appendices

The Appendices includes student learning objectives, assignment details, and Revised SOLE survey and Blue Survey.

Appendix 7A. Student Learning Objectives and Honors College Seminar Guidelines

Honors College Seminar Guidelines: Students are expected to master the following General Education Natural Sciences, Course Specific, Global Diversity and Writing Intensive student learning outcomes (Natural Sciences Competency = "NSC," Course Specific Competency Outcomes = "CSC," Global Diversity Competency = "GDC," and Writing Intensive Competency = "WIC"). Ocean Literacy Principle (OLP) and the corresponding number of the principle.

We designed the course so students of any major, from Biology to Business, could benefit. Below is a list of our student learning objectives with aligned competencies. We have also indicated related Ocean Literacy principles alongside each objective.

Students who have successfully complete this course are able to:

1. Explain the causes and impacts of climate change on the NC and Baja marine ecosystems. (NSC1) (OL-1,2,3,4,5,6,7).
2. Use historical data and modeling techniques to analyze and predict impacts of global climate change on the NC and Baja coastlines. (NSC2) (OL-1,2,3,4,5,6,7).
3. Critique scientific data, including climate fluctuations on the Atlantic and Pacific

Oceans and migratory patterns of animals' species for both areas, and identify limitations (NSC3) (OL- 1,2,3,5,6).

4. Accurately apply scientific data to advocate for equitable protection from the impacts of global climate change. (NSC4) (OL-3,5,6).
5. Evaluate through conversations with stakeholders and historical records, challenges of working on protection and management of threatened species. (CSC) (OL-5,6).
6. Evaluate various approaches to mitigating climate change and its impact on the blue economy of NC and Baja. (CSC) (OL-1,2,3,4,5,6,7).
7. Apply critical thinking skills to evaluate the issue and impacts of global climate change on both NC and Baja from multiple perspectives including the inequitable impacts of climate change. (GDC2) (OL- 1,2,3,4,5,6,7).
8. Demonstrate in their journal writing the complex, relevant topic of climate change and its impacts to migratory species using the coastlines of NC and Baja, addressing significant questions through engagement with and effective use of credible sources. (WIC1) (OL-3,5,6).
9. Create an "Ocean Advocacy Paper" to promote agency for climate change through the selection of an advocacy program and create an argument for how to promote awareness for the whale species. (WIC2) (OL-2,3,6).
10. Demonstrate that they understand writing as a process that can be made more effective as they draft and revise their advocacy papers. (WIC3).
11. Proofread and edit their own writing, avoiding grammatical and mechanical errors in their stakeholder project drafts. (WIC4).
12. Assess and explain the choices they made as they reflect on their final advocacy papers. (WIC5).

Appendix 7B. Ocean Advocacy Paper

Part I: Ocean Advocacy (3–4 pages). For this section, students research and selected an existing environmental initiative for either the NC or Baja coast (could be the same initiative they presented with their partner) then wrote a paper advocating for the support of this project. In your paper, students were to:

- Explain this project and the rationale behind it.
- Identify the stakeholders impacted by this project.
- Consider who might resist this project and so identify the audience with whom you will share your argument.
- Create an argument to support this project.
- Address the complexities surrounding this project.
- Report any preliminary results for this project.
- Summarize reasons audience should support this project.

Part II: Reflection on Ocean Literacy (1–2 pages). For this section, students explained Ocean Literacy and how their ideas about Ocean Literacy have been impacted by this course. They explained their rationale for selecting this initiative and how this project connects to Ocean Literacy Principles.

Part III- Personal Reflection (no word limits or minimum). Reflect on this experience, responding to the following:

- Has this experience changed you in any way? If so, how?
- How did the field experience within this course impact your ideas, feelings, or interest about the Ocean?
- What was your connection to the natural world prior to the course? How do you feel your connection to the natural world has been impacted by this course?
- Reread your personal reflection above. How could you summarize the impact of your time of learning and exploring in Baja? Provide a 1–2 sentence blurb that encapsulates your experience.

Appendix 7C. Journal Prompts from Field Experience

- In the airport: What is something I want to work on (a personal goal) during this trip? How will I manage my expectations (for whale watching, for example)? What would make this trip great?
- On Day 1: Distributed Ocean Literacy Quotes. Pick one quote and reflect on how this quote relates to you personally. Return to the quote throughout the week and reflect on how your relationship to the quote changes.
- Day 4 (Travel Day): Reflect on your time in Loreto. What were some personal highlights? What challenges did you face? What connections can you make from this experience to something back home?
- Day 5/6: Anthony Bourdain said, "Travel isn't always pretty. It isn't always comfortable. Sometimes it hurts, it even breaks your heart. But that's okay. The journey changes you; it should change you. It leaves marks on your memory, on your consciousness, on your heart, and on your body. You take something with you. Hopefully, you leave something good behind." What will you take with you from this trip and what will you leave behind?
- Day 7 (Return Day): Revisit your Ocean Literacy quote and your thoughts about it from the beginning of the week. Have your thoughts/feelings changed throughout the course/field experience? How might you bring an Ocean ethic back home with you and what might that look like for you?

Appendix 7D. Revised Ocean Literacy and Experience (SOLE) Instrument

Correct answers are identified in bold.

1. Approximately how much of the Earth is covered by Ocean? (OL1). 1
   (a) 30%
   (b) 50%
   (c) 60%
   (d) **70%**
   (e) 90%
   (f) 97%
2. Rivers supply most of the salt to the Ocean, which comes from (OL1). 3
   (a) Seafloor
   (b) **Eroding land**
   (c) Volcanic emissions
   (d) Atmosphere
   (e) All of these
   (f) None of these
3. What processes cause sea level changes? (OL1). 6
   (a) Plate tectonics
   (b) Ice caps melt & grow
   (c) Seawater expands & contracts
   (d) Sea level does not change
   (e) **Answer a, b, & c**
4. Approximately how much of the Earth's water is fresh and unfrozen (neither ice nor Ocean)? (OL1). 7
   (a) >50%
   (b) 40–50%
   (c) 20–30%
   (d) 10–20%
   (e) **3%**
   (f) 1%
5. The Ocean is large and finite. Its resources are (OL1). 13
   (a) Unlimited
   (b) All renewable
   (c) All non-renewable
   (d) **Limited**
   (e) Answer a and b
6. Which of the following are transported by rivers from watersheds to estuaries and to the Ocean? (OL2). 15
   (a) Nutrients
   (b) Salts
   (c) Sediments
   (d) Pollutants
   (e) **All of these**
   (f) Answer a and b
7. Climatic conditions constantly change and erode the landscape of barrier islands (beaches). Climatic changes occur in the form of (OL1). 19
   (a) Heavy winds
   (b) Wave action
   (c) Tidal surges
   (d) Coastal storms
   (e) **All of these**
   (f) None of these

8. Which sources put the most oil in the Ocean? (OL3). 23
   (a) Oil spills from ships
   (b) Leaks from refineries and pipelines
   (c) **Used motor oils washed into storm drains**
   (d) Leaks from offshore oil rigs
   (e) None of these sources put oil in the Ocean
9. All but one if the following decompose in Ocean water (OL3). 27
   (a) Sewage
   (b) Tin cans
   (c) **Plastic bags**
   (d) Chemical fertilizers
10. The Ocean controls weather and climate by dominating which of the Earth's systems? (OL3). 28
    (a) Energy
    (b) Plants
    (c) Water
    (d) Carbon
    (e) **Answer a, c, and d**
    (f) None of these systems
11. Most rain that falls on land originally evapo rated from the (OL3). 30
    (a) **Ocean**
    (b) Lakes
    (c) Rivers
    (d) Rain does not begin in Ocean
    (e) None of these
12. The Ocean dominates the Earth's carbon cycle. Approximately how much primary production on Earth takes place in the sunlit areas of the Ocean? (OL3). 32
    (a) 30%
    (b) **50%**
    (c) 60%
    (d) 70%
    (e) 90%
    (f) 97%
13. The Ocean has and will continue to have a significant influence on climate change by storing, absorbing, and moving (OL5). 33
    (a) Salts
    (b) Carbon
    (c) Heat
    (d) Water
    (e) Plants
    (f) **Answer b, c, and d.**
14. Where is the greater diversity of living organisms found? (OL5). 36
    (a) On the land
    (b) **In the Ocean**
    (c) Both equally
15. What produces most of the Earth's oxygen? (OL5). 39
    (a) Forests
    (b) **Ocean**
    (c) Both Equally
    (d) None of these
16. Ocean life ranges in size from the smallest virus to the largest animal that has lived on Earth, called the (OL5). 41
    (a) Giant squid
    (b) Basking shark
    (c) **Blue whale**
    (d) Sperm whale
17. Marine habitat that have brackish water and provide productive nursery areas for many marine species are (OL5) 46
    (a) seas
    (b) **estuaries**
    (c) rivers
    (d) open Ocean
    (e) lagoons
18. Which Ocean ecosystem provides habitat for one-third of all marine species? (OL7) 48
    (a) **Coral reef**
    (b) Seagrass meadow
    (c) Mangrove forest
    (d) Open Ocean
    (e) Estuary
19. Which of the following statements are true about the Ocean? (OL6) 52
    (a) It provides food and medicine.
    (b) It provides mineral and energy resources.
    (c) It provides transportation and jobs.
    (d) It benefits our economy and national security.
    (e) **All of these**
    (f) Both c and d
20. What does the statement, the Ocean and humans are inextricably connected mean? Humans need the Ocean. (OL6) 54
    (a) For freshwater
    (b) For oxygen
    (c) To regulate the temperature
    (d) For new health cures
    (e) **All of these**

21. Humans affect the Ocean in a variety of ways. Human development and activity often leads to (OL6) 55
    (a) Pollution (point, non-point, noise)
    (b) Physical changes to beaches
    (c) Removal of most large vertebrates
    (d) **Answer a, b, & c**
    (e) Humans do not affect the Ocean
22. Which of the following statements is most relevant to Ocean Literacy? Much of the world's population lives. (OL6) 57
    (a) Near rivers
    (b) In rural areas
    (c) **In coastal areas**
    (d) In mountain areas
    (e) In wooded areas

Appendix 7E: Blue Survey Questions

Below are a number of statements that may or may not describe your beliefs about protecting the Ocean (Cudaback 2006). Rate each statement on a scale of 1 to 5, where:

1 = Strongly Disagree; 2 = Disagree; 3 = Neutral; 4 = Agree; 5 = Strongly agree.

Attitudes—Ocean stewardship (7items)

Strongly agree, agree, neutral, disagree, strongly disagree.

The health of the Ocean is important to human survival.

My actions can have a significant effect on the health of Ocean and coastal areas.

I have a personal responsibility to work for the health of Ocean and coastal areas.

Business and industry should be responsible for Ocean sustainability.

I understand the issues facing the global Ocean.

Individual citizens should be responsible for Ocean sustainability.

Ocean sustainability is more important than economic growth.

Ocean as economic resource (3 items)

Strongly agree, agree, neutral, disagree, strongly disagree.

It is all right for humans to use the Ocean as a resource for economic purposes.

We should no longer use the Ocean as a resource for economic purposes.

Maritime economic activities are compatible with Ocean sustainability.

Ocean-friendly behaviour (4 items)

Strongly agree, agree, neutral, disagree, strongly disagree.

I opt for plastic-free alternatives.

I avoid products with ingredients that are toxic for the marine environment or that are derived from endangered marine organisms.

I reduce my energy consumption at home.

I take short showers.

Willingness to act (4 items)

Strongly agree, agree, neutral, disagree, strongly disagree.

I would be willing to reduce my energy consumption at home.

I would be willing to avoid products with ingredients that are toxic for the marine environment or that are derived from endangered marine organisms.

I would be willing to opt for plastic-free alternatives.

I would be willing to take short showers.

Ocean and Environmental Literacy (10 items)

Strongly agree, agree, neutral, disagree, strongly disagree.

1.____ I know some specific things I could do to help the Ocean.

2.____ I am familiar with the issues facing the global Ocean.

3.____ I have enough background knowledge to write a substantive letter to my congressional representative about an issue affecting the Ocean.

4.____ The Ocean and coastal regions overall are so vast and healthy that they can continue to absorb pollution and other kinds of man-made stresses for the foreseeable future.

5.____ Human-made stresses are endangering coastal regions and the Ocean's ability to sustain itself and may well be leading to long-term damage and serious problems.

6.____ We do not need to worry about the health of the Ocean, because we will develop new technologies to keep them clean.

7.____ What I do in my life doesn't impact the Ocean at all?

8.____ Government should be responsible for protecting marine environments.

References

Brewster K (2004) Whales, they give themselves: conversations with Harry Brower, Sr. University of Alaska Press

Canada (1993) *Marine mammal regulations (SOR/93–56)*. Retrieved August 10, 2024, from https://laws-lois.justice.gc.ca/eng/regulations/sor-93-56/

Cavas B, Acık S, Koc S, Kolac M (2023) Research trends and content analysis of Ocean Literacy studies between 2017 and 2021. Front Mar Sci 10:1200181. https://doi.org/10.3389/fmars.2023.1200181

Coleman KM (1983) Manilius' monster. Hermes 111(2):226–232. http://www.jstor.org/stable/4476314

Corkeron PJ, Connor RC (1999) Why do baleen whales migrate? Mar Mamm Sci 15(4):1228–1245. https://doi.org/10.1111/j.1748-7692.1999.tb00887.x

Frank M (2000) Engineering systems thinking and systems thinking. Syst Eng 3:63–168

Greely, T.M., & Lodge, A. (2009) Measuring Ocean Literacy: What teens understand about the Ocean using the Survey of Ocean Literacy and Engagement (SOLE). Conference proceeding

Hildebrand L, Bernard KS, Torres LG (2021) Do gray whales count calories? Comparing energetic values of gray whale prey across two different feeding grounds in the eastern North Pacific. Front Marine Sci 8:683634. https://doi.org/10.3389/fmars.2021.683634

Hoare P (2023, July 6) Grey whales seen seeking human help to remove parasites. *The Guardian*. https://www.theguardian.com/environment/2023/jul/06/grey-whales-seen-seeking-human-help-to-remove-parasites

Jones ML, Swartz SL, Leatherwood S (eds) (1984) The Gray whale: Eschrichtius robustus. Academic Press

Kuhnlein HV, Humphries MM (2017) Gray whales. In: *Traditional animal foods of indigenous peoples of Northern North America*. McGill University. http://traditionalanimalfoods.org/mammals/whales-dolphins-porpoisess/page.aspx?id=6398

Laist DW (2017) North Atlantic right whales: From hunted leviathan to conservation icon (Kindle edition). Johns Hopkins University Press

Lee T, Jones MG, Chesnut K (2019) Teaching systems thinking in the context of the water cycle. Res Sci Educ 49(1):137–172

Marine Mammal Commission (2019, March) *Marine Mammal Protection Act*. Retrieved August 10, 2024, from https://www.mmc.gov/wp-content/uploads/MMPA_March2019.pdf

National Oceanic and Atmospheric Administration (2013) Ocean Literacy: The essential principles and fundamental concepts of Ocean sciences for learners of all ages. https://repository.library.noaa.gov/view/noaa/39086

NOAA (2023, April 13) *Gray whale*. Retrieved August 10, 2024, from https://www.fisheries.noaa.gov/species/gray-whale

NOAA (2024a, June 7) *North Atlantic right whale*. Retrieved August 10, 2024, from https://www.fisheries.noaa.gov/species/north-atlantic-right-whale

NOAA (2024b, May 1) *2019–2023 Eastern North Pacific gray whale UME closed*. Retrieved August 10, 2024, from https://www.fisheries.noaa.gov/national/marine-life-distress/2019-2023-eastern-north-pacific-gray-whale-ume-closed

NOAA. (2013). Ocean Literacy: The essential principles and fundamental concepts of Ocean sciences for learners of all ages version 2. Brochure resulting from the 2-week on-line workshop on Ocean Literacy through science standards. National Oceanic and Atmospheric Administration. First published June 2005, revised March 2013

Organisation for Economic Co-operation and Development. (2016). OECD factbook 2015–2016: Economic, environmental and social statistics. OECD Publishing. https://doi.org/10.1787/factbook-2015-en

Papadopoulos JK, Ruscillo D (2002) A ketos in early Athens: An archaeology of whales and sea monsters in the Greek world. Am J Archaeol 106(2):187–227. http://www.jstor.org/stable/4126243

Paredes Coral E, Deprez T, Mokos M, Vanreusel A, Roose H (2022) The blue survey: Validation of an instrument to measure Ocean Literacy among adults. Mediterr Mar Sci 23(2):321–326. https://doi.org/10.12681/mms.26608

Parsons ECM (2004) Sea monsters and mermaids in Scottish folklore: Can these tales give us information on the historic occurrence of marine animals in Scotland? Anthrozoös 17(1):73–80. https://doi.org/10.2752/089279304786991936

Scammon CM (2007) Marine mammals of the northwestern coast of North America. Sheridan Books

Swartz SL (2016) Why do gray whales appear 'curious' about whale-watchers? Laguna San Ignacio Ecosystem Science Program. https://www.sanignaciograywhales.org/wp-content/uploads/2016/10/Why-Friendly-FINAL-2-OCT-2016.pdf

Van Ginkel R (2004) The Makah whale hunt and leviathan's death: Reinventing tradition and disputing authenticity in the age of modernity. Etnofoor 17(1/2):58–89. http://www.jstor.org/stable/25758069

Velazco FM (2022a) The leviathan's playing: Retrospective on mediations with gray whales in the Ojo de Liebre Lagoon (Mexico). Leonardo 55(5):530–534. https://doi.org/10.1162/leon_a_02260

Velazco FM (2022b) Northwest traveler: A story of gray whales and humans. Noro Originals. https://www.researchgate.net/publication/376170802_Northwest_Traveler_A_story_of_gray_whales_and_humans

Author Contributions and Biographical Profiles

[1]*Tammy D. Lee: Co-Conceptualization; Writing—Original Draft, Review, and Editing.

[2]Bonnie Glass: Co-Conceptualization, Writing—Original Draft, Review, and Editing.

[3]Gail Lemiec: Writing—Review and Editing.

[4]Fernando Martín Velazco: Co-Conceptualization; Writing—Original Draft, Review, and Editing.

*Corresponding author at leeta@ecu.edu

Tammy D. Lee, PhD, Associate Professor in Science Education in the Department of Mathematics, Science, and Instructional Technology Education at East Carolina University. Dr. Lee's research focuses on science educator development using the framing of science teacher identity and connectiveness to nature. As part of teacher development, she researches the impact of local and global climate science and Ocean Literacy curriculum on climate change awareness on educators and students' identity. She has designed and implemented global climate science and Ocean Literacy curriculum implementing place-based experiences, which entail study abroad experiences for educators in five different countries.

Bonnie B. Glass, MA, Master Teaching Instructor in Science Education in the Department of Mathematics, Science, and Instructional Technology Education at East Carolina University. Ms. Glass' research focuses on science educator development through impactful field experiences which connect students to nature. She researches the impact of Ocean Literacy curriculum on climate change awareness, identity, and agency. She has designed and implemented global climate science and Ocean Literacy curriculum implementing place-based experiences which entail in-country and study abroad experiences for educators.

Gail Lemiec, Unique Experiences Coordinator at the NC Aquarium at Fort Fisher. Ms. Lemiec has worked for the NC Aquariums for over 10 years, serving in the Education Department, mainly offering professional development workshops and immersive opportunities for educators. She has been an informal environmental educator for over 20 years in the state of North Carolina and has lived and taught in all three major geographical regions in the state. Ms. Lemiec earned her bachelor's degree from Northland College and a master's degree in Zoology from Miami University of Ohio. Her Master's capstone focused on Ocean Literacy knowledge and attitudes in students across the state. She works to connect people with NC Aquarium conservation projects and leads hands-on courses to empower educators to bring conservation into the classroom.

Fernando Martín Velazco is a multidisciplinary artist and storyteller based in Baja, Mexico. His work focuses on the field of expanded arts and performative literature, generally based on research processes in wild areas. He has written for cultural publications in Mexico and for the US arbitrated magazines, Media N-Journal and Leonardo. He has also participated in-artist residency programs in Mexico, Austria, Saudi Arabia, and the United States, including the Djerassi Foundation program. He recently published Viajera del Noroeste (The Northwest Traveler), a transmedia about the cultural history of gray whales, on the digital platform Noro.mx, and collaborates with the international art and technology project, Domestic Light.

Ocean Literacy in Science Curriculum: A Quantitative Content Analysis in Sweden

15

Géraldine Fauville, Craig Strang, Ewa Skantz, Jens Bjelvenmark, and Anne Solli

Abstract

In 2023, the United Nations Secretary-General's Special Envoy for the Ocean, Peter Thomson sent a letter to all the UN member states to recommend that they correct the under-representation or absence of Ocean-related subjects in their school curricula. While Sweden has identified Ocean Literacy as one of four focus areas concerning its contribution to the UN Decade of Ocean Science for Sustainable Development, it has not yet taken action to address Thomson's letter. As a first step in this direction, we analyzed the subjects of *Geography* and *Science Studies,* including biology, physics, and chemistry for the compulsory school (age 7 to 16), and *Science Studies* and *Biology* for the upper secondary school (age 16 to 19), in relation to Ocean Literacy and Ocean-related content. The 182 learning goals were individually coded according to five variables. Results demonstrate that (1) Ocean Literacy concepts are important to attaining over 75% of the learning goals, while (2) Ocean Literacy concepts, or even aquatic-focused words are nearly absent due mostly to a general vagueness in the formulation of the learning goals. We argue that the learning goals contribute to an Ocean blind spot for Swedish students that might inhibit their general understanding of science and challenge their understandings of the Ocean planet on which they live. This study offers a concrete way to engage in discussion with Swedish policymakers and educators concerning the importance, and the absence of, the Ocean in the Swedish curriculum. We hope the methodology developed in this study will be adapted and used in other countries to foster international collaboration toward addressing Thomson's call for action for a bluer education.

Keywords

Content analysis · Formal education · School curriculum · Marine environmental education · Ocean Literacy (OL) · SDG14 · Ocean Decade · Ocean blindness · Terrestrial bias

G. Fauville (✉) · E. Skantz · A. Solli
University of Gothenburg, Gothenburg, Sweden
e-mail: geraldine.fauville@gu.se; ewa.skantz@gu.se; anne.solli@gu.se

C. Strang
The Lawrence Hall of Science, University of California, Berkeley, CA, USA
e-mail: cstrang@berkeley.edu

J. Bjelvenmark
Gullmarsgymnasiet, Lysekil, Sweden
e-mail: jens.bjelvenmark@lysekil.se

T. J. Kennedy (ed.), *Ocean Literacy: The Foundation for the Success of the Ocean Decade, Volume I*, Sustainable Development Goals Series, https://doi.org/10.1007/978-3-031-98145-6_15

15.1 Introduction

In this chapter, we highlight formal education as the key to improving young people's knowledge of the Ocean and its importance to all life on Earth, and to accelerating societal transformation for a sustainable future. This study analyzes the Swedish science curriculum to identify the extent of its alignment with the Ocean Literacy (OL) framework. We hope the method and codebook developed in this study will be widely adapted and used to accelerate the research and international collaboration on Ocean Literacy in curricula around the world.

Sweden is a strong actor in the UN Decade of Ocean Science for Sustainable Development (Ocean Decade), launched by the United Nations in 2021 to tackle the declining health of the Ocean. A Swedish national committee was commissioned to help the government prioritize and implement actions within the Ocean Decade framework (FORMAS 2020). OL emerged as one of the key focus areas to be addressed during the Ocean Decade. This focus on OL translated into the Swedish Agency for Marine and Water Management bringing together various organizations and actors to strengthen the work with marine education. Furthermore, in 2022, Sweden contributed 400,000 USD to support the work of the Ocean Decade and launched a national research program for the Ocean and water. Given the Swedish commitment to OL and support for the Ocean Decade, it is particularly timely to scrutinize the Swedish science curriculum in hopes of future actions to ensure a national curriculum that is rich in Ocean-related content.

In 2023, the United Nations Secretary-General's Special Envoy for the Ocean, Peter Thomson sent a letter to all the UN member states alerting them to the under-representation or absence of Ocean concepts in school curricula. He stated that the "inclusion of the Ocean in school curricula around the world is essential if we are going achieve Ocean Literacy and lay better foundations for Ocean science," and claimed that "embracing Ocean Literacy through a whole-school approach will ultimately lead to universal recognition that the Ocean is crucial for the survival of humankind on Planet Earth." (Thomson 2023, para 4). This call positioned OL as an essential pillar of the Ocean Decade. It also positioned formal education as the lynchpin for achieving OL.

OL plays a central role in the Ocean Decade. It encompasses what every person should know about the Ocean to make informed, responsible decisions regarding the health of the Ocean (Cava et al. 2005). Research findings in many countries consistently indicate low levels of Ocean Literacy among the public (see Eddy 2014), and as Thomson points out in his letter, this shortfall of OL is attributable to the limited extent to which Ocean concepts are addressed in formal science education curriculum, as shown in several studies over the last 15 years.

Hoffman and Barstow (2007) analyzed the Earth science curriculum across the U.S. and found that the OL principles and concepts were mostly absent. Gough (2017) came to a similar conclusion by counting key words connected to OL in the Australian curriculum. She found that the word "Ocean" appears twice, while the word "marine" is absent. The words "coast" and "coastal" are present four and five times respectively. Castle et al. (2010) analyzed the extent to which marine topics are taught in English schools. They concluded that the Ocean-related concepts were mainly taught as part of geography through description of coastal issues, and that marine-related topics are seldom taught, as there is no requirement for them to be taught in the national curriculum. After analyzing the Brazilian curricula at the regional and federal levels, Pazoto et al. (2022) argued that while OL is better represented in the Brazilian curriculum than in other countries, "the representativeness of OL in Brazilian documents is below the recommended level for a person to be considered Ocean literate" (ibid., p. 1). In Canada, McPherson et al. (2018) studied the high school science curriculum and revealed the limited inclusion of OL principles and concepts. In their second study, they interviewed high school science teachers about the challenges to integrating Ocean-related concepts in science courses. Several interviewees mentioned that "without the inclusion of Ocean education in the curriculum outcomes, there was "not enough additional time to include these concepts" (McPherson et al. 2020, p. 134). The invisibility of the Ocean in national curricula is also noted as a key challenge for teachers in a large European study, which identified barriers to teaching students about the Ocean (Fauville et al. 2018). A study from the Azores also showed the limited extent to which 426 participating teachers address Ocean science in their teaching because of the "scarcity in the curricula and [their] insufficient knowledge on the subject" (Costa et al. 2024, p. 8).

The National Marine Educators Association (NMEA n.d.) developed a rating scale and measured the alignment of OL to the U.S. Next Generation Science Standards that govern the science curriculum in nearly all states for children ages 5–18. They found that while OL principles and concepts are explicitly included in only 6% of the standards, the same ideas are either essential or provide useful examples to understanding 27% of the standards. Based on the intrinsic importance of understanding Ocean ideas to understanding general science concepts, Strang et al. (2007) argues that it is not possible for students to be science literate without being Ocean literate. Pocze et al. (2020) analyzed the primary and secondary school curricula in nine European countries to find possible entry points or opportunities to include Ocean concepts into existing subjects of study. They found plentiful opportunities to teach about the Ocean at both the primary (natural science, environmental education, nature and plants, animals and humans) and secondary (biology, chemistry, physics, geography) levels. However, as described above, evidence from Canada, the Azores, the U.S., and across Europe demonstrates the imperative of Thomson's call to UN members: if Ocean Literacy is not explicitly included in national curriculum frameworks, these plentiful opportunities have always been and will continue to be unrealized. Ryabinin et al. (2019) supported this conclusion, arguing that without adding more Ocean science to the formal curriculum, students cannot be reached fairly, equitably and consistently.

While these studies use various methods, together they draw a dire picture of the inadequacy of Ocean-related content in the curriculum in every country studied. OL provides a coherent and widely agreed upon framework to correct this situation. The journey to incorporate OL into the routine learning experience of every child must begin with a uniform analysis of existing national curriculum frameworks to identify specific gaps in relation to learning about the Ocean and the opportunities for filling them.

Since the early 2000s, several collaborative efforts in the U.S., led by many organizations under the umbrella of the NMEA, have developed an OL Framework comprised of the OL principles and concepts, an Ocean Literacy Scope and Sequence for Grades K–12, the Alignment of Ocean Literacy and the Next Generation Science Standards, and the International Ocean Literacy Survey. These tools are freely available to be adapted and used to remedy the lack of OL included in the national curriculum of any country. We think national ministries of education will be more susceptible to the remedy once the specific contours of the challenge they are trying to address have been clearly described. We hope that the methodology and codebook described in this chapter will provide the international Ocean Literacy community with another important tool for systematically identifying the presence and absence of Ocean concepts in national curricula, and for comparing findings and progress across countries.

15.2 Material and Method

15.2.1 Material

This study analyzes the Learning Goals (LGs) in the two Swedish curricula in the formal education system:

- The Curriculum for Compulsory School, Preschool Class and the Leisure Time Center (Lgr 22), addressing education for children aged 7–16.
- The Curriculum for Upper Secondary Education (Gy11), focusing on voluntary education for 16–19-year-olds.

The scope of the study is limited to the science-related courses; Geography, Biology, Physics, and Chemistry for compulsory school, and Natural Science Studies and Biology for the upper-secondary school, as every upper secondary student will have to study at least one of them (except for Engineering students). The LGs are organized around four grade bands (Table 15.1).

The selected courses include 214 LGs. Several LGs are repeated multiple times for different courses (e.g., the LGs of physics, chemistry and biology for grade 1–3 are identical). We removed 32 duplicates to avoid coding the same content several times. Our final dataset includes 182 LGs. At the beginning of this project, the Swedish curricula were available only in Swedish. We translated the LGs to English for analysis.

15.2.2 Method

We used a content analysis approach to investigate the content of the 182 LGs. Content analysis is defined as "the systematic, objective, quantitative analysis of message characteristics," (Neuendorf 2017, Chap. 1, p. 2) and is used to make inferences about specific characteristics within a given body of text. Content analysis can be human- or computer-coded. To ensure scientific rigor, in the case of human-coded content analysis, it is essential for the coders to co-create the procedure and to calibrate their coding until they achieve high agreement. This method has been used in various research contexts such as analyses of school textbooks, or the tone of social network comments (see Neuendorf 2017, Chap. 1). For example, Smith et al. (2008) used content analysis to comb through 142 anti-bullying school policies in England. Based on existing documents, the research teams identified categories of interest, tested each category on a subset of anti-bullying policies and revised the categories. This process was repeated once more to fine-tune their scheme. Each policy was then independently coded by 2 coders, and they measured their percentage agreement. The authors found out that most policies addressed improving school climate and a statement concerning contacting parents when bullying incidents occur. However, homophobic bullying and cyberbullying were often left out of the schools' policies.

Table 15.1 Equivalent between Swedish grade bands, students' ages and international appellation

Grade band	Students' age	Appellation
1–3	7–10	Primary
4–6	10–13	Primary
7–9	13–16	Lower secondary
10–12	16–19	Upper secondary

Each coder in this study was selected for a particular expertise essential to successfully conducting this content analysis. Their collective expertise includes marine science, Ocean Literacy, marine education, and Swedish science education at different grade levels. The method included two phases: the creation of the codebook, and the use of the codebook to analyze the 182 LGs.

The creation of the codebook (phase 1) was highly iterative and similar to the one described in Smith et al. 2008. Reading and discussing the LGs several times led to a series of variables. These variables were created based on previous research in the field of marine education and curriculum (see Introduction) and on characteristics of the Swedish LGs themselves. Once a set of variables was created, a subset of LGs was randomly selected. Each coder individually coded these LGs using the variables. The coding of the subset of LGs was compared across coders. All disagreements were discussed at length leading to modification of the variables and re-wording of their descriptions. When a revised version of the variables was finalized, the coders again received a subset of LGs and used the variables to code individually, leading to more challenges to be resolved. The variables were slowly refined through iterative group discussions followed by individual coding of a subset of LGs. This resulted in a slow and thorough refinement of the variables. This process revealed itself to be difficult due to the nature of the Swedish LGs, which tend to be vague and broad, making their precise interpretation and coding challenging. The broad nature of most of the LGs is illustrated with these examples:

- Seasonal changes in nature. Some animal and plant life cycles and adaptations to different habitats and seasons. (grades 1–3)
- Critical review and use of information related to biology. (grades 4–6)
- Applications of cell and molecular biology. Opportunities, risks and ethical issues. (grade 10–12)

The final codebook includes the following five variables we used for coding each LG (see the Recommended Resources section for more information about the codebook created for this project).

Aquatic Focus

Similar to Gough's (2017) analysis searching for words with aquatic focus (e.g., "coastal" or "Ocean"), our first variable consisted of coding with "Yes" or "No" if aquatic words are included. The presence of words such as "Ocean", "water", "precipitation" constituted an aquatic focus.

Terrestrial Bias

This refers to our familiarity, as land-living, air breathing organisms, with terrestrial ecosystems and phenomena. This bias translates into exclusion of fresh and saltwater ecosystems from various fields such as conservation or marine education (Castello 2021; National Science Teaching Association & National Marine Educators Association 2025). The National Marine Education Association pointed out this bias in a draft of the U.S. Next Generation Science Standards published by the National Academy of Sciences. This variable coded for the absence or presence of terrestrial bias—where terrestrial examples were provided while marine ones could have been provided but were not– in LGs.

Type of LG

The Swedish curriculum is divided into two types of LGs:

- Knowledge of scientific concepts and explanatory models.
- Ability to use science, for example, to evaluate information, communicate, and make decisions. Additionally, it is described that students should develop the ability to conduct systematic investigations in the subject.

To capture this distinction, we coded each LG as "content LG" focusing on students' *knowledge* or "method LG" focusing on *doing science.*

Ocean Literacy Principle Role (OLP Role)

This variable was informed by the alignment conducted in the United States where they created a 4-point scale coding for the relationship between the NGSS and the OLP (for additional details describing the alignment of OL to NGSS, see the Recommended Resources section). We adapted this pre-existing variable to the Swedish context by coding each LG according to the role that the OL framework (principles and concepts) plays in understanding it. The OLP role was coded as:

- Essential: understanding at least one OL principle or concept is essential to fully attain the LG.
- Example: when at least one OL principle or concept serves as a useful example (but is not essential) to attain the LG.
- No role: when none of the OL principles or concepts play a significant role in attaining the LG.
- Place-based: The Swedish curricula include some LGs focusing on the local environment, making them impossible to code for OLP role, since the role would change depending on the location of the student.

OLP and LG Alignment

For each LG coded as "essential" or "example" in the OLP role variable, we then identified and recorded the OLP(s) and Fundamental Concept(s) the LG aligns with. We used both the OL Guide and the OL Scope and Sequence for Grades K–12 (available in the Recommended Resources section) for this coding. The OLPs are:

- OLP1: Earth has one big Ocean with many features
- OLP2: The Ocean and life in the Ocean shape the features of Earth
- OLP3: The Ocean is a major influence on weather and climate
- OLP4: The Ocean makes Earth habitable
- OLP5: The Ocean supports a great diversity of life and ecosystems
- OLP6: The Ocean and humans are inextricably interconnected
- OLP7: The Ocean is largely unexplored.

In the second phase of this method, the final version of the codebook was used to analyze each of the 182 LGs. The lead author coded all the LGs while each of the other four researchers coded approximately 25% of the LGs. In this way, each LG was coded independently by two researchers. Once both coders completed their portion of the analysis individually, the two sets of coding for each LG were compared. The number of disagreements were counted. The reliability between coders was measured through percent agreement (McHugh 2012). For all variables, the percentage agreement was at least 80% except in one occurrence where it was 73% (for the OLP role variable in one of the four coding pairs). Each coding disagreement was first discussed by the pair of coders and, in most cases, resolved. A few disagreements were brought back to the whole research team for additional discussion and resolution. Once the 182 LGs were coded, and codes recorded on a spreadsheet by our four pairs of coders (each pair coding a fourth of the LGs) and all the disagreements were resolved, the analysis was conducted in the statistical software R.

15.3 Results

The dataset, including 182 LG, was coded to investigate the extent of the inclusion of Ocean-related content in the Swedish science and geography curriculum. The results from the five variables analyzed are presented below (see the Recommended Resources section for a link to the additional table).

15.3.1 Aquatic Focus

Throughout the 182 LGs, only 13 LGs include an aquatic focus. While there are five LGs each in grades 1–3 and grades 4–6 that include an aquatic focus, only 3 LGs include aquatic words in grades 7–9. Grades 10–12 are distinguished by an absence of any aquatic words. In the 13 LGs that include an aquatic focus (Table 15.2), only two of them use the word "Oceans" (sic). Most of the others address concepts related to water that are not directly related to Ocean Literacy.

15.3.2 Terrestrial Bias

Out of 182 LGs, five present a terrestrial bias. One LG in grade 1–3 and two in grade 4–6 specifically mention "animals, plants and fungi" thus excluding significant forms of Ocean life such as algae and photosynthetic microbes. A similar bias is present in grades 10–12 with a LG focusing on "plants and fungi." In grades 7–9, one LG mentions water as a solvent and transporter in "soil and plants" which excludes transport through Ocean circulation and in Ocean organisms such as algae. Since there is no soil and almost no plants in the Ocean, the focus of the LG solely on water as a solvent and transporter in soil and plants gives an incorrect notion that the water cycle and the role of water in the life cycle of photosynthetic organisms occurs only on land. In fact, water is critically important as a solvent and transporter on the entire planet, notably on the 71% of the planet covered by Ocean.

Table 15.2 List of the LGs with aquatic focus. The words underlined constitute the aquatic reference

Grade	Learning goals with aquatic focus
1–3	• How materials can be sorted by some properties, such as appearance, whether they are magnetic, and whether they **float or sink in water**. How the materials can be recycled.
	• Some mixtures and how they can be separated into their various constituents, for example by **evaporation and filtration**.
	• The **different forms of water**: solid, liquid and gas. Evaporation, boiling, condensation, melting and solidification.
	• Spatial conditions in nature and environment for human population and settlement, for example land, **water** and climate.
	• Names and locations of the world's continents and **oceans** as well as of countries and places that are significant to the student.
4–6	• Common weather phenomena and their causes, such as how winds and **precipitation** occur.
	• Classification of substances and materials based on the properties **solubility**, conductivity, acidic or basic. The properties and **cycle of water**.
	• Earth's natural resources, such as **water**, farmland, forests and minerals. Where on earth different natural resources are found and how people's use of the resources affects the landscape and human habitats.
	• Name and location of geographical objects in Sweden, Europe and the world. A selection of **oceans, lakes, rivers**, mountains, deserts, regions, countries and cities.
	• Some basic causes and consequences of poverty and ill health, for example lack of access to education, health care and **water**. Work to improve people's living conditions, for example through the United Nations (UN)
7–9	• **Water** as a solvent and transporter of substances, for example in soil, plants and the human body.
	• Some chemical processes in soil, air and **water** and their connection to questions about the environment and health, for example the greenhouse effect, **water purification** and the spread of environmental toxins.
	• Spatial distribution of vulnerable places on earth, for example areas at risk of recurring **floods**, droughts or earthquakes. How social planning and human activities affect vulnerability and how risks can be prevented.

15.3.3 Type of LG

About two-thirds of the LGs across grade bands address scientific content (content LGs) while a third address the scientific practices (method LGs). The proportion of these two types of LGs changes across grade bands (Fig. 15.1). In the early grade bands, the focus is mainly on scientific content with few method LGs. As the students advance, the percentage of method LGs increases to constitute almost half of the LGs in grades 10–12.

15.3.4 OLP Roles

In this variable, for each LG, we identify the role that the OLP plays in attaining the LG. Table 15.3 presents the distribution of the OLP roles across all the LGs. About half of all LGs were coded as "example" in terms of the OLP role (e.g., "Some stories about how scientific knowledge has developed"). This means that in 50% of the LGs, the OLP could serve to exemplify these LGs in useful ways. In more than 20% of the LG, addressing the OLP is "essential," (e.g., "Natural scientific theories about the origin and development of life"). Seven LGs were specifically place-based (e.g., "Food chains and cycles in the local environment"). Twenty-three percent of the LGs address topics that are not substantively related to the OLPs. In this way the OLPs do not play a role in attaining the specific LG (e.g., in the LG "Some of the human organs, their names and general function").

When exploring the OLP roles in relation to the types of LG, we noticed some differences. In most of the method LGs (81%), the OLP can be a useful example (Fig. 15.2). However, in only a few method LGs (5%) are the OLP seen as essential. Also, the OLPs play no role in only 10% of the method LGs. In content LGs, this strong difference between example and essential does not exist. The OLP roles of essential, example and no role for content LGs are all situated around 30% with a slight dominance of example with 37%.

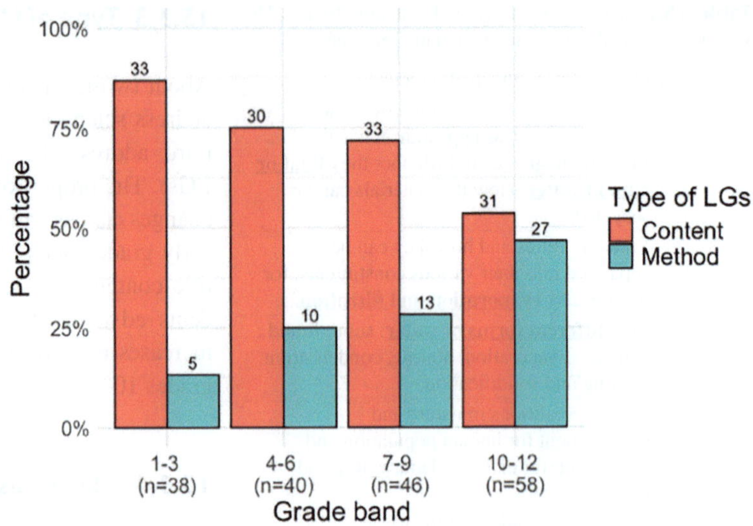

Fig. 15.1 Percentage of LG by type and grade band

Table 15.3 Number (and percentage) of LGs per grade band for the different OLP roles

	OLP role			
Grade band	Essential	Example	No role	Place-based
1–3 (n = 38)	5 (13.16%)	15 (39.47%)	12 (31.58%)	6 (15.79%)
4–6 (n = 40)	13 (32.50%)	19 (47.50%)	7 (17.50%)	1 (2.50%)
7–9 (n = 46)	14 (30.43%)	23 (50.00%)	9 (19.56%)	0 (0.00%)
10–12 (n = 58)	9 (15.52%)	35 (60.35%)	14 (24.14%)	0 (0.00%)
Total (n = 182)	41 (22.53%)	92 (50.55%)	42 (23.08%)	7 (3.85%)

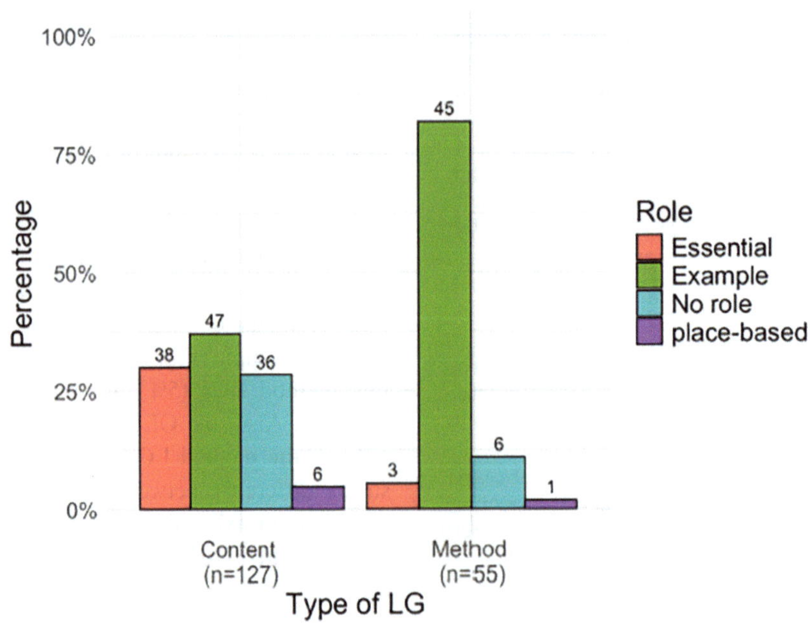

Fig. 15.2 Percentage of LGs by OLP role and LG type

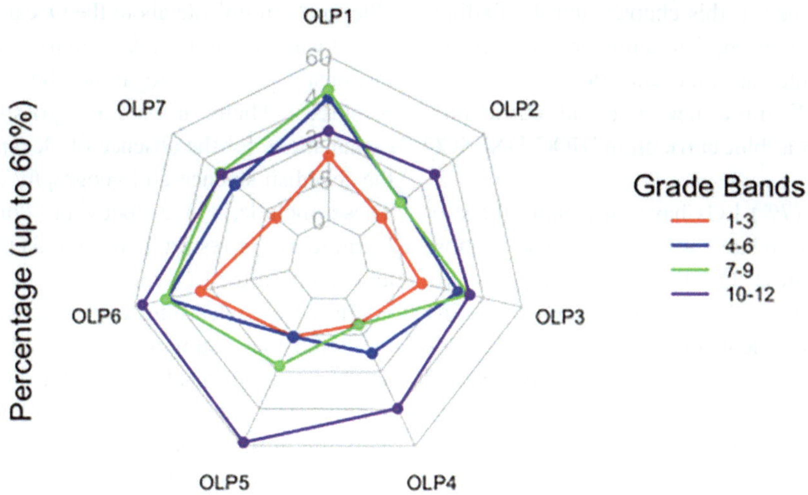

Fig. 15.3 Percentage of LGs aligning with each of the seven OLP

15.3.5 OLPs and LGs Alignment

When exploring the alignment of the LGs with the OLP, our results reveal that the OLP ideas are relevant across grade bands (Fig. 15.3). All the OLPs align with at least 10% of the LGs in any given grade band. The percentage of OLPs aligning with LGs goes all the way to almost 60% with OLP5 in grades 10–12 having the highest proportion. Figure 15.3 illustrates that the higher the grand band, the more the LGs require inclusion of the content of each OLP.

15.4 Discussion

The letter from UN Secretary General's Special Envoy for the Ocean, Peter Thomson (2023) to the IOC/UNESCO Executive Secretary addressed,

> *...the need for inclusion of Ocean Literacy in the curriculum frameworks of UN Member States. Despite the fact that the Ocean is inextricably connected to human health, provides the basis of climate regulation and our food systems, and consistently inspires art, literature and cultural practices across the world, it is currently remarkably misunderstood, under-represented or absent from school curricula... I believe we must acknowledge that the Ocean is not sufficiently emphasized in formal education...(and) the need for urgent attention to be given to the inclusion in our academic institutions of the Ocean's central place in human and planetary well-being.* (Thomson 2023, paras. 1, 3)

Around the world, public schools represent a key mechanism to ensure equitable education (UNESCO 2017) across racial, gender, economic, and geographic boundaries. If understanding the importance of the health of the Ocean, and its influence on planetary health, is not included in school curriculum, it will likely not be taken up at scale in the general population, except in indigenous coastal communities where this deeply held understanding is pre-existing (Glithero et al. 2024).

With its 3218 km of coastline and hundreds of years of rich fishing and maritime history, Sweden is among the Nordic countries with a common vision of becoming "the most sustainable and integrated region in the world" by 2030 (Sepponen et al. 2021, p. 3). The Nordic countries acknowledge that securing Ocean sustainability is a critical milestone toward achieving this vision. However, the nearly complete absence of Ocean-related content from the Swedish science and geography curriculum is counterintuitive to the Nordic countries vision around sustainability, and is a dramatic proof point for Thomson's call for urgent attention. The methodology described in this chapter and the findings resulting from its application in the Swedish context constitute an important steppingstone for including OL in a new national curriculum, referred to as a "blue curriculum" (IOC-UNESCO 2022).

Only 13 (7%) LGs have an aquatic focus or include water-related words; only two of those directly address Ocean-related concepts (see Sect. 15.3.1.). The aquatic focus steadily decreases as students move up through the grade bands until there are only three aquatic-focused LGs in grades 7–9, and not one in any discipline of science or geography in grades 10–12. Of the LGs with an aquatic focus, only two specifically mention the Ocean or any aquatic environment. Those two, both at primary grades, refer to "Oceans" (sic) simply as the label for a geographical feature, which only addresses a small part of one of the 45 fundamental concepts of Ocean Literacy (OLP 1, Fundamental Concept a.), and represents a low level of declarative knowledge (recall of simple facts). The other aquatic focused LGs address concepts related to water that are indirectly or not at all relevant to Ocean Literacy. It is also notable that the two LGs that refer to the Ocean use the plural form (Oceans), which is in contradiction to language of the Thomson letter and the accepted usage established in OLP 1: Earth has one big Ocean. This grammatical detail is foundational to the idea that there is one interconnected body of water on Earth that all organisms, including humans, depend on and share. This leads to students' understanding of the global movement of water, carbon, energy, and pollutants through the water cycle and Ocean circulation.

In line with previous research (e.g., Castle et al. 2010; Costa et al. 2024; Gough 2017; McPherson et al. 2018), the Ocean is underrepresented and somewhat misrepresented in the Swedish curriculum, making it highly unlikely, as stated by Thomson (2023), that the vast majority of students will ever attain the ultimate goals of OL: to understand the Ocean's influence on you and your influence on the Ocean; to understand the Essential Principles and Fundamental Concepts about the Ocean; to be able to communicate about the Ocean in a meaningful way; and to be able to make informed and responsible decisions regarding the Ocean and its resources. There are several possible, partial explanations for the absence of Ocean content in the Swedish science and geography curriculum. These include, (1) a focus on "place-based" learning, (2) terrestrial bias, and (3) the structure and nature of the LGs themselves.

Place-based learning experiences emphasize the belief that young students should learn first and foremost in and about their local place so that learning is contextualized and relevant to the daily life of the learner. When curriculum frameworks and learning experiences are designed to be mostly place-based, only students living on or near the coastline or other aquatic bodies will have the opportunity to become Ocean literate (Castle et al. 2010). We anticipated that this could be an unintended deterrent to Ocean Literacy in the Swedish science and geography curriculum, and so our codebook included a variable to identify "place-based" LGs. However, only seven LGs (3.85%) are place-based (see Sect. 15.3.4.).

Terrestrial bias in our "modern," formal education system, unlike in many indigenous coastal, seafaring and maritime cultures, has for generations focused exclusively on the science of the land. In analyses of the U.S. Next Generation Science Standards, many examples of terrestrial bias were identified, such as:

- Referring to living things or organisms as "plants and animals" or in the case of the Swedish Learning Goals, "plants, animals and fungi" (excludes algae and photosynthetic microbes, the most abundant life in most of Earth's living space);
- Describing the carbon cycle and/or climate change without referring to carbon sequestration in the Ocean (the largest reservoir of carbon on Earth is in the deep Ocean).

To our surprise, only five of the analyzed LGs included terrestrial bias (see Sect. 15.3.2). While those five could be improved to be more accurate, they do not account for the nearly complete absence of Ocean concepts in the remaining LGs.

We did not initially anticipate that the structure and nature of the LGs would be the greatest factor resulting in the absence of Ocean concepts—and the absence of terrestrial bias. Most of the analyzed LGs are broad, vague and focus on general competences having to do with using scientific knowledge (see Sect. 15.2.2). Such broad and vague LGs that include few or no examples, do not explicitly include or exclude the Ocean nor do they display terrestrial bias. The terrestrial bias is not actually absent, but rather, implicit. When a learning goal refers broadly to "ecosystems," it is left up to teachers (and instructional materials developers) to select which ecosystems to teach about. As research in various countries has shown, teachers acknowledge their own lack of Ocean knowledge and feel ill-equipped to teach about Ocean ecosystems, leading to leaving the Ocean out of their teaching practice if it is not clearly specified in the curriculum (Fauville et al. 2018; McPherson et al. 2020; Costa et al. 2024). Teachers need to be provided with adequate, specific training, to equip them with the appropriate content and pedagogical knowledge needed to teach about the Ocean. The importance of teacher preparation and ongoing professional learning about the Ocean has already been argued in the Ocean Literacy community (e.g., Freitas et al. 2022; Glithero et al. 2024).

In summary, the broader and more vague the LGs, the less likely it is that the Ocean will be explicitly included. The less explicit the Ocean concepts are, the less likely it is that teachers and materials developers will include them in learning experiences. To change the status quo absence of Ocean concepts in the enacted science and geography curriculum, Ocean concepts must be explicit in the intended curriculum.

The broad and vague nature of the LGs provides teachers and materials developers with expansive freedom to interpret how and what aspects of them should be taught to students. Empowering teachers with this autonomy and creative license also inevitably results in inequity and a lack of quality control. This is potentially in contradiction with Swedish goal of equitable, high quality learning opportunities for all students. Because the LGs are broad, vague and subject to individual interpretation (by educators and researchers), it is unclear if the intent is to teach these ideas in a comprehensive way or to provide learners with any relevant example of them as an introduction. We found that many of the LGs could easily be divided into several discrete and more specific concepts, and that most could, in their current form, be taught in a wide variety of ways.

The combined results from the variables OLP roles and aquatic focus, demonstrates the importance of understanding the Ocean to understand general science concepts, and the degree to which Swedish students are not provided the opportunity to learn about the Ocean. At grades 1–3, Ocean concepts are essential or useful examples in 53% of the LG, and are explicitly mentioned in only 13% of the LG. At grades 4–6, Ocean concepts are essential or useful examples in 80% of the LG, and are explicitly mentioned in only 13% of the LG. At grades 7–9, Ocean concepts are essential or useful examples in 80% of the LG, and are explicitly mentioned in only 7% of the LG. Most dramatically in grades 10–12, Ocean concepts are essential or useful examples in 76% of the LG, and are explicitly mentioned in none of the LG. Well over half, and from grades 4–12, over three-fourths of the Swedish science and geography curriculum require some understanding of the Ocean, and yet the Ocean is not mentioned at all, except as a labels for certain features on a map.

We also note that as described in Sect. 15.3.5, all seven OLPs are essential to or useful examples of many LGs at all grade bands. We combine the roles of "essential" and useful "example" here to illuminate the importance of Ocean concepts in the general science and geography curriculum (e.g., concepts about the nature and structure of ecosystems, the evolution and adaptation of whole phyla of organisms, carbon sequestration). When Ocean examples are categorically excluded from the teaching of science and geography LGs, Swedish students will have an "Ocean blind spot" that challenges their understandings of the Ocean planet on which they live. Our findings support the claim that learners cannot be science literate without being Ocean literate (Strang et al. 2007). Students cannot, for example, understand primary productivity on Earth unless they learn about chemosynthesis in deep sea vents and the photosynthesis taking place in the Ocean, along with the typical instruction about trees. The importance of OL to be science literate was illustrated in Fauville and colleagues' study (2017) where Swedish high-school students learned and talked about Ocean acidification. The students' questions were analyzed and revealed that students had an incomplete understanding of photosynthesis. Their grasp of this phenomenon was limited to the terrestrial environment while they wondered if and how photosynthesis took place in the Ocean.

15.5 Conclusion

In addition to contributing to the dialogue around the Swedish science curriculum, we hope that our work may inspire the international community of Ocean educators and researchers to analyze and compare the presence of Ocean Literacy in the curriculum of many countries. It is important to note that the inclusion of Ocean-related content in national curricula is equally relevant for landlocked countries or regions. As argued by IOC-UNESCO (2022), it is important for "landlocked, rural and remote regions with little or no contact with the sea, but still influenced by the Ocean" (p. 100). The weather, climate, availability of fresh water and conditions for life to exist in every country are influenced by the Ocean. We all live on an Ocean planet.

We recognize that educational standards, curriculum frameworks and learning goals vary widely between countries (Pocze et al. 2020). We are convinced, however, that the codebook and methodology we have developed and tested can be adapted, improved, and used in a wide variety of settings, especially with the aid of international collaboration and thought partnership. Through collaboration, we hope that during and beyond the Ocean Decade, we can shine a bright light on the Ocean blind spot that has dimmed understanding around the world of the critical importance of the health of our one Ocean. We anticipate that useful comparisons and best practices will emerge that can accelerate our response to Thomson's call and the growing momentum toward global Ocean Literacy.

15.5.1 Summary

Key concepts explored in this chapter include:

- Despite Sweden's commitment to the Ocean Decade, an analysis of 182 learning goals in the national science curriculum found that Ocean-related content is nearly nonexistent, creating an "Ocean blind spot" for students.
- Formal education plays a critical role in advancing Ocean Literacy. The study reinforces the argument that integrating Ocean Literacy into national curricula is essential for ensuring that students develop a foundational understanding of the Ocean's role in climate, biodiversity, and planetary health.
- Broad and vague learning goals limit the inclusion of Ocean Literacy. Many learning goals lack specificity, allowing for interpretation by educators and materials developers, which often leads to the exclusion of Ocean-related content in instruction.
- Terrestrial bias contributes to the Ocean's omission in science education. The curriculum often prioritizes terrestrial ecosystems over marine environments, reinforcing a worldview that neglects the Ocean's critical role in sustaining life on Earth.
- A global approach to curriculum analysis can strengthen Ocean Literacy. The methodology developed in this study offers a scalable approach for evaluating Ocean Literacy in curricula worldwide, supporting international collaboration to promote a "blue curriculum" in education systems across different countries.

15.5.2 Recommended Resources

The following resources provide further insights and support on this topic:

- The final codebook, the dataset, and additional results: https://osf.io/3cd86/?view_only=ee5123446cf5487280e640750a7fb431
- The Ocean Literacy Scope and Sequence for Grades K–12: https://www.marine-ed.org/ocean-literacy/scope-and-sequence
- The Ocean Literacy Guide: https://www.marine-ed.org/ocean-literacy/overview
- Ocean Literacy and Next Generation Science Standards (NGSS): https://www.marine-ed.org/ocean-literacy/ngss-alignment

Acknowledgements Thank you to the Swedish Institute for the Marine Environment (Havsmiljöinstitutet) for its financial support. Thank you to the Centre for Sea & Society, at the University of Gothenburg for financially supporting our team's participation in the European Marine Science Educators Association Conference in 2024.

References

Castello L (2021) Science for conserving Amazon freshwater ecosystems. Aquat Conserv Mar Freshwat Ecosyst 31(5):999–1004. https://doi.org/10.1002/aqc.3615

Castle Z, Fletcher S, McKinley E (2010) Coastal and marine education in schools: Constraints and opportunities created by the curriculum, schools, and teachers in England. In: *Ocean yearbook 24*, pp 425–444

Cava F, Schoedinger S, Strang C, Tuddenham P (2005) Science content and standards for Ocean Literacy: A report on Ocean Literacy. https://coexploration.org/Oceanliteracy/documents/OLit2004-05_Final_Report.pdf

Costa AC, Freitas D, Santos AI, Botelho AZ, Parente MI, Behr A, Rodrigues AM, Guerra H, Cascalho J, Mendes A (2024) Where is Ocean Literacy in Oceanic islands' schools? The Azores case. Mar Policy 163:106062. https://doi.org/10.1016/j.marpol.2024.106062

Eddy TD (2014) One hundred-fold difference between perceived and actual levels of marine protection in New Zealand. Mar Policy 46:61–67. https://doi.org/10.1016/j.marpol.2014.01.004

Fauville G (2017) Questions as indicators of Ocean Literacy: Students' online asynchronous discussion with a marine scientist. Int J Sci Educ 39(16):2151–2170. https://doi.org/10.1080/09500693.2017.1365184

Fauville G, McHugh P, Domegan C, Mäkitalo Å, Friis Møller L, Papathanassiou M, Alvarez Chicote C, Lincoln S, Batista V, Copejans E, Crouch F, Gotensparre S (2018) Using collective intelligence to identify barriers to teaching 12–19 year olds about the Ocean in Europe. Mar Policy 91:85–96. https://doi.org/10.1016/j.marpol.2018.01.034

FORMAS (2020) *Ett svenskt bidrag till FN:s årtionde för havsforskning för hållbar.* https://formas.se/download/18.23ea31d174bf101a2f136fc/1601381628002/r3-2020-ett-svenskt-bidrag-till-fns-artioende-havsforskning.pdf

Freitas C, Bellgrove A, Venzo P, Francis P (2022) Towards a 2025 National Ocean Literacy Strategy: Current status and future needs in primary education. Front Mar Sci 9:883524. https://doi.org/10.3389/fmars.2022.883524

Glithero LD, Bridge N, Hart N, Mann-Lang J, McPhie R, Paul K, Peebler A, Wiener C, Yen C, Kelly R, McRuer J, Hodgins D, Curtin F (2024) Ocean Decade vision 2030 white papers—Challenge 10: Restoring society's relationship with the Ocean. UNESCO-IOC. (The Ocean Decade Series, 51.10.), Paris. https://doi.org/10.25607/ekwn-wh61

Gough A (2017) Educating for the marine environment: Challenges for schools and scientists. Mar Pollut Bull 124(2):633–638. https://doi.org/10.1016/j.marpolbul.2017.06.069

Hoffman M, Barstow D (2007) Revolutionizing Earth system science education for the 21st century: Report and recommendations from a 50 state analysis of Earth science education standards. TERC, Cambridge, MA

IOC-UNESCO (2022) A new blue curriculum: A toolkit for policy-makers. UNESCO. https://unesdoc.unesco.org/ark:/48223/pf0000380544

McHugh ML (2012) Interrater reliability: The kappa statistic. Biochem Med 22(3):276–282. https://doi.org/10.11613/BM.2012.031

McPherson K, Wright T, Tyedmers P (2018) Examining the Nova Scotia science curriculum for international Ocean Literacy principle inclusion. Int J Learn Teach Educ Res 17(11):1–16. https://doi.org/10.26803/ijlter.17.11.1

McPherson K, Wright T, Tyedmers P (2020) Challenges and prospects to the integration of Ocean education into high school science courses in Nova Scotia. Appl Environ Educ Commun 19(2):129–140. https://doi.org/10.1080/1533015X.2018.1533439

National Marine Educators Association (n.d.) Introduction to the Alignment of Ocean Literacy to the Next Generation Science Standards (NGSS). https://www.marine-ed.org/Ocean-literacy/ngss-alignment

National Science Teaching Association & National Marine Educators Association (2025) The importance of teaching and learning about the Ocean: Joint position statement. https://www.nsta.org/sites/default/files/NSTA-NMEA_Joint-Position-Statement_The-importance-of-teaching-learning-about-the-ocean_FINAL.pdf

Neuendorf KA (2017) The content analysis guidebook. SAGE Publications, Inc. https://doi.org/10.4135/9781071802878

Pazoto CE, Silva EP, Duarte MR (2022) Ocean Literacy in Brazilian school curricula: An opportunity to improve coastal management and address coastal risks? Ocean & Coastal Management 219:106047. https://doi.org/10.1016/j.ocecoaman.2022.106047

Pocze B, Tasiopoulou E, Copejans E (2020) Ocean Literacy for all – curriculum analysis. European Schoolnet, Brussels. https://files.eun.org/SciEduDept/EU4Ocean/EU4Ocean-curriculum-analysis-EUN-FINAL.pdf

Ryabinin V, Barbière J, Haugan P, Kullenberg G, Smith N, McLean C, Troisi A, Fischer A, Aricò S, Aarup T, Pissierssens P, Visbeck M, Enevoldsen HO, Rigaud J (2019) The UN Decade of Ocean Science for Sustainable Development. Front Mar Sci 6:470. https://doi.org/10.3389/fmars.2019.00470

Sepponen S, Hjelt M, Kuusela-Opas E, Pakkala A, Suominen F, Värre U (2021) *Sustainable Ocean economy: Mapping of Nordic strongholds.* https://www.nordicinnovation.org/Oceaneconomy mapping

Smith PK, Smith C, Osborn R, Samara M (2008) A content analysis of school anti-bullying policies: Progress and limitations. Educ Psychol Pract 24(1):1–12. https://doi.org/10.1080/02667360701661165

Strang C, deCharon A, Schoedinger S (2007) Can you be science literate without being Ocean literate? Current: J Mar Educ 23(1):7–9. https://doi.org/10.5281/ZENODO.30563

Thomson P (2023) Call to include Ocean Literacy in school curricula by 2025 as a part of the education for sustainable development. https://unesdoc.unesco.org/ark:/48223/pf0000386655

UNESCO (2017) A guide for ensuring inclusion and equity in education. UNESCO. https://doi.org/10.54675/MHHZ2237

Author Contributions and Biographical Profiles

[1]*Géraldine Fauville: Co-conceptualization; Writing—Original Draft, Review, and Editing; Methodology—Investigation, Validation, Formal Analysis, Data Curation, and Visualization; Project Administration—Supervision; Funding Acquisition.

[2]Craig Strang: Co-conceptualization; Writing—Original Draft; Methodology—Investigation.

[3]Ewa Skantz: Co-conceptualization; Writing—Original Draft; Methodology—Investigation.

[4]Jens Bjelvenmark: Co-conceptualization; Writing—Review and Editing; Methodology—Investigation.

[5]Anne Solli: Co-Conceptualization; Writing—Review and Editing; Methodology—Investigation.

*Corresponding Author at geraldine.fauville@gu.se

Géraldine Fauville, PhD, is an Associate Professor and Senior Lecturer at the Department of Education, Communication and Learning, University of Gothenburg, Sweden. With a double background in marine biology and in education and IT, her research focuses on how immersive media, such as virtual reality, contribute to marine environmental education and promote Ocean and environmental literacies.

Craig Strang is Associate Director Emeritus of the Lawrence Hall of Science at the University of California, Berkeley. He co-led the United States Ocean Literacy Campaign and the development of the Ocean Literacy Framework, including the Alignment of Ocean Literacy with the Next Generation Science Standards. He is currently a consultant co-leading the statewide California Campaign for Outdoor Learning.

Ewa Skantz, PhD, is a senior lecturer at the Department of Education, Communication and Learning, University of Gothenburg. An overall research interest is how digitalization changes the conditions for social interaction, communication, teaching, and learning in school. Her research interest is Ocean Literacy and teachers' professional digital competence. Currently, she is working in a project: Advancing teacher professional digital competence (funded by the Swedish Research Council).

Jens Bjelvenmark, PhD, is a marine biologist, certified biology and chemistry teacher, and head of biology and the science program at Gullmarsgymnasiet, a High School with a marine profile. He has been awarded for his teaching by the Swedish National Academy of Sciences and selected by the Swedish government as a member of a STEM delegation aiming to enhance interest in science, technology, engineering and mathematics among young people.

Anne Solli is Senior Lecturer in Science education at the Department of Pedagogical, Curricular and Professional Studies at the Faculty of Education, University of Gothenburg. She is involved in the science teacher training program. Her research interest includes socio-scientific literacies such as young citizen agency and how teachers engage with global climate change in a networked society.

16 Development of an Aquaponics STEAM Program for Improving Ocean Literacy

Tsuyoshi Sasaki and Kota Sawada

Abstract

Proposed solutions to Ocean-related issues are linked to social values, environmental behavior, and marine governance, all of which necessitate individuals to possess Ocean Literacy (OL). The different frameworks and approaches to OL have been proposed. However, this diversity can cause ambiguity and complexity, and a comprehensive synthesis of the progress made is needed. Therefore, it is essential to provide specific examples that illustrate the relationship between OL and particular fields. This study examines an Aquaponics STEAM Program that cultivates plants and fish in a sustainable circular model for middle school students aged 13–14, aiming to enhance scientific inquiry skills and relational values (RVs) related to the OL components based on the theory of change. Results show significant differences in scientific inquiry skills between Group A, engaged in aquaponics activities, and Group B, which was not. Pre-survey analyses revealed a conflation of intrinsic and instrumental values, while post-survey results indicated that Group A could distinguish RVs, instrumental values, and intrinsic values, suggesting that hands-on experiences fostered this differentiation. Aquaponics activities can effectively cultivate OL frameworks and competencies aligned with school curriculum goals.

Keywords

Ocean Literacy (OL) · Scientific inquiry skills · Theory of change · Nature-connectedness · Ocean Decade · Teacher training program · National curriculum · Integrated study · VUCA

16.1 Introduction: Program for Improving Ocean Literacy

The natural environment is undergoing significant changes on a global scale. Abnormal weather occurs worldwide, including increased frequency of wildfires, larger typhoons and hurricanes, and sea temperatures exceeding 30 degrees Celsius (86 degrees Fahrenheit) in certain regions. These changes, in turn, affect the distribution of marine life. The primary drivers of these Ocean changes are land-based and include population growth, increasing urbanization, rising individual consumption, and climate change (Nash et al. 2017). Modern society is often described as being in an era of VUCA (Volatility, Uncertainty, Complexity,

T. Sasaki (✉) · K. Sawada
Tokyo University of Marine Science and Technology, Tokyo, Japan
e-mail: t-sasaki@kaiyodai.ac.jp; s184013@edu.kaiyodai.ac.jp

T. J. Kennedy (ed.), *Ocean Literacy: The Foundation for the Success of the Ocean Decade, Volume I*, Sustainable Development Goals Series, https://doi.org/10.1007/978-3-031-98145-6_16

Ambiguity), where numerous unresolved issues are accumulating. The causes of these challenges have become intricate, leading us into a challenging period where straightforward scientific and technological solutions are inadequate to address them. Proposed solutions to many of the problems facing the Ocean are closely linked to social values, socio-environmental actions, and social and Ocean governance (United Nations Environment Programme 2024), and to achieve these solutions, it is essential that many people around the world have Ocean Literacy (OL). The ability to comprehend and respond to the interactions between the Ocean and humans is referred to as OL. In the coming era, every individual needs to become Ocean literate, capable of understanding and acting on these interactions to preserve the Earth's environment. Currently, OL has emerged as one of the key objectives of the United Nations Decade of Ocean Science for Sustainable Development (UN Ocean Decade), actively supported by UNESCO's Intergovernmental Oceanographic Commission (IOC) (Guan et al. 2023).

OL education and research have developed since the 2000s, following the establishment of OL in the USA (Plankis and Marrero 2010). Since then, OL has transcended the borders of the United States and is increasingly recognized globally. Various associations have been established to promote and enhance OL in society, including the European Marine Science Educators Association (EMSEA), the Canadian Network for Ocean Education (CaNOE), the International Pacific Marine Educators Network (IPMEN), the Australian Association for Environmental Education (AAEE), RELATO (Latin American Education Network for the Ocean), and the Asia Marine Educators Association (AMEA) (Santoro et al. 2024). During the first decade of OL, significant research has concentrated on knowledge and awareness. Additionally, educational materials were created that emphasize OL principles. It arose from the recognition of the deficiency of Ocean-related subjects in formal education and the necessity for a more comprehensive framework to promote the integration of Ocean sciences into national and state curricula. The growing acknowledgment of OL's potential practical role and value has established OL as a central element of international frameworks and policy objectives (McKinley and Burdon 2020).

OL is increasingly regarded as a solution for Ocean sustainability policy, with the potential to drive societal behavior change. For instance, "OL as a concept and approach is radically evolving from being a tool to be applied in formal education and training contexts to a tool and an approach for society as a whole, aimed at triggering actions towards Ocean sustainability" (Shellock et al. 2024). Recently, it has become evident that a connection with the Ocean plays a crucial role in fostering positive attitudes and enacting change to conserve the Ocean environment (Sasaki 2022). Consequently, research has concentrated on attitudes, communication, behavior, and activism, in addition to knowledge and awareness (Brennan et al. 2019). Additionally, Santoro et al. (2024) define four skills based on the theory of change and advocate for implementing OL education in public and social education to cultivate these skills for collective well-being and sustainable development, as outlined in the following.

- Knowledge and awareness: Through problem-based, action-based, and wonder-based knowledge, provides a foundation for young people to engage with and debate Ocean topics and issues.
- Nature-connectedness: Fostered by spending time and learning in nature, strengthens positive attitudes and emotional connection with the marine environment.
- Values and attitudes: Towards the Ocean, developed alongside peers and teachers in the school environment, guide students to make informed and sustainable choices to safeguard the marine environment, and influence others for collective change.
- Competencies: Enable and empower students to take sustainable action.

Recent studies have proposed several additional dimensions of OL: (i) awareness, (ii) attitude, (iii) behavior, (iv) activism, (v) communication, (vi) emotional connections, (vii) access and experience, (viii) adaptive capacity, and (ix) trust and transparency McKinley et al. 2023). OL is now recognized as both a process and an outcome that fosters a society that understands, values, and cares for the Ocean (Shellock et al. 2024).

Despite this progress, limitations to the evolution of OL remain (McKinley et al. 2023). First, due to the origin and interdisciplinary nature of OL, there are disparities in its conceptualization and application. In just the last 5 years, a range of papers has outlined different frameworks and approaches to OL. This diversity can create ambiguity and complexity for those seeking to operationalize OL. Second, this rapidly evolving field lacks a comprehensive synthesis of the work and progress achieved to date (Shellock et al. 2024). Additionally, there is a lack of specific information regarding the contents of OL skills, and it remains unclear how activities in research or industrial fields influence OL skills, indicating that further discussion is needed. The contents of Challenges 1 to 8 in the UN Ocean Decade Vision 2030 White Papers, published by IOC, were reviewed. However, there was little specific information regarding OL education, leading us to conclude that the previous analysis remains valid. Consequently, it is essential to provide specific examples that illustrate the relationship between OL and particular research and industrial fields.

This research presents a concrete effort to link OL concept with Challenge 3: Sustainably Nourish the Global Population, as outlined in the UN Ocean Decade Vision 2030 White Papers. We developed teaching materials called the "Aquaponics STEAM Program" and conducted classes to assess changes in the OL components based on the theory of change, especially "nature-connectedness", "values and attitudes", "competencies". This educational program is expected to enable learners to master the OL components while relating them to "a circular food system that integrates aquatic and terrestrial life" of Challenge 3, in accordance with OL Basic Principle 6.g: "to develop OL individuals who are considerate of natural resources and the marine environment" (Intergovernmental Oceanographic Commission, 2024).

16.2 Purpose and Preparation for the Study

This study examined changes in participants' OL before and after the implementation of the Aquaponics STEAM Program, and between Group A which engaged in aquaponics activities and Group B which did not, using a newly developed method to evaluate OL components related to nature-connectedness, values and attitudes, and competencies.

The goal of this program is to cultivate individuals who, in accordance with the fundamental principles of aquatic marine environmental literacy, can observe their local aquatic environment (thinking skills), consider various issues collaboratively (think skills), understand the comprehensive knowledge of aquatic environmental literacy (knowledge and skills), take responsible actions based on a broad perspective (judgment skill), and communicate these insights clearly to a wider audience (expression skill) (Sasaki 2022). As a result of the Canal Study, it has been demonstrated that participating students not only enhance their skills to think, make judgments, and express themselves, but also deepen their connection with the canal, increase their sense of competence, and show a greater willingness to improve the water quality of the canal (Sasaki 2022).

16.3 Method

16.3.1 Study Site

Konan Middle School (ages 13–15) in Minato Ward, Tokyo has been conducting the Canal Study since 2010 in collaboration with Tokyo University of Marine Science and Technology to enhance OL. The Canal Study is part of the formal curriculum under the subject "Integrated Studies". Japanese national curriculum aims to develop three competencies: "knowledge and skills", "skills to think, make judgments, and express themselves", and "motivation to learn and humanity" (Ministry of Education, Culture, Sports, Science and Technology (MEXT) 2024). In Integrated Studies, there is a particular emphasis on fostering "skills of thinking, judgment, and expression".

16.3.2 Development of the Aquaponics STEAM Program

In this study, we developed teaching materials of the aquaponics system. Aquaponics is a technology that cultivates plants alongside fish, where microorganisms break down ammonia contained in fish waste, supplying nitrates that serve as nutrients for plant growth. This process creates a circular aquaculture system. The hotel's food waste is quickly decomposed using BUIK, a high-temperature fermentation bacteria to create eco-feed for aquaponics. The name BUIK is an acronym that stands for Bacteria and the three donors: Mr. Uchishiro, Mr. Ichinose, and Mr. Kondo. This aquaponics system is anticipated to play a significant role in "promoting sustainable fish farming while considering the marine environment" (SDG 14: Life Below Water), avoiding the "protein crisis" (SDG 2: Zero Hunger), "minimizing and effectively utilizing food waste" (SDG 12: Responsible Consumption and Production), and "a circular food system that reduces loss and waste and integrates aquatic and terrestrial life" (UN Ocean Decade: Challenge 3).

The STEAM program aims to develop skills in Science, Technology, Engineering, and Mathematics, as well as OL (nature connectedness, values and attitudes, competencies), which is related to scientific inquiry skills and relational values (described below) that are positioned as part of the liberal arts. The purpose of this study was to determine whether middle school students' scientific inquiry skills and relational values related to OL components improve through participation in aquaponics activities that utilizes eco-feed. The target species for the aquaponics were yellowfin goby "MAHAZE" (*Acanthogobius flavimanus*) for the fish component and shoreline purslane (*Sesuvium portulacastrum*) for plants component. Goby fish inhabit the nearby Takahama Canal and are one of the most familiar fish species among students. The students selected Goby fish for cultivation and conducted the collection of the fish.

16.3.3 Aquaponics STEAM Program Activities and Survey

The classroom activity and survey were conducted from July 6 to November 9, 2023, by dividing 2 grades and 7 classes into two groups: Group A and Group B. Group A consisted of classes A and B of Grades 7 and 8 (ages 13–14), while Group B consisted of classes C and D of Grades 7 and 8 (ages 13–14) (Table 16.1).

Table 16.1 Aquaponics STEAM program activities and survey

	Group A	Group B
	Classes A and B of Grades 7 and 8 (ages 13–14) including 109 students (59 male, 50 female)	Classes C and D of Grades 7 and 8 (ages 13–14) including 75 students (40 male and 35 female)
1st session	Pre-survey of Ocean Literacy, Lecture: the Cross Model Worksheet method (Sasaki 2022), food issues related to the SDGs	
2nd session	Lecture: Nitrogen circulation in aquaponics	
3rd session	Aquaponics activities	Post-survey of Ocean Literacy
4th session	Post-survey of Ocean Literacy	Aquaponics activities

University students in a teacher training program, who were expected to obtain a teaching license, were assigned to each classroom in groups of two to four. During the first questionnaire survey on OL, they read questions aloud to middle school students and provided explanations.

16.3.4 Evaluation Methods for Scientific Inquiry Skills

Competence in OL components based on the theory of change provide a framework for developing skills that empower young people to take effective action. This aligns with UN Sustainable Development Goal 14: Life Below Water, the Competence in Education for Sustainable Development or the OECD PISA Global Competence Framework. Our research focused on scientific inquiry skills as a part of OL and aimed to clarify methods for developing these skills by evaluating the effects of Aquaponics STEAM Program before and after the implementation, as well as the presence (Group A) or absence (Group B) of aquaponics activities (Table 16.1).

Previous studies on scientific inquiry skills often used questionnaires from the researcher's perspective to replicate real inquiry situations. In this study, we conducted aquaponics activities using the Cross Model Worksheet (Sasaki 2022) to enhance these skills. The scientific inquiry process is structured around the following components: (1). Understanding the current situation, (2). Identifying problems, (3). Formulating hypotheses, (4). Methods for hypothesis verification, (5). Results, (6). Discussion, and (7). Future challenges. The Japanese curriculum guidelines (MEXT 2024) specify three competencies, with "skills to think, make judgments, and express themselves" representing scientific inquiry skills. The evaluation criteria for these skills, based on the scientific inquiry process, are as follows.

- Thinking skills: (1). The method for verifying the hypothesis to address the problem and hypothesis formulation is documented. (2). The hypothesis can be established by focusing on the changing variables. (3). The method for verifying the hypothesis includes the conditions outlined in the problem and hypothesis formulation. (4). There is an anticipation of the potential results that can be obtained from the verification of the hypothesis.
- Judgment skills: (1). The consideration is directly related to the hypothesis formulation. (2). The validity of the hypothesis formulation is assessed based on the results.
- Expression skill: (1). The facts derived from the results are expressed in alignment with the hypothesis.

We developed seven items and an assessment method to evaluate scientific inquiry skills using a cross-model worksheet, measuring changes in scientific inquiry skills before and after the STEAM program using the Mann-Whitney *U* test by js-STAR XR+ (Table 16.2).

16.3.5 Evaluation Methods for Relational Values

The global environmental crisis is fundamentally attributed to a loss of connection between humans and nature. Recently, the importance of nature management grounded in values that emphasize this connection has been proposed. These values, which focus on the relationship between humans and nature, are referred to as relational values (hereafter abbreviated as "RVs"). RVs are defined as "preferences, principles, and virtues based on relationships between people and between people and nature, connected to human relationships, policies, and social norms, and including eudaimonic values related to a good life" (Chan et al. 2016). RVs are a crucial concept in Nature's Contributions to People, adopted by the IPBES (Intergovernmental Platform on Biodiversity and Ecosystem Services) as a new approach for maintaining ecosystems (Pascual et al. 2017). We have decided to replace and evaluate the concept of nature-connectedness from OL's Theory of Change, which states that "fostering positive attitudes and emotional connections with the marine environment is achieved by spending time and learning in nature," with RVs.

Additionally, as shown in Table 16.3, RVs are classified into "individual RVs" and "collective

Table 16.2 Seven items to assess three categories of scientific inquiry skills and their evaluation methods

Categories of skills	Contents of skills	Evaluation method
Thinking skills	1. Can identify investigable issues.	Check the description about 1. Understanding the current situation.
	2. Can hypothesize focusing on changing variables.	Check the description about 2. Identifying problems, 3. Formulating hypotheses.
	3. The method for observation/experimentation to verify the issue and hypothesis is entered.	Check the description about 4. Methods for hypothesis verification.
	4. Has a foresight of what results will be obtained through observation/experimentation.	Check the description about 3. Formulating hypotheses and 4. Methods for hypothesis verification.
Judgment skills	1. The consideration responds directly to the hypothesis.	Check the description about 3. Formulating hypotheses, 6. Discussion.
	2. The validity of the hypothesis is examined based on the results.	Check the description about 3. Formulating hypotheses, 5. Results, 6. Discussion.
Expression skills	1. Can illustrate "results" using diagrams or tables. Check the description about "5. Results.	Check the description about 5. Results.

Table 16.3 Relational values are classified into primarily individuals and involving the human collective (Chan et al. 2016)

	Elements	Examples
Relational values (primarily individual)	Individual identity	This place is important to me, to who I am as a person.
	Stewardship eudaimonic	My care for this land fulfills me, helps me lead a good life.
	Stewardship principle/ virtue	Keeping the land healthy is the right thing to do.
Relational values (involving the human collective)	Cultural identity	Place is important to my people, to who we are as a people.
	Social cohesion	Being in nature provides a vehicle for me to connect with people.
	Social responsibility	Caring for ecosystems is crucial to caring for my fellow humans, present and future.
	Moral responsibility to non-humans	Caring for all lifeforms and physical forms is a moral necessity.

RVs". RVs have gained attention due to the perceived limitations of traditional nature conservation, which primarily focused on instrumental and intrinsic values. Instrumental values refer to the joy and satisfaction that individuals derive from experiencing or observing nature and are often assessed in monetary terms. In contrast, intrinsic values represent the worth that nature holds independently of human perception (Chan et al. 2016).

Previous studies suggest that RVs may change with experience (Britto dos Santos and Gould 2018). This study aims to determine whether RVs change with experience in aquaponics. We conducted a comparative analysis using the Mann-Whitney *U* test by js-STAR XR+ to examine changes in RVs, before and after the experience, based on the presence or absence of aquaponics experience. Additionally, to explore changes in the structure of students' values toward nature, we conducted an exploratory factor analysis regarding intrinsic values, instrumental values, and RVs. The number of factors was determined based on the Kaiser criterion (eigenvalue >1). Factors were extracted using the generalized least squares method, and rotation was performed using direct oblimin rotation. In cases of missing values, list-wise deletion was applied for factor analysis. To evaluate internal consistency, we measured Cronbach's alpha, and to examine the validity of the exploratory factor analysis, we conducted the Kaiser-Meyer-Olkin (KMO) sampling adequacy test and Bartlett's test of sphericity. Each analysis was performed using SPSS version 16.0.

Table 16.4 Questionnaire items as a part of OL to find out how learning about canals changes middle school students' awareness of values

Is the canal valuable? (intrinsic values)
Q1: All creatures living in the canal have the right to live. (intrinsic value 1)
Q2: The canal should be safeguarded to preserve nature.(intrinsic value 2)
Does the canal provide advantages? (instrumental values)
Q1: The canal ecosystem is vital for providing food like fish and shellfish. (instrumental value 1)
Q2: The canal ecosystem is vital for preventing and mitigating disasters such as floods, tsunamis, and typhoons. (instrumental value 2)
Q3: The canal ecosystem is vital for water in the canal. (instrumental value 3)
Q4: The canal ecosystem is important in maintaining the food chain for plants and animals. (instrumental value 4)
Are we connected to the canal? (relational values)
Q1: The canal is an important place to me. (individual identity: RV1)
Q2: The canal is an important place for residents. (cultural identity: RV2)
Q3: My work with the canal has connected me with others. (social cohesion: RV3)
Q4: Caring for the canal ensures the Well-being of current and future generations. (stewardship eudaimonic: RV4)
Q5: We have a moral duty to protect it and its wildlife. (moral responsibility to non-humans: RV5)
Q6: Protecting the canal enhances our lives. (social responsibility: RV6)
Q7: Keeping the canal healthy is important. (stewardship principle/virtue: RV7)

To clarify changes in RVs between the local canal and students, as well as shifts in their values regarding nature, the questionnaire items included items related to the RVs in Table 16.3, as well as "Instrumental Values" and "Intrinsic Values" (Table 16.4). Based on Uehara et al. (2020), we used a Likert scale (1 = not at all applicable, 5 = very applicable) for the questions. We conducted the questionnaire survey before and after the implementation, as well as the presence (Group A) or absence (Group B) of aquaponics activities previously described in Table 16.1.

16.4 Results

16.4.1 Changing Scientific Inquiry Skills

The total scores of Group A and Group B on the pre- and post-questionnaires for scientific inquiry skills (thinking, judgment, and expression) are presented in Table 16.5.

A Mann-Whitney *U* test compared the scientific inquiry skills scores of Group A and Group B, focusing on skills of thinking, judgment, and expression as well as the total score across these three areas in the pre- and post-questionnaire surveys. The analysis revealed no significant difference in the pre-questionnaire scores (n.s.), and between the pre-questionnaire and post-questionnaire scores for Group A (n.s.). However, Group A's scores in scientific inquiry, thinking and judgment skills are significantly higher than Group B's in the post-questionnaire, with *p*-values of 0.0751, 0.0040 and 0.038, respectively (Table 16.6).

16.4.2 Changing Relational Values

Results of the Mann-Whitney *U* Test

For Groups A and B, the Mann-Whitney *U* test conducted on the pre-test and post-test results for RVs indicated no significant differences across any of the items(n.s.).

Results of Exploratory Factor Analysis

For Group A in the pre-survey, the Cronbach's alpha coefficient was 0.91, the KMO measure of sampling adequacy was 0.89, and the Barlett test of sphericity was $p < .001$. An exploratory factor analysis extracted three factors (Table 16.7). The first factor was composed of items related to RVs, so it was named the "RVs" factor. The second factor primarily consisted of items related to intrinsic values but also included items 3 and 4 of instrumental values; therefore, it was labeled the "Intrinsic Values/Instrumental

Table 16.5 Number of correct answers to 7 questions on scientific inquiry skills (Thinking skills 1–4, Judgment skills 1–2, Expression skill 1)

	Number of valid responses	Score of thinking skill1	Score of thinking skill2	Score of thinking skill3	Score of thinking skill4	Total score of thinking skills	Score of Judgment skill1	Score of Judgment skill2	Total Score of judgment skills	Expression skill	Total score
Pre-survey of Group A	118	87	81	66	56	290	78	53	131	16	437
(%)		73.7%	68.6%	55.9%	47.5%		66.1%	44.9%		13.6%	
Post-survey of Group A	109	92	76	67	63	298	78	55	133	7	438
(%)		84.4%	69.7%	61.5%	57.8%		71.6%	50.5%		6.4%	
Pre-survey of Group B	77	57	51	47	35	190	53	38	91	7	288
(%)		74.0%	66.2%	61.0%	45.5%		68.8%	49.4%		9.1%	
Post-survey of Group B	78	57	43	35	34	169	48	24	72	8	249
(%)		73.1%	55.1%	44.9%	43.6%		61.5%	30.8%		10.3%	

Table 16.6 Comparison among scientific inquiry skills-thinking, judgment, and expression-between Post-Group A and Group B using the Mann-Whitney *U* test

Index	Post-Group A	Post-Group B	Statistics Z	Significant of criterion
Scientific inquiry skills	4.11 ± 1.77	3.45 ± 1.67	2.6	0.0093**
Thinking skills	2.76 ± 1.12	2.29 ± 1.12	2.64	0.0083**
Judgment skills	1.29 ± 0.90	1.04 ± 0.85	4.774	0.0601+
Expression skill	0.06 ± 0.23	0.12 ± 0.32	0.7	0.4839

+.05 < p < .10, +p < .05, ** p < .01

Table 16.7 Results of exploratory factor analysis for Group A (pre-survey)

	Relational values	Instrumental values/ Intrinsic values	Instrumental values
Cultural identity: RV2	**0.87**	−0.02	0.01
Individual identity: RV1	**0.86**	−0.02	−0.04
Social cohesion: RV3	**0.75**	−0.14	−0.06
Stewardship eudaimonic: RV4	**0.73**	0.05	−0.08
Social responsibility: RV6	**0.63**	0.24	0.11
Moral responsibility to non-humans: RV5	**0.59**	0.31	−0.02
Intrinsic value2	0.05	**0.85**	0.08
Intrinsic value1	0.01	**0.76**	0.01
Instrumental value 4	−0.01	**0.65**	−0.27
Stewardship principle/virtue: RV7	0.12	**0.65**	0.03
Intrinsic value2	0.14	−0.1	**−0.86**
Instrumental value 3	−0.04	**0.5**	**−0.54**
Intrinsic value1	0.15	0.28	**−0.41**
Inter-factor correlation			
Relational values			−0.43
Intrinsic values: Maintaining the health of the food chain			−0.37

Values" factor. The third factor was entirely composed of items related to instrumental values, leading to its designation as the "Instrumental Values" factor. For Group A in the post-survey, the Cronbach's alpha coefficient was 0.91, the KMO measure of sampling adequacy was 0.86, and the Bartlett test of sphericity was p < .001. Exploratory factor analysis identified three distinct factors (Table 16.8). The first factor in the post-survey was labeled "Intrinsic Values" as it comprised items related to intrinsic values. The second factor, composed of items related to RVs, was designated as the "RVs" factor. The third factor, primarily composed of items related to instrumental values, was designated as the "Instrumental Values" factor.

For Group B in the pre-survey, the Cronbach's alpha coefficient was 0.90, indicating high internal consistency. The KMO measure of sampling adequacy was 0.84, and the Barlett test of sphericity was $p < .001$. Exploratory factor analysis revealed three factors (Table 16.9). The first factor was identified as "the Health of the Food Chain", as it included items RV7, RV6, and instrumental value 4, which exhibited cross-loadings. The second factor was labeled the "RVs" factor, comprising items specifically related to RVs. The third factor was termed the "Instrumental values/Intrinsic values" factor, as it primarily consisted of items associated with instrumental values and intrinsic values. For Group B in the post-survey, the Cronbach's alpha coefficient was 0.93, the KMO measure of sampling adequacy was 0.90, and the Barlett test of sphericity was $p < .001$. As a result of the exploratory factor analysis, two factors were extracted (Table 16.10). The first factor primarily comprised items related to instrumental values and intrinsic values, but it also included items associated with RV7 and RV5. Saito et al. (2022) noted that the principles and virtues of stewardship and moral responsibility toward nonhumans are best considered as a single component when examining Japanese ethics regarding nature. Consequently, these elements were combined into the first factor designated as the "Instrumental Values, Intrinsic Values, and Worldview". The second factor was designated as the "RVs" factor, as it is composed of items related to RVs.

Table 16.8 Results of exploratory factor analysis for Group A (post-survey)

	Intrinsic values	Relational values	Instrumental values
Intrinsic value2	**0.99**	−0.04	0.06
Intrinsic value1	**0.74**	0.06	0.01
Cultural identity: RV2	0.05	**0.91**	−0.08
Individual identity: RV1	−0.06	**0.91**	−0.03
Social cohesion: RV3	0.09	**0.72**	0.07
Stewardship eudaimonic: RV4	−0.1	**0.71**	0.17
Moral responsibility to non-humans: RV5	0.08	**0.61**	0.19
Social responsibility: RV6	0.27	**0.50**	0.23
Instrumental value 3	−0.05	−0.1	**0.95**
Instrumental value 4	0.15	−0.02	**0.79**
Instrumental value 1	0.06	0.07	**0.56**
Instrumental value 2	−0.13	0.37	**0.46**
Stewardship principle/virtue: RV7	0.11	0.14	0.39
Inter-factor correlation			
Intrinsic values		0.36	0.42
Relational values			0.61

Table 16.9 Results of exploratory factor analysis for Group B (pre-survey)

	Social responsibility for maintaining the health of the food chain	Relational values	Instrumental values/ Intrinsic values
Stewardship principle/virtue: RV7	**0.53**	−0.18	0.06
Social responsibility: RV6	**0.38**	−0.29	0.27
Social cohesion: RV3	−0.28	**−0.91**	0.05
Individual identity: RV1	0.09	**−0.88**	−0.04
Cultural identity: RV2	0.22	**−0.78**	0
Stewardship eudaimonic: RV4	0.29	**−0.66**	0.16
Instrumental value1	−0.33	0	**0.89**
Instrumental value4	**0.41**	0.21	**0.75**
Instrumental value2	−0.04	−0.03	**0.65**
Intrinsic value2	0.23	−0.07	**0.57**
Intrinsic value3	0.09	−0.13	**0.56**
Intrinsic value1	0.31	−0.11	**0.42**
Moral responsibility to non-humans: RV5	0.18	−0.17	**0.40**
Inter-factor correlation			
Social responsibility for maintaining the health of the food chain		−0.33	0.36
Relational values			−0.43

Table 16.10 Results of exploratory factor analysis for Group B (post-survey)

	Worldviews of instrumental values and intrinsic values	Relational values
Intrinsic value 2	**0.95**	0.13
Instrumental value 3	**0.83**	0.11
Instrumental value 4	**0.74**	−0.19
Intrinsic value1	**0.71**	0
Stewardship principle/virtue: RV7	**0.69**	−0.16
Instrumental value 1	**0.54**	−0.01
Moral responsibility to non-humans: RV5	**0.47**	−0.44
Instrumental value 2	**0.4**	−0.26
Social cohesion: RV3	−0.11	**−0.96**
Stewardship eudaimonic: RV4	0.08	**−0.86**
Individual identity: RV1	0.02	**−0.85**
Cultural identity: RV2	**0.41**	**−0.52**
Social responsibility: RV6	**0.44**	**−0.45**
Inter-factor correlation		
Worldviews of instrumental values and intrinsic values		−0.62

16.5 Discussion

16.5.1 Changing Scientific Inquiry Skills

In this research, we aimed to analyze the changes in scientific inquiry skills that occurred before and after the implementation of the aquaponics activities by utilizing the scientific inquiry framework outlined in the evaluation indicators of the Aquaponics STEAM Program.

As a result of the Mann-Whitney *U* test, no significant difference was found in the scientific inquiry skills of Group A and Group B in the pre-survey. However, a significant difference was observed in the thinking and judgment skills of the scientific inquiry skills of Group A more than Group B in the post-survey. In the aquaponics activities conducted by Group A, middle school students completed the inquiry process for each item on a cross-model worksheet, while university students facilitated the inquiry activities of the middle school students, as illustrated below.

> University Student A: *"Now, let's observe as we fill out the cross-model worksheet. Did you find something?"*

Middle School Student A: *"The plants are growing larger."*
University Student A: *"I understand. Please write it down on the sheet."*
Middle School Student A: *"This isn't a result; it's more of a hypothesis."*

Several studies have revealed challenges in cultivating problem-setting skills in Japan, which is a crucial aspect of scientific inquiry skills (Ogura and Moriyama 2021). However, Sakamoto et al. (2016) assert that a combination of inquiry activities grounded in scientific principles and laws, along with interventions in question generation, constitutes an effective teaching strategy for fostering scientific thinking. This approach not only encourages the development of high-quality scientific questions but also incorporates prospects for further inquiry. Additionally, Miyamoto (2016) suggests that sharing and refining hypotheses among students facilitates the hypothesis-setting process. Middle school students used cross-model worksheets to externalize their inquiry processes, while university students facilitated their engagement in inquiry activities and question generation based on scientific principles. It is believed that these factors contributed to an enhancement of scientific inquiry skills, especially thinking skills. Both Groups A and B likely scored high in the pre-survey because they had support from university students, unlike in the post-survey, which they completed independently.

16.5.2 Changing Views on Nature

Another purpose of this study was to clarify the changes in RVs before and after educational practice, as well as the changes in the relationships among RVs, instrumental values, and intrinsic values. The Cronbach alpha coefficient of this survey was high, ranging from 0.90 to 0.93. Based on the Cronbach alpha coefficient, it can be said that the survey items in this study are suitable for investigating values regarding nature. In addition, the KMO measurement of sampling adequacy was 0.84 to 0.90, and Barlett's test of sphericity was $p < .001$, so it can be said that the response data in this study were suitable for performing factor analysis.

As a result of an exploratory factor analysis conducted on the pre-survey of Group A, three factors were identified: the "RVs" factor, the "Intrinsic Values/Instrumental Values" factor, and the "Instrumental Values" factor. Additionally, three factors were extracted from the post-survey of Group A: the "Intrinsic Values" factor, the "RVs" factor, and the "Instrumental Values" factor. In the pre-survey, the "Intrinsic Values/Instrumental Values" factor comprised a blend of intrinsic values and instrumental values. However, in the post-survey, intrinsic values were distinctly associated with the "Intrinsic Values" factor, and only items related to intrinsic values included. This indicates that during the pre-survey stage, intrinsic values and instrumental values could not be differentiated. On the other hand, the post-survey revealed three extracted factors: RVs, instrumental values, and intrinsic values. This indicates that the experience of raising fish and plants in aquaponics may have allowed the participants to differentiate and identify these three values.

Exploratory factor analysis was conducted on the response data from the pre-survey of Group B, resulting in the extraction of three factors: social responsibility for maintaining the health of the food chain, RVs, and Instrumental Values/Intrinsic Values. During the pre-survey stage, instrumental values and intrinsic values could not be distinctly classified or recognized, suggesting that they were partially conflated with the RVs items. In the post-survey, two factors emerged: RVs and Instrumental Values/Intrinsic Values/Worldview. The Instrumental Values/Intrinsic and Values/Worldview factor was influenced by the instrumental values and intrinsic values, as well as RV3 and RV7, which were integrated as worldviews by Saito et al. (2022). Consequently, it became evident that in Group B, RVs, instrumental values, and intrinsic values were not distinctly classified or recognized in either the pre- or post-surveys.

Comparing the results of the exploratory factor analysis of Group A and Group B during the pre-survey stage, neither group was able to distinguish between intrinsic values and instrumental values. However, in the post-survey stage, Group A was able to classify and recognize RVs, instrumental values, and intrinsic values, while Group B was unable to classify RVs, instrumental values, and intrinsic values. In this study, Group A's post-survey demonstrated an ability to differentiate among these values. Himes and Muraca (2018) argue that excluding one of the three values or reducing them to a single type in value assessment can lead to a negative bias in the evaluation process. When we focus solely on the instrumental values of an object, it may be perceived as something quantifiable, commercializable, and replaceable through monetary evaluation (Muraca 2016). By recognizing all three values through the experience of aquaponics activities, students can evaluate nature more holistically, considering a range of values rather than relying on a singular perspective.

16.5.3 Limitations of This Research

To investigate students' values regarding the canal, we conducted a survey using the specific word "canal", but many previous studies have investigated and analyzed RVs in general nature (Uehara et al. 2020; Saito et al. 2022). Although it has been reported that there is no significant difference between RVs based on a specific place and RVs based on a general nature (Saito et al. 2022), no research has been conducted that focuses on pre- and post-recognition changes like this study and investigating changes in RVs based on general nature may provide new insights. In addition, because this study focused on quantitative surveys, it was difficult to evaluate what specific parts of the canal students' value. In the future, it is necessary to analyze students' values in more detail by conducting RVs analysis using qualitative surveys as well as quantitative surveys. This study was based on a questionnaire survey of middle school students, and in order to generalize, it is necessary to expand the scope of the study.

16.6 Conclusion

In this study, we clarified the Aquaponics STEAM Program effectively enhances scientific inquiry skills. Competencies in the OL framework based on the theory of change are skills that enable students to act for collective well-being and sustainable development. These competencies reflect the social-emotional and behavioral aspects of the learning objectives in the United Nations Sustainable Development Goal 14 "Life Below Water", are vital for in Education for Sustainable Development (ESD), and the ability to effect change through knowledge, skills, and higher-level cognitive skills centered on critical thinking and learning, as defined in the OECD PISA Global Competence Framework. These competencies are essential twenty-first-century skills for effective intercultural exchange, and collective well-being (Santoro et al. 2024). An Aquaponics STEAM Program that promotes scientific inquiry can greatly enhance these skills.

Middle school students were able to recognize the nearby canal from the perspective of RVs. This indicates increased awareness of nature-connectedness in the OL framework based on the theory of change. A nature-connectedness fosters pro-environmental attitudes and behaviors. When individuals view environmental harm as self-harm, they are less likely to damage nature. Interventions like nature experiences, virtual reality, and multimedia can strengthen this connection (Santoro et al. 2024). Our study results indicate that classroom aquaponics activities with peers and university students effectively raise awareness of the nature-connectedness (the nearby canal) in the OL framework. In addition, aquaponics activities are expected to foster the canal that surrounds their lives, encourage young people to take the steps of the theory of change model, and play a role in creating an opportunity for sustainable action. Valuing and improving the canal environment and choosing actions to transform it into a better state will influence others and become an opportunity for collective change. This program enhances competencies such as scientific inquiry skills, fosters awareness of nature-connectedness, and cultivates values and attitudes. As described above, the Aquaponics STEAM Program is proposed to be effective as an education initiative of OL based on the theory of change model.

The OECD has established a goal of fostering students' Global Competence by 2030. Global Competence is defined as the ability to critically analyze global and intercultural issues from multiple perspectives, to understand how differences influence one's own and others' perceptions, judgments, and ways of thinking, and to interact openly, appropriately, and effectively with individuals who speak different languages, all while respecting human dignity. Matsuo (2017) examined the curricula of nine OECD member countries and identified three distinct categories: (i) basic literacies for managing language, numbers, and information, including literacy, numeracy, and information and communication technology (ICT); (ii) higher-order cognitive skills, such as critical thinking and metacognitive strategies; and (iii) social skills, encompassing social competence and self-management, which pertain to interpersonal relationships and independence within those contexts. In the 2017 revision of the curriculum guidelines, the Japanese national curriculum aims to develop three competencies: "Ensure that knowledge and skills", "Develop students' skills to think, make judgments, and express themselves", and "Cultivate motivation to learn and humanity" (MEXT 2024). The Aquaponics STEAM Program not only enhances competencies, nature-connectedness, attitudes, and values within an OL framework but also serves as an effective educational program for cultivating the competencies that align with the educational goals of the OECD Education 2030 initiative and its member countries, including Japan.

16.6.1 Summary

Key concepts explored in this chapter include:

- The Aquaponics STEAM Program effectively enhances middle school students' scientific inquiry skills by integrating hands-on learning with sustainable, circular food systems that mirror real-world Ocean-related challenges.
- The program supports the development of Ocean Literacy competencies—such as nature-connectedness, values and attitudes, and action-oriented competencies—aligned with the Theory of Change framework and the UN Ocean Decade's Challenge 3.
- Through comparative analysis, students who participated in aquaponics activities demonstrated a clearer understanding and distinction between relational, instrumental, and intrinsic values, indicating the program's impact on shaping environmental ethics.
- Aquaponics education connects students to their local environment—in this case, the canal ecosystem—promoting a sense of responsibility and encouraging pro-environmental behaviors rooted in relational values.
- The study affirms that experiential, interdisciplinary programs like aquaponics can bridge national education goals and international sustainability frameworks, offering a scalable model for advancing Ocean Literacy and global competency.

16.6.2 Recommended Resources

The following resources provide further insights and support on this topic:

- Chan, K. M. A., Balvanera, P., Benessaiah, K., Chapman, M., Díaz, S., Gómez-Baggethun, E., … Turner, N. (2016). Why protect nature? Rethinking values and the environment, *PNAS, 113*(6), 1462–1465.
- Intergovernmental Oceanographic Commission. (2024, November 20). Ocean Decade Vision 2030 White Papers: Challenge 3: Sustainably nourish the global population. https://Oceandecade.org/publications/Ocean-decade-vision-2030-white-papers/
- Kluczkovski A., Ehgartner U., Pugh E., Hockenhull I., Heaps-Page, A. Williams, Thomas J. M. H., … Denby K. (2024). Aquaponics in schools: Hands-on learning about healthy eating and a healthy planet, *Nutrition Bulletin, 49*(3), 327–344.
- Santoro, F., Tereza de Magalhães, A. V., Middleton, F., & Buchanan-Dunlop, J. (2024, November 20). A new Blue Curriculum: a toolkit for policy-makers.

Acknowledgements We thank all contributors to this research, particularly the teachers, staff, and students at Konan Middle School and the Tokyo University of Marine Science and Technology. We also appreciate the publisher and editors for the opportunity to publish. This research was funded by the Japan Society for the Promotion of Science (JSPS) under Grant No. 22 K02515 and is part of the Decade Action initiative "Establishment of Circulatory Aquaculture, Improvement of Shallow Waters, and Evaluation System Aiming for Blue Economy, ID 217", linked to the UN Decade of Ocean Science for Sustainable Development 2021–2030.

References

Brennan C, Ashley M, Molloy O (2019) A system dynamics approach to increasing Ocean Literacy. Front Mar Sci 6:360

Britto dos Santos N, Gould RK (2018) Can relational values be developed and changed? Investigating relational values in the environmental education literature. Curr Opin Environ Sustain 35:124–131

Chan KMA, Balvanera P, Benessaiah K, Chapman M, Díaz S, Gómez-Baggethun E et al (2016) Why protect nature? Rethinking values and the environment. *PNAS* 113(6):1462–1465

Guan S, Qu F, Qiao F (2023) United Nations decade of Ocean science for sustainable development (2021–2030): from innovation of Ocean science to science-based Ocean governance. Front Mar Sci 9:1091598

Himes A, Muraca B (2018) Relational values: the key to pluralistic valuation of ecosystem services. Curr Opin Environ Sustain 35:1–7

Intergovernmental Oceanographic Commission (2024, November 20) Ocean decade vision 2030 white papers: challenge 3: sustainably nourish the global population. https://unesdoc.unesco.org/ark:/48223/pf0000390117?posInSet=3&queryId=96df7063-986b-430f-9760-1296d144c398

Matsuo T (2017) Competencies required in the 21st century and curriculum reforms in Japan and abroad. J Natl Inst Educ Policy Res 146:9–22

McKinley E, Burdon D (2020) Understanding Ocean Literacy and Ocean climate-related behaviour change in the UK-work package 1: evidence synthesis. Daryl Burdon Ltd, Hull. https://oceanconservationtrust.org/wp-content/uploads/Review-of-Ocean-Literacy-1.pdf

McKinley E, Burdon D, Shellock RJ (2023) The evolution of Ocean Literacy: a new framework for the United Nations Ocean Decade and beyond. Mar Pollut Bull 186:114467

Ministry of Education, Culture, Sports, Science and Technology (2024, November 20) Japanese national curriculum guideline for middle school. https://www.mext.go.jp/content/20201008-mxt_kyoiku02-000005242_1.pdf

Miyamoto N (2016) The relationship between hypothesis formulation and data interpretation in junior high school science: focusing on sharing and refining hypotheses based on causal relationships. Sci Educ Res 55(3):341–350

Muraca B (2016) Relational values: a Whiteheadian alternative for environmental philosophy and global environmental justice. BalK J Philos 8(1):19–38

Nash N, Whitmarsh L, Capstick S, Hargreaves T, Poortinga W, Thomas G et al (2017) Climate-relevant behavioral spillover and the potential contribution of social practice theory. Wiley Interdiscip Rev Clim Chang 8(6):e481

Ogura M, Moriyama J (2021) A perspective on research topics aimed at developing problem-finding and problem-setting skills in junior high school technology classes. Sch Educ Res Hyogo Univ Teacher Educ 34:343–356

Pascual U, Balvanera P, Díaz S, Pataki G, Roth E, Stenseke M et al (2017) Valuing nature's contributions to people: the IPBES approach. Curr Opin Environ Sustain 26:7–16

Plankis B, Marrero ME (2010) Recent Ocean Literacy research in United States public schools: results and implications. Int Electron J Environ Educ 1(1):21–51

Saito T, Hashimoto S, Basu M (2022) Measuring relational values: do people in greater Tokyo appreciate place-based nature and general nature differently? Sustain Sci 17:837–848

Santoro F, Tereza de Magalhães AV, Middleton F, Buchanan-Dunlop J (2024, November 20) A new Blue Curriculum: a toolkit for policy-makers

Sasaki T (2022) Toward achievement of the UN Ocean Decade: does "CANAL STUDY" in Central Tokyo Bay Area Enhance Ocean Literacy? Mar Technol Soc J 56(3):91–100

Sasaki T (2023) Research on problems through independent theme setting. Phys Educ 71(2):142–145. https://www.ingentaconnect.com/content/mts/mtsj/2022/00000056/00000003/art00015;jsessionid=1xqkt5trslwfd.xic-live-02#:~:text=Abstract-,Abstract, water%20quality%20of%20Tokyo%20Bay

Shellock RJ, Fullbrook L, McKinley E, Cvitanovic C, Kelly R, Martin V (2024) The nature and use of Ocean Literacy in achieving sustainable Ocean futures: a systematic map. Ocean Coast Manag 257:107325. https://doi.org/10.1016/J.OCECOAMAN.2024.107325

Uehara T, Sakurai R, Tsuge T (2020) Cultivating relational values and sustaining socio-ecological production landscapes through Ocean Literacy: a study on Satoumi. Environ Dev Sustain 22(2):1599–1616

United Nations Environment Programme (2024, November 20) Making peace with nature: a scientific blueprint to tackle the climate, biodiversity and pollution emergencies. https://wedocs.unep.org/20.500.11822/34948

Author Contributions and Biographical Profiles

[1]*Tsuyoshi Sasaki: Co-conceptualization; Writing—Original Draft, Review, and Editing; Methodology—Data Collection and Analysis; Project Administration and Supervision; Grant Writing and Funding Acquisition.

[2]Kota Sawada: Co-conceptualization; Methodology—Data Collection and Analysis; Writing—Original Draft.

*Corresponding author at t-sasaki@kaiyodai.ac.jp

Tsuyoshi Sasaki, PhD, is professor at the Tokyo University of Marine Science and Technology, where he is dedicated to advancing marine education and fostering international collaboration. His career in education began in 1990 as a high school teacher at Iwate Prefectural Miyako Fisheries High School. In 2006, he joined Tokyo University of Marine Science and Technology as an Associate Professor and became a member of the National Marine Educators Association at the same year. A recognized leader in marine education across Asia, Dr. Sasaki has served as President of the Aquatic Marine Environmental Educators Association since 2012 and the Asia Marine Educators Association since 2015. He was promoted to Professor in 2018. In 2023, he played a key role in establishing the Ocean Town Community School, a pioneering initiative to integrate Ocean Literacy into community-based learning.

Kota Sawada, is a graduate student at the Tokyo University of Marine Science and Technology. He is currently engaged in research and educational activities focused on marine sustainability and Ocean Literacy. Working closely with Dr. Tsuyoshi Sasaki, he contributes to initiatives that strengthen the connection between academic research and community-based Ocean education, including efforts related to the recently launched Ocean Town Community School.

17 Investigating Ocean Literacy Levels of Pre-Service Science and Social Studies Teachers in Türkiye: The Influence of Gender, Ocean Experience, and Environmental Activities

Bülent Çavaş, Şermin Açık, Simge Koç, and Mısra Kolaç

Abstract

Ocean Literacy, a crucial component of environmental education, encompasses understanding the Ocean's influence on people and the impact of human activities on the Ocean. This chapter reports an investigation that is related to Ocean Literacy levels of pre-service science and social studies teachers, focusing on the variables of gender, Ocean experience, and participation in environmental activities. The study employs a survey to gather data from a representative sample of pre service teachers from a faculty of education located in seaside city in Türkiye. According to the quantitative analysis, it is found that there were significant differences in two items with female participants demonstrating higher levels of awareness and understanding. However, there was no significant difference in terms of participation in environmental activities, participants' sea experience showed a significant difference in their Ocean Literacy levels regarding the actual source of rainwater that falls on the land. The results underscore the importance of integrating comprehensive Ocean education into teacher training programs and highlight the need for targeted initiatives to bridge gaps in Ocean Literacy, ultimately fostering a more informed and environmentally conscious teaching workforce.

Keywords

Ocean Literacy (OL) · Pre-service teachers · Gender · Ocean experience · Environmental activities

17.1 Introduction

The Ocean, a fundamental ecosystem of our Planet, serves as the source of life, the regulator of climate, and is indispensable for our existence. Covering 70% of the Earth, this vast body of water not only sustains life but also hosts countless species, absorbs carbon dioxide, produces a significant portion of the oxygen, and is a major source of food and employment. However, various factors, such as climate change, pollution, the unregulated and excessive exploitation of marine resources, and ecosystem degradation, are increasingly stressing this ecosystem and driving

B. Çavaş (✉) · Ş. Açık · S. Koç · M. Kolaç
Dokuz Eylül University, Izmir, Türkiye
e-mail: bulent.cavas@deu.edu.tr;
sermin.acikcinar@deu.edu.tr; simge.koc@deu.edu.tr

T. J. Kennedy (ed.), *Ocean Literacy: The Foundation for the Success of the Ocean Decade, Volume I*, Sustainable Development Goals Series, https://doi.org/10.1007/978-3-031-98145-6_17

it towards a critical existential threat (Cava et al. 2005; Brennan et al. 2019). Ensuring the sustainable use of the Ocean while preserving its health remains one of the most significant contemporary challenges (Ryabinin et al. 2019).

This threat facing our Planet is growing more serious by the day. The WWF Living Planet Report, based on the most comprehensive findings to date, indicates that between 1970 and 2018, wildlife populations have declined by an average of 69%, with freshwater species experiencing the greatest global decline at 83%. Our Planet has already warmed by 1.2 °C due to widespread industrial activities. Currently, 50% of warm-water coral reefs have been lost due to various causes, and it is feared that over 99% could be wiped out as global warming continues to escalate (WWF 2022).

In the face of these serious risks that endanger our future, the most crucial solution is to raise awareness among all people living on our Planet, regardless of their age, to reduce the stress factors on the ecosystem. To this aim, the term Ocean Literacy (OL), coined in 2004 by a group of Ocean scientists and educators in the United States, was emphasized to ensure that people become informed about the Ocean. This group recognized the lack of necessary and sufficient content in formal education to foster Ocean awareness and initiated efforts to promote and encourage the inclusion of Ocean sciences. Similarly, the Pew Ocean Commission Report, also published in 2004, argued that public awareness must be enhanced to enable the Ocean's recovery. In this context, they called for the beginning of an era of Ocean Literacy that would connect people to the marine environment (Steel et al. 2005; Santoro et al. 2017). One of the most notable developments, regarded as a success for the Ocean community, is the United Nations' declaration of the Sustainable Development Goals (SDGs) for the period 2021–2030, which includes a dedicated goal related to the Ocean among a total of 17 goals. The 14th of these goals focuses on life below water. In the same year, the United Nations also organized a conference titled "Our Ocean, Our Future" to support the 14th Sustainable Development Goal. As a result of these developments, the visibility of OL increased, bringing attention to the importance of OL by promoting Ocean-related education, supporting OL, and addressing issues such as Ocean conservation (Santoro et al. 2017; Mogias et al. 2019). Bibliometric studies indicate that while the initial increase in research on OL was inconsistent, the past 6 years have seen a notable rise in its popularity. The data also supports the notion that OL has become a globally researched topic (Çavaş et al. 2023; Shellock et al. 2024; Costa and Caldeira 2018).

Ocean Literacy is not merely a theoretical term or a collection of information; it is a movement aimed at enhancing our understanding of the Ocean. It encompasses the ability to comprehend both the impact of the Ocean on humans and the influence of human activities on the Ocean. One of the most critical steps toward ensuring a sustainable future is to promote OL among individuals, helping them to recognize the vital role of the Ocean in our lives. An Ocean-literate person understands the fundamental concepts of how the Ocean functions, can communicate about it effectively, and is able to make informed and responsible decisions regarding Ocean resources (Cava et al. 2005). Ocean Literacy has seven essential principles and 45 fundamental concepts. The seven fundamental principles (NOAA 2013) are as follows:

(1) The Earth has one big Ocean with many features.
(2) The Ocean and life in the Ocean shape the features of Earth.
(3) The Ocean is a major influence on weather and climate.
(4) The Ocean made Earth habitable.
(5) The Ocean supports a great diversity of life and ecosystems.
(6) The Ocean and humans are inextricably interconnected.
(7) The Ocean is largely unexplored.

Recent developments in OL research have emphasized the need for a more in-depth exploration of the term, leading to an evolution in its practical concept. It now emphasizes not just raising public awareness of Ocean-related issues but also equipping people with the tools and strategies to translate Ocean knowledge into actions and behaviors that support sustainable Ocean use. Recent research has identified several more aspects of OL, including awareness, attitude, behavior, activism, communication, emotional connections, access and experience, adaptive capacity, and trust and transparency (Brennan et al. 2019; McKinley et al. 2023).

Due to its geographical position, with three sides surrounded by seas, Türkiye is a nation with well-developed maritime relations. However, despite this, it is evident that the learning objectives related to the Ocean and seas in the curriculum are extremely limited. Similarly, it has been found that topics related to OL are addressed at very minimal levels in education faculties. Encouraging pre-service teachers in terms of OL is of significant importance for the protection of our country's seas. To take steps toward identifying and addressing societal gaps on this increasingly important topic, it is essential first to understand the level of OL. There are globally developed surveys designed to measure OL levels. In this study, the OL Survey, developed by Mogias et al. (2019) was used to measure the OL levels of middle school students (age of 13–14) and pre-service science and social studies teachers.

17.2 Literature Review

Supporting OL in primary and secondary education is of significant importance. Unfortunately, school curricula are deficient in topics related to OL, and universities also lack educational programs dedicated to it (Visbeck 2018). To address this gap, it is essential to assess the OL levels of students and teacher candidates and update educational programs to better suit their needs.

The study conducted by Mogias et al. (2019) assessed the OL of elementary students (grades 3–6) from Italy, Croatia, and Greece, finding a moderate overall knowledge of Ocean sciences with some national variations. The study highlighted common misconceptions across all countries, particularly regarding the origin of atmospheric oxygen and the global interconnectedness of seas. Students engaged in nature-related activities or non-formal education, such as visits to aquariums, exhibited higher knowledge levels. The study emphasizes the need for curriculum enhancements to improve OL and protect the Mediterranean Sea.

Greely (2008) developed an assessment tool to measure the OL of high school students in the United States. The results revealed that participants' attitudes towards the environment significantly positively influenced their OL. In the study by Steel et al. (2005) aimed at assessing the public's OL levels, data were collected from 1233 individuals. The study found that the public lacked sufficient knowledge of environmental terms and Ocean-related issues. No significant difference was observed between individuals living near the coast and those living farther inland in terms of understanding key terms and answering Ocean-related questions; both groups struggled with these tasks. Another study by Tsai and Chang (2019), conducted with high school students aged 16–18, showed that students had difficulty answering questions about information not included in their textbooks. The results indicated that students' knowledge of ecosystems and climate was inadequate.

Chang et al. (2014) conducted a study focusing on vocational high school students by employing both a questionnaire survey and a qualitative analysis of concept maps. The study included 285 students, aiming to assess their OL and provide insights for marine science and literacy education in vocational high schools. The findings indicated that the students' understanding of OL concepts needed significant improvement. Another study conducted by Lwo et al. (2013) utilized concept mapping to create open-ended questions designed to evaluate Taiwanese high school students' grasp of Ocean concepts and their overall OL. The results showed that the students had an average pass rate of about 50% in assessing their misconceptions about marine science, with more than half of the questions achieving a pass rate below 50%.

In light of this information, it is evident that the level of OL worldwide is insufficient, and efforts to improve it are necessary. Measuring the OL levels of students and pre-service teachers in Türkiye is crucial for recognizing this deficiency and raising awareness about the need for curriculum revisions.

17.3 Problem Statement

What are the Ocean Literacy levels of pre-service teachers?

17.3.1 Sub-problems

1. How do pre-service teachers' Ocean Literacy levels vary according to gender?
2. How do pre-service teachers' Ocean Literacy levels vary according to their experiences at sea?
3. How do pre-service teachers' Ocean Literacy levels vary according to Whether They participate in environmental and nature activities in schools?

17.4 Materials and Methods

17.4.1 Participants

The participants in this study are third and fourth year pre-service teachers who are required to attend formal education courses with mandatory attendance in the faculty of education in Izmir-Türkiye (see Fig. 17.1). The study consisted of 100 pre-service teachers; seventy-six percent of the participants were females while twenty-four of them were males.

17.4.2 Data Collection Tool

The questions in the scale are presented in Appendix 17.1. In this study, the validity and reliability of the Ocean Literacy Scale were tested after being adapted into Turkish. The scale was translated into Turkish by a team of experts in science education, marine sciences, and foreign languages. During the translation process, the original English version of the scale was independently translated into Turkish by three experts. These translations were then compared to create a unified Turkish form. To ensure linguistic equivalence, the Turkish version was back-translated into English, and the compatibility between the two versions was confirmed. The reliability of the scale was tested by calculating the Cronbach's alpha internal consistency coefficient. Cronbach's alpha values for the overall scale and each sub-

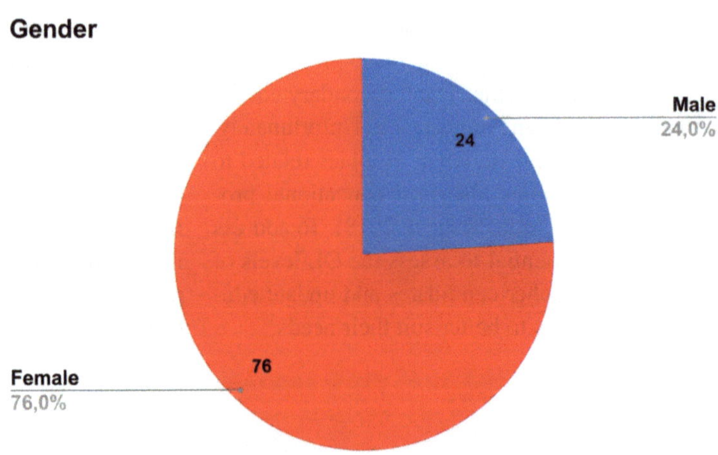

Fig. 17.1 Distribution by gender

Table 17.1 Alignment of the survey questions with the seven essential principles of the OL framework

Basic principles of OL	Questions
1. The earth has an Ocean with many features	1, 7, 13
2. The Ocean and life in the Ocean shape the characteristics of the earth	2, 10
3. The Ocean has a great influence on climate and weather	5, 11
4. The Ocean has made the earth habitable	3, 8
5. The Ocean supports the diversity of life and ecosystems.	4, 9
6. The Ocean and people are inextricably linked	12, 14, 16
7. Oceans are largely unexplored	6, 15

dimension were found to be above.70, indicating high reliability. The distribution of the questions in the scale according to the seven basic principles of the OL Framework is as follows (see Table 17.1).

17.4.3 Data Analysis

The level of OL of education faculty pre-service teachers is measured objectively through a test and explained with quantitative data.

16 multiple-choice questions were prepared in accordance with the basic principles of OL and presented to the students. At the same time, the gender of the students, their participation in activities related to nature and the environment, and whether they watch documentaries about nature were also included in the questions and taken into consideration.

Students scored 1 point for each correct answer and 0 point for each incorrect answer. The score varied between 0–16 points. The average score was 8.5, indicating the balance point of the scale.

The appropriateness of the content to the level of the students, the language used and the degree of difficulty were evaluated by experts examined the degree to which the scales represented the basic principles of OL and their scientific adequacy.

17.5 Findings

17.5.1 How Do Pre-service Teachers' OL Levels Vary According to Gender?

Table 17.2 shows the number and percentages of correct scores of the pre-service teachers according to their gender in the OL test. It is understood that there is significant gender difference in item Q5. In Q5, 37.5% of the male pre-service teachers gave correct answers, while this percentage was found to be 17.1% for the female pre-service teachers. In Q5 "Rainwater falling on land originally came from..." it is understood that male and female pre- service teachers have different levels of knowledge about the source of water.

In the scale consisting of a total of 16 questions, questions 14, 15 and 16 are graded in terms of agreement unlike the first 13 questions and are explained separately in the Table 17.3 below.

Table 17.3 shows that there is difference between the pre-service teachers' agreement scores in question 16. While it is understood that there is general agreement with the question posed as "What I throw in the sink has an influence on the sea", it is seen that the participation rate of female pre-service teachers is significantly different from that of male pre-service teachers. A gap influenced by gender-based experiences has been identified in many researches. Numerous studies on environmental issues reveal that female students tend to have higher levels of awareness. Kaya et al. (2009) attribute this to the generally greater environmental sensitivity observed among female students. Boeve-de Pauw and Van Petegem (2010) suggest that this difference may stem from societal expectations for women to adopt more nurturing and protective roles.

It is understood that the differences between the groups in terms of gender in the question of where the rainwater falling on the land actually comes from and the effect of everything thrown in the sink on the sea are statistically significant. It was found that there was no significant differ-

Table 17.2 Differences in the number and percentages of correct scores of pre-service teachers according to their gender in the OL test

Questions	Gender Female N (%)	Male N (%)	P-value
Q1. If you were a fish, could you reach by swimming every Ocean?	6 (25)	10 (13.2)	0.168
Q2. If you walk in the mountains and see a rock containing fossil fish, it means that:	63 (82.9)	17 (70.8)	0.198
Q3. First living organisms on earth lived:	65 (85.5)	20 (83.3)	0.793
Q4. Marine environment is home to:	72 (94.7)	21 (87.5)	0.226
Q5. Rainwater falling on land originally came from…	**13 (17.1)**	**9 (37.5)**	**0.030**[a]
Q6. Which of the following environments is the least explored?	67 (88.2)	22 (91.7)	0.632
Q7. The Aegean Sea is connected with:	21 (27.6)	9 (37.5)	0.358
Q8. Which of the following are the main source of oxygen that living beings breathe?	36 (47.4)	15 (62.5)	0.196
Q9. The largest animal ever living on earth is:	19 (27.5)	50 (72.5)	0.217
Q10. The coastline is shaped by:	71 (93.4)	20 (83.3)	0.132
Q11. How would the climate of your town be if there were not the Aegean Sea?	44 (57.9)	14 (58.3)	0.970
Q12. Most of world goods (e.g. your toys) are transported…	67 (88.2)	22 (91.7)	0.632
Q13. Most of the water on the earth is:	42 (55.3)	16 (66.7)	0.324

[a] It was found that the differences between the groups in terms of the emphasized feature were statistically significant

ence between female and male pre-service teachers in the questions of the OL test other than the two questions mentioned above.

17.5.2 How Do Pre-service Teachers' OL Levels Vary According to Their Experiences at Sea?

Table 17.4 shows the number and percentages of answers obtained from the OL test according to the pre-service teachers' sea experiences. The statistical data obtained shows that almost all of the pre-service teachers did not have a significant difference in the number of correct answers they got from the OL test according to their sea experiences such as going to the beach, going swimming, and engaging in water sports. However, it was only observed that there was a significant difference only in Q5. The participants' sea experience showed a significant difference in their Ocean Literacy levels regarding the actual source of rainwater that falls on the land.

According to the results of the research (see Table 17.5), it was observed that there were no significant differences between the answers given by the pre-service teachers to the test in terms of sea experiences on question basis.

17.5.3 How Pre-service Teachers' OL Levels Change According to Whether They Participate in Environmental and Nature Activities in Schools

Table 17.6 shows the number and percentages of correct scores of pre-service teachers from the OL test according to the frequency of participation in activities on environment and nature. The statistical data obtained show that there is no significant difference in the number of correct answers obtained from the OL test according to the frequency of pre-service teachers' participation in activities on environment and nature.

Table 17.7 shows the agreement scores of answers on the OL test according to pre-service teachers' participation in activities on environment and nature. Similarly, the statistical data obtained show that there is no significant difference in the number of correct answers obtained from the OL test according to the frequency of pre-service teachers' participation in activities on environment and nature. The answers given for the facts about, the impact of living far from the coast on our lives, needs to be done to protect the sea and the impact of everything that goes down the sink on the sea did not have a significant difference in the agreement scores of answers on the OL test according to pre-service teachers' participation in activities on environment and nature.

Table 17.3 Differences in the agreement scores of pre-service teachers according to their gender in the OL test

	Gender						
	Female			Male			
	N (%)			N (%)			
Questions	1	2	3	1	2	3	P-value
Q14. The sea influences my life, even if I live far from the coast.	67(88.2)	6(7.9)	3(3.9)	24(100)	0	0	0.210
Q15. We need to study and know the sea to protect it.	75(98.7)	1(1.3)	–	23(95.8)	1(4.2)	–	0.384
Q16. What I throw in the sink has an influence on the sea.	**73(96.1)**	**3(3.9)**	**0**	**19(79.2)**	**2(8.3)**	**3(12.5)**	**0.005***

* ≤ 0.05 (5 percent significance level)

Table 17.4 Differences in the number and percentages from the OL test according to pre-service teachers' sea experiences

	Number (N) and percentages (%) from the OL test according to sea experience					
	1	2	3	4	5	
Question	N (%)	N (%)	N (%)	N (%)	N (%)	P-value
Q1	3 (18.8)	3 (18.8)	3 (18.8)	2 (12.5)	5 (31.3)	0.532
Q2	10 (12.5)	19 (23.8)	23 (28.7)	12 (15)	16 (20)	0.654
Q3	12 (14.1)	18 (21.2)	27 (31.8)	12 (14.1)	16 (18.8)	0.946
Q4	12 (12.9)	22 (23.7)	29 (31.2)	12 (12.9)	18 (19.4)	0.529
Q5	**2 (9.1)**	**10 (45.5)**	**7 (31.8)**	**3 (13.6)**	**0 (0)**	**0.013[a]**
Q6	13 (14.6)	20 (22.5)	25 (28.1)	13 (14.6)	18 (20.2)	0.171
Q7	4 (13.3)	8 (26.7)	7 (23.3)	8 (26.7)	3 (10)	0.085
Q8	6 (11.8)	12 (23.5)	16 (31.4)	7 (13.7)	10 (19.6)	0.992
Q9	11 (15.9)	15 (21.7)	23 (33.3)	8 (11.6)	12 (17.4)	0.586
Q10	11 (12.1)	21 (23.1)	29 (31.9)	13 (14.3)	17 (18.7)	0.862
Q11	6 (10.3)	10 (17.2)	23 (39.7)	9 (15.5)	10 (17.2)	0.272
Q12	12 (13.5)	20 (22.5)	28 (31.5)	13 (14.6)	16 (18)	0.916
Q13	8 (13.8)	15 (25.9)	20 (34.5)	5 (8.6)	10 (17.2)	0.361

[a] It was found that the differences between the groups in terms of the emphasized feature were statistically significant

Table 17.5 Differences in the agreement scores of answers from the OL test according to pre-service teachers' sea experiences

		Differences in the agreement scores from the OL test according to sea experience					
Question		1.00	2.00	3.00	4.00	5.00	P-value
Q14	Agree	12 (13.2)	20 (22)	28 (30.8)	13 (14.3)	18 (19.8)	0.882
	Don't know	0 (0)	1 (16.7)	3 (50)	1 (16.7)	1 (16.7)	
	Disagree	1 (33.3)	1 (33.3)	1 (33.3)	0 (0)	0 (0)	
Q15	Agree	12 (12.2)	22 (22.4)	32 (32.7)	13 (13.3)	19 (19.4)	0.237
	Don't know	1 (50)	0 (0)	0 (0)	1 (50)	0 (0)	
	Disagree	0 (0)	0 (0)	0 (0)	0 (0)	0 (0)	
Q16	Agree	13 (14.1)	18 (19.6)	29 (31.5)	13 (14.1)	19 (20.7)	0.521
	Don't know	0 (0)	2 (40)	2 (40)	1 (20)	0 (0)	
	Disagree	0 (0)	2 (66.7)	1 (33.3)	0 (0)	0 (0)	

Table 17.6 Differences in the number and percentages of correct scores on the OL test according to pre-service teachers' participation in activities on environment and nature

	Number (N) and percentages (%) from the OL test according to participation in activities on environment and nature				
	1	2	3	4	
Question	N (%)	N (%)	N (%)	N (%)	P-value
Q1	3 (18.8)	2 (12.5)	7 (43.8)	4 (25)	0.278
Q2	20 (25)	22 (27.5)	26 (32.5)	12 (15)	0.630
Q3	19 (22.4)	24 (28.2)	29 (34.1)	13 (15.3)	0.821
Q4	23 (24.7)	26 (28)	31 (33.3)	13 (14)	0.426
Q5	7 (31.8)	6 (27.3)	7 (31.8)	2 (9.1)	0.676
Q6	22 (24.7)	23 (25.8)	31 (34.8)	13 (14.6)	0.457
Q7	6 (20)	12 (40)	8 (26.7)	4 (13.3)	0.357
Q8	16 (31.4)	10 (19.6)	18 (35.3)	7 (13.7)	0.122
Q9	14 (20.3)	18 (26.1)	27 (39.1)	10 (14.5)	0.547
Q10	23 (25.3)	25 (27.5)	30 (33)	13 (14.3)	0.305
Q11	11 (19)	17 (29.3)	22 (37.9)	8 (13.8)	0.704
Q12	22 (24.7)	26 (29.2)	29 (32.6)	12 (13.5)	0.396
Q13	17 (29.3)	16 (27.6)	20 (34.5)	5 (8.6)	0.153

17.6 Conclusion

17.6.1 Discussion

In examining the differences in the number and percentages of correct scores of pre-service teachers on the OL test according to their gender, the findings reveal that there is only one item with statistically significant differences: "Where does rainwater that falls on land actually come from?" This outcome suggests that while overall OL may not differ significantly between male and female pre-service teachers, their understanding of specific concepts related to the water cycle and its connection to oceanic processes does.

One of the findings that female pre-service teachers are more likely than their male counterparts to agree with the statement "Everything I put in the sink has an effect on the sea" is indicative of a broader trend in environmental awareness and responsibility that has been observed in various studies. This gender difference in environmental attitudes is a well-documented phenomenon, often reflecting deeper social and psychological factors. Women, particularly in educational settings, may feel a greater sense of duty to model environmentally responsible behavior, especially as future educators. This sense of responsibility may be heightened by the recognition of the cumulative effects of seemingly small actions, like what is poured down the

Table 17.7 Differences in the agreement scores of answers on the OL test according to pre-service teachers' participation in activities on environment and nature

		Differences in the agreement scores from the OL test according to participation in activities on environment and nature				
Question		1.00	2.00	3.00	4.00	P-value
Q14	I agree	23 (25.3)	25 (27.5)	30 (33)	13 (14.3)	0.652
	I don't know	0 (0)	2 (33.3)	3 (50)	1 (16.7)	
	I don't agree	0 (0)	1 (33.3)	2 (66.7)	0 (0)	
Q15	I agree	23 (23.5)	27 (27.6)	34 (34.7)	14 (14.3)	0.744
	I don't know	0 (0)	1 (50)	1 (50)	0 (0)	
	I don't agree	0 (0)	0 (0)	0 (0)	0 (0)	
Q16	I agree	20 (21.7)	26 (28.3)	32 (34.8)	14 (15.2)	0.897
	I don't know	2 (40)	1 (20)	2 (40)	0 (0)	
	I don't agree	1 (33.3)	1 (33.3)	1 (33.3)	0 (0)	

sink, on larger environmental systems. Men, on the other hand, may perceive less direct control or responsibility over these outcomes, leading to a lower level of agreement with the statement.

The stronger awareness shown by female pre-service teachers towards the statement "Everything I put in the sink has an effect on the sea" can be considered as a practical reflection of the principle that "The Ocean and humans are inextricably linked," one of the core principles announced by NOAA in 2013. The awareness can be addressed in the course content of experimental science lessons that chemical wastes should not be thrown into sink drains. The course content on chemical waste management can be expanded. In order to direct students to more environmentally sensitive instructional behaviors, practices that support the development of awareness of experimental materials that have toxic and long-term effects on aquatic life can be included in the teacher training system. Thus, in the near future, while the pre-service teachers are interacting with their students, this can be effective in spreading the level of awareness on OL issues among students.

Another important finding is that significant differences in OL levels exist based on pre-service teachers' experiences at sea, particularly in response to the item "Where does rainwater that falls on land actually come from?" suggests that direct interaction with marine environments plays a critical role in understanding the water cycle's connection to the Ocean. Those who have spent time at sea are more likely to directly observe the vastness of the Ocean, the patterns of evaporation, and the subsequent condensation that forms clouds and leads to precipitation over land. This firsthand experience reinforces the concept that rainwater, while falling on land, often originates from oceanic evaporation, making the connection between Ocean and land more tangible and comprehensible.

The finding that there were no significant differences in OL levels among pre-service teachers based on their participation in environmental and nature activities in schools is both surprising and thought-provoking. This result suggests that mere participation in such activities may not be sufficient to significantly enhance OL, pointing to the need for a deeper examination of the content and delivery of these activities.

17.6.2 Recommendations

It is recommended that teacher education programs revise their curricula to incorporate the dimensions of Ocean Literacy and include courses with practical applications focused on Ocean awareness and conservation. This curriculum enhancement will equip teachers with the knowledge and skills needed to integrate Ocean Literacy into their educational practice and inspire Ocean-conscious behaviors among students.

To foster greater Ocean Literacy, awareness, and conservation efforts among pre-service teachers, we, as authors suggest implementing Ocean and marine-based active teaching methods through out-of-school learning experiences as part of the teacher education process. By integrating experiential, hands-on learning activities in natural marine environments, future teachers can gain a deeper understanding of Ocean ecosystems and conservation challenges. This approach aims to strengthen their commitment to Ocean stewardship and empower them to convey these principles effectively in their future classrooms. Out-of-school learning settings, such as coastal field trips, marine research stations, and interactive museum exhibits, will provide immersive educational opportunities that enhance understanding and inspire proactive engagement in Ocean conservation.

As a final remark, it is recommended to establish collaborations between faculties of education and marine science institutes which the research results supports this partnerships (Çavaş et al. 2021; Fadzil et al. 2019). Such partnerships would provide pre-service teachers with opportunities to participate in projects and academic research led by scientists in the field of marine sciences. Through these collaborations, teacher candidates could engage directly in scientific work, gain valuable insights into marine ecosystems, and deepen their understanding of Ocean conservation efforts. This hands-on experience with real-world research would not only enhance their knowledge of Ocean Literacy but also inspire them to bring scientifically informed perspectives to their future teaching practices.

17.6.3 Summary

Key concepts explored in this chapter include:

- Future research could explore why "water cycle and its connection to oceanic processes" item displayed a gender difference, potentially examining factors such as previous educational experiences, instructional methods, or even socio-cultural influences that might contribute to different levels of understanding in specific areas of OL. Additionally, targeted educational interventions could be developed to address these gaps, ensuring that all pre-service teachers, regardless of gender, have a comprehensive understanding of key concepts in OL.
- The gender differences observed in this study highlight the importance of tailored environmental education strategies. For female pre-service teachers, who already demonstrate a higher level of environmental concern, educational programs might focus on empowering them with strategies to influence others and integrate these values into their future teaching. For male pre-service teachers, it could be beneficial to emphasize the tangible connections between everyday actions and environmental outcomes, fostering a more personal connection to environmental responsibility.
- The significant difference in OL levels, particularly regarding the understanding of where rainwater comes from, underscores the value of sea experiences in shaping a more informed and connected view of the Ocean's role in global systems. By promoting direct engagement with the marine environment, educational programs can foster a deeper, more accurate comprehension of critical oceanic processes, contributing to a more Ocean-literate population.
- The absence of significant differences in OL levels among pre-service teachers based on their participation in environmental and nature activities suggests that current approaches to environmental education may not be adequately addressing Ocean-related topics. To foster a more Ocean-literate teaching workforce, it is essential to ensure that environmental education in schools includes robust, targeted OL components. By doing so, pre-service teachers can be better prepared to understand and convey the vital role of the Ocean in global environmental systems.

17.6.4 Recommended Resources

The following resources provide further insights and support on this topic:

Books

- Koutsopoulos, K., & Stel, J. H. (Eds.). (2021). OL: Understanding the Ocean. Springer.
- Santoro, F et al (Eds.) (2017) OL for All—A Tool Kit. IOC/UNESCO & UNESCO Venice Office, Paris (IOC Manuals and Guides, 80)
- Greely, T. (2008). OL and reasoning about Ocean issues: The influence of content, experience and morality. University of South Florida.

Journals

- *Frontiers in Marine Science:* https://www.frontiersin.org/whats-new/marine-science/home
- *Marine Policy:* https://www.sciencedirect.com/journal/marine-policy
- *Environmental Education Research:* https://www.tandfonline.com/journals/ceer20
- *Sustainability:* https://www.mdpi.com/journal/sustainability
- *Ocean and Coastal Research:* https://www.ocr-journal.org/

Websites

- Intergovernmental Oceanographic Commission: https://www.ioc.unesco.org/en
- OL Portal: https://Oceanliteracy.unesco.org/
- European Union—OL: https://maritime--forum.ec.europa.eu/theme/Ocean-literacy-and-blue-skills/Ocean-literacy_en

Acknowledgement This chapter has been supported by the Scientific Research Coordination Unit of Dokuz Eylul University under the project number SBA-2022-2691.

References

Boeve-de Pauw J, Van Petegem P (2010) A cross-national perspective on youth environmental attitudes. Environmentalist 30:133–144

Brennan C, Ashley M, Molloy O (2019) A system dynamics approach to increasing OL. Front Mar Sci 6(360). https://doi.org/10.3389/fmars.2019.00360

Cava F, Schoedinger S, Strang C, Tuddenham P (2005) Science content and standards for OL: A report on OL. http://coexploration.org/Oceanliteracy/documents/OLit200405_Final_Report.pdf

Çavaş B, Çapar S, Çavaş L, Yahşi Ö (2021) Turkish STEM teachers' opinions about the scientist-teacher-student partnership. J Turk Sci Educ (TUSED) 18(4):622–637

Çavaş B, Açık S, Koç S, Kolaç M (2023) Research trends and content analysis of OL studies between 2017 and 2021. Front Mar Sci 10:1200181

Chang CC, Yang WC, Lwo LS (2014) Evaluating vocational high school students' marine science literacy and their misconceptions. Sci Educ Monthly 371:2–17

Costa S, Caldeira R (2018) Bibliometric analysis of OL: An underrated term in the scientific literature. Mar Pollut 87:149–157. https://doi.org/10.1016/j.marpol.2017.10.022

Fadzil HM, Saat RM, Awang K, Adli DSH (2019) Students' perception of learning STEM related subjects through scientist-teacher-student partnership (STSP). J Balt Sci Educ 18(4):537–548

Greely T (2008) OL and reasoning about Ocean issues: The influence of content, experience and morality. Ph.D. dissertation. University of South Florida, Tampa, FL

Kaya E, Akıllı M, Sezek F (2009) An analysis of high school students' attitudes towards the environment in terms of gender. Mehmet Akif Ersoy Univ J Fac Educ 18:43–54

Lwo LS, Chang CC, Tung YP, Yang WC (2013) Marine science literacy and misconceptions among senior high school students. J Res Educ Sci 58:51–83. https://doi.org/10.6209/jories.2013.58(3).03

McKinley E, Burdon D, Shellock RJ (2023) The evolution of OL: A new framework for the United Nations Ocean Decade and beyond. Mar Pollut Bull 186:114467. https://doi.org/10.1016/j.marpolbul.2022.114467

Mogias A, Boubonari T, Realdon G, Previati M, Mokos M, Koulouri P, Cheimonopoulou MT (2019) Evaluating OL of elementary school students: Preliminary results of a cross-cultural study in the Mediterranean region. Front Mar Sci 6:396. https://doi.org/10.3389/fmars.2019.00396

National Marine Educators Association (2010) Ocean Literacy scope and sequence for grades K-12. National Marine Educators Association, College Park, MD

NOAA (National Oceanic and Atmospheric Administration) (2013) OL: The essential principles and fundamental concepts of Ocean sciences for learners of all ages. NOAA, Washington, DC

Ryabinin V, Barbière J, Haugan P, Kullenberg G, Smith N, McLean C, Troisi A, Fischer A, Arico S, Aarup T, Pissierssens P, Visbeck M, Enevoldsen HO, Rigaud J (2019) The UN Decade of Ocean Science for Sustainable Development. Front Mar Sci 6:470. https://doi.org/10.3389/fmars.2019.00470

Santoro F, Santin S, Scowcroft G, Fauville G, Tuddenham P (2017) OL for all—a toolkit. UNESCO Venice Office, Paris

Shellock RJ, Fullbrook L, McKinley E, Cvitanovic C, Kelly R, Martin V (2024) The nature and use of OL in achieving sustainable Ocean futures: A systematic map. Ocean Coast Manag 257:107325

Steel BS, Smith C, Opsommer L, Curiel S, Warner-Steel R (2005) Public OL in the United States. Ocean Coast Manag 48:97–114. https://doi.org/10.1016/j.ocecoaman.2005.01.002

Tsai LT, Chang CC (2019) Measuring OL of high school students: Psychometric properties of a Chinese version of the OL scale. Environ Educ Res 25:264–279. https://doi.org/10.1080/13504622.2018.1542487

Visbeck M (2018) Ocean science research is key for a sustainable future. Nat Commun 9:4. https://doi.org/10.1038/s41467-018-03158-3

WWF (2022) Living planet report 2022—building a nature-positive society. (eds: Almond REA, Grooten M, Juffe Bignoli D, Petersen T). WWF, Gland

Author Contributions and Biographical Profiles

[1]*Bülent Çavaş: Co-conceptualization; Writing—Original Draft, Review, and Editing; Project Administration.

[2]Şermin Açık: Co-conceptualization; Writing—Original Draft, Review, and Editing.

[3]Simge Koç: Co-conceptualization; Writing—Original Draft, Review, and Editing.

[4]Mısra Kolaç: Co-conceptualization; Writing—Original Draft, Review, and Editing.

*Corresponding Author at bulent.cavas@deu.edu.tr

Bülent Çavaş, PhD, is a distinguished professor and scholar in the field of science education. He further honed his expertise with a Post-Doctoral fellowship at Middle East Technical University in Ankara, Turkey, focusing on Career Choices in STEM fields. Bülent served as the President (2017–2010) of the International Council of Associations for Science Education (ICASE), where he continues to influence global science education policies and practices. Additionally, he acts as an external expert for the European Commission, evaluating various projects in Brussels, Belgium. In addition, Bülent contributes to the global dialogue on Ocean sustainability as a member of the UNESCO–Ocean Decade Strategic Communication Group (SCG) and serves as a consultant for the Inter-Islamic Network on Oceanography, advising on scientific and educational initiatives to promote Oceanographic research and knowledge-sharing among Islamic countries.

Simge Koç, PhD, is an Assistant Professor. She completed her undergraduate (2008), graduate (2011), and doctoral (2017) studies in Science Education at Dokuz Eylül University. As a doctoral researcher in 2014–2015, she studied at the Laboratory Safety Institute in Boston, Massachusetts, USA. She has published numerous national and international papers and articles in the field of science education and has authored three books and several book chapters on Laboratory Safety and Teacher Professional Development Education. She also worked as a researcher for a national project on Ocean Literacy, further expanding her contributions to science education. Since 2021, she has been serving as a Lecturer in the Department of Science Education at Dokuz Eylül University.

Şermin Açık, PhD, is the Dean of Maritime Faculty at Dokuz Eylül University since 2023. She graduated from the Faculty of Fisheries at Ege University in 1999, completed her master's degree in 2002, and her PhD in 2007. She was appointed as associate professor in 2012 and professor in 2017. She served as the director of the Institute of Marine Sciences and Technology between 2019–2021 and has published 35 research papers in journals indexed by the Science Citation Index (SCI), 60 national-international papers, 4 book chapters and 1 international book. She has worked in nearly 35 projects supported by various institutions and is a member of the Marine Commission of Inter-University Council Presidency.

Mısra Kolaç, completed her undergraduate degree in the Science Education Department at Dokuz Eylül University Faculty of Education in 2021. After graduation, she began her master's degree in the Marine Living Resources Department at Dokuz Eylül University Institute of Marine Sciences and Technology. Her research focuses on the taxonomy of Hydrozoa. She also participated in a national project on Ocean Literacy. Through the Erasmus+ Student Exchange program, she studied for one semester in 2024 at the National and Kapodistrian University of Athens in Greece. She serves as a Black Sea Young Ambassador within the BRIDGE-Black Sea project, supported by the European Union's Horizon 2020 research and innovation programme.

18 Insights from Project SEA: Building Ocean Literacy and Enhancing Marine Science Education Through Teacher Professional Development

Lauren Madden, Louise Ammentorp, Nathan Magee, Graceanne Taylor, Gabriela Rivera-Peña, and Alexia Schianodicola

Abstract

The United Nations Educational, Scientific, and Cultural Organization (UNESCO) declared 2021–2030 as the Decade of Ocean Science for Sustainable Development, aiming to establish "Transformative Ocean science solutions for sustainable development, connecting people and our Ocean." A key strategy to achieve this goal is to enhance p-12 instructional strategies related to marine science. Building knowledge and skills in children addresses Ocean Literacy upstream, increasing the likelihood that students exposed to marine science instruction become a well-informed populace committed to protecting the world's Ocean. We share lessons learned from *Project SEA: Science Education and Action,* a New Jersey-SeaGrant-funded initiative that prepared 70 teachers in New Jersey to integrate marine science and climate change into NGSS-aligned lessons between 2020–2024. *Project SEA* represents a partnership between higher education and a nonprofit. Teachers completed surveys to measure their confidence in teaching, and knowledge of marine science and related subjects. Earlier analyses reveal that through our series of workshops, teachers' content knowledge and confidence in teaching marine science improved, as demonstrated by increased scores on items selected from the Survey of Ocean Literacy, written teaching reflections, and comments from focus group discussions. These teachers described the personal and professional benefits of building a community of like-minded educators committed to bringing Ocean science into preschool–12 learning settings. We will describe our experiences, share selected research findings from across our dataset, and offer recommendations for purposeful p–12 marine science education.

Keywords

Ocean Literacy (OL) · Marine science curricula · p–12 classroom learning · Informal learning

L. Madden (✉) · L. Ammentorp · N. Magee · A. Schianodicola
The College of New Jersey, Ewing, NJ, USA
e-mail: maddenL@tcnj.edu; ammentor@tcnj.edu; magee@tcnj.edu; schiana1@tcnj.edu

G. Taylor
Save Barnegat Bay, Toms River, NJ, USA
e-mail: education@savebarnegatbay.org

G. Rivera-Peña
Trenton Public Schools, Trenton, NJ, USA
e-mail: griverapena@trenton.k12.nj.us

18.1 Introduction

We are in the midst of what the United Nations Educational, Scientific, and Cultural Organization (UNESCO) has coined the *Ocean*

T. J. Kennedy (ed.), *Ocean Literacy: The Foundation for the Success of the Ocean Decade, Volume I*, Sustainable Development Goals Series, https://doi.org/10.1007/978-3-031-98145-6_18

Decade (2021–2030). This moment in history has called attention to the critical importance of our Ocean. The Ocean regulates weather and climate and provide food for much of the planet's human population. The Ocean serves as a sink for atmospheric carbon and adjust to human impacts such as pollution and overfishing. The complex marine systems that regulate so much of our planet's activity are still poorly understood by the general public, a gap we seek to address through supporting explicit marine science education in preschool–12 (p–12) learning settings.

Through our work, we seek to create tools and strategies for marine science teaching and learning in primary and secondary school settings by supporting teachers' content and pedagogical knowledge in this critical area. In an extensive literature review on development of environmental scientific literacy, Kumar et al. (2024) noted that, "teachers' knowledge and practices can impact the coming generation's attitude toward the environment and play an important role in producing environmentally literate students." (p. 8) Therefore, we assert that building teachers' knowledge and pedagogical skills related to marine science can support the development of students' Ocean Literacy.

18.2 Marine Science Education

Though 71% of Earth's surface is covered in water, and nearly 97% of that water is found in the Ocean, people hold a multitude of misconceptions related to the Ocean. In a focus-group study of 10–11-year-old children who reported learning about the Ocean in school and on television, Ballantyne (2004) found that children held a number of incorrect or partially correct conceptions related to the Ocean. These ideas ranged from a misunderstanding of the reasons why the sea is salty (e.g., that people throw salt in it) to explanations for currents (e.g., whales and other large animals swimming could cause currents). Wardlaw et al. (2022) also found that without explicit connections, students often neglect to recognize marine science efforts by scientists (in this case monitoring seaweed) as science. However, these incomplete understandings are not held by children alone. Fletcher et al. (2009) found that though the general public in the UK reported an interest in marine environments, there was a general lack of understanding of a variety of issues related to marine science. These reports echoed earlier ones in the United States (e.g., Steel et al. 2005) which showed that citizens held varied amounts of knowledge related to the Ocean. The variation that existed was based on geography (distance from the coast) and socioeconomic status.

When we move past citizens alone, and consider teachers specifically, it is critical to include marine science education in order to build a scientifically literate populace. Markos et al. (2017) used a quantitative survey instrument, the Survey of Ocean Literacy and Experience (SOLE), to study Greek preservice teachers' Ocean Literacy. This study revealed that teachers held a wide variety of conceptions related to marine science. Furthermore, the study showed that the instrument was useful for identifying teachers' conceptions and planning pre- and in-service teacher education to meet their needs purposefully. Payne and Zimmerman (2010) reported on the dearth of information in K–12 science textbooks, curricula, standards, and assessments related to marine environments while making a strong case for preparing current and future teachers to address the Ocean in their instruction. These authors contrast this lack of coverage of marine science content with the systemic global environmental issues directly related to the Ocean (e.g., global climate change, ozone depletion, decreasing biodiversity), and offer an instructional vision framework with direct connections to Ocean and aquatic education.

Given these findings, it is clear that addressing the gap in marine science education is crucial for enhancing marine science literacy, and science literacy more broadly. Science literacy involves not only understanding scientific concepts, but also applying that knowledge to make informed decisions in everyday life. (National Academies of Sciences, Engineering, and Medicine 2016). Ke et al. (2021) argue that individuals need interactions with multiple scientific models to develop the science literacy skills needed to understand complex systems. This argument aligns with the environmental literacy guidelines created by the North American Association for Environmental Education (NAAEE 2019), which emphasize the importance of considering the cultural and relational factors that drive humans' understanding and interaction with the environment. For marine science education, this means equipping teachers with both content expertise and effective pedagogical strategies. Given the nuances of the world's Ocean, it is critical to fill teachers' toolboxes with scientific content, resources and strategies that are suited to support their students in understanding the world's Ocean.

18.3 Theoretical Grounding: Ocean Literacy

Much like scientifically literate people use science to guide decision-making in day-to-day life, our work is grounded in the idea that Ocean-literate people can make informed and responsible decisions regarding the Ocean and its resources (NOAA 2024). The *Ocean Literacy Framework* comprises seven key ideas, the *Essential Principles* of Ocean science. These are:

1. Earth has one big Ocean with many features
2. The Ocean and life in the Ocean shape the features of Earth.
3. The Ocean is a major influence on weather and climate.
4. The Ocean makes Earth habitable.
5. The Ocean supports a great diversity of life and ecosystems.
6. The Ocean and humans are inextricably connected.
7. The Ocean is largely unexplored.

We assert that building Ocean Literacy begins with purposeful classroom instruction starting in the earliest years. We believe that grounding the development of this Ocean Literacy in the local context through place-based examples is critical (Kumar et al. 2024). To aid teachers in building their own Ocean Literacy and that of their students, we launched *Project SEA: Science Education, and Action,* a curricular and professional development project that took place over the 2020–2024 academic years.

18.4 Project SEA: Science, Education, and Action

With generous funding from the New Jersey SeaGrant Consortium, we developed a comprehensive program for enhancing place-based Ocean Literacy in New Jersey through Project SEA. The funding was secured over two consecutive two-year cycles (spanning 2020–2024) and allowed us to develop marine science curricular materials and a series of workshops for six cohorts of both current and future teachers (practicing and preservice teachers working collaboratively), in an effort to build Ocean Literacy in p–12 students across the state. A total of 70 teachers participated in our Project SEA workshops over the past 4 years. If we were to conservatively assume that each teacher-participant shared what they learned with just one colleague, and each of these teachers worked with just 25, p–12 students, our work had the potential to impact as many as 3500 children in each school year.

18.4.1 The Project SEA Team

With a vision for inclusivity that values multiple perspectives, we assembled a team of formal and informal educators to implement Project SEA. The Project SEA principal investigator and team leader is Dr. Lauren Madden, a professor of elementary science education with extensive experience preparing current and future teachers to teach science and an academic background in marine science. The project has two co-PIs: Drs. Louise Ammentorp and Nathan Magee. Dr. Ammentorp is a professor of early childhood and elementary teacher education with research interests in nature-based learning and collaboration between practicing and future teachers. Dr. Magee is an atmospheric physicist whose research focuses on the relationship between clouds and climate. His work includes research and outreach to support the development of high-quality science teachers. The team's informal educator was Ms. Graceanne Taylor, who is the education director of *Save Barnegat Bay,* a coastal New Jersey nonprofit organization. Throughout Project SEA, Ms. Taylor played a critical role in all aspects of the project. Over the past 4 years, several undergraduate student assistants have served as key members of our project team, most notably Gabriela Rivera-Peña and Alexia Schianodicola, who are co-authors on this chapter. These student assistants contributed in many ways ranging from logistical support to curriculum development assistance.

18.4.2 Our Places

Our work utilizes a place-based approach (Knapp 2005), and as such, centers on New Jersey. Place-based instruction require learners to consider the landscape and their connection to it as they learn and grow. New Jersey is a densely populated state in the mid-Atlantic region of the US with over 130 miles of coastline (See Fig. 18.1). We sought to engage with teachers across the entire state, but most participants lived and worked in the central part of New Jersey. The project's PI team is based at The College of New Jersey (TCNJ), a small, public, primarily undergraduate institution in central New Jersey near the Pennsylvania border. Faculty from TCNJ's School of Education and School of Science worked together to provide a clear focus on marine science content and pedagogy. TCNJ's facilities include state of the art model classrooms and laboratories including a Scanning Electron Microscope.

TCNJ's School of Education recently created an outdoor classroom designed to facilitate hands-on learning experiences in a natural setting. The area includes many native plants (planted with students) and is located next to the two campus lakes. Our work included a partnership with *Save Barnegat Bay,* a nonprofit organization based in coastal New Jersey committed to restoring and protecting the Barnegat Bay ecosystem through advocacy and education. Save Barnegat Bay is situated on 40 acres of woods and includes a large outdoor classroom with an amphitheater, demonstration gardens, a compost area, a greenhouse, rain barrels and more. It also has easy, walkable access to both fresh and brackish water.

18.4.3 Project Timeline

Project SEA was first awarded funding to support curriculum development and teacher professional learning for the period of 2020–2022 with a focus on elementary and middle school teachers (grades K–8). Soon after the award was made, the COVID-19 pandemic began. As a result, the team elected to switch focus and conduct preliminary research about New Jersey teachers' preparedness to teach about marine environments and climate change. Our early research (see Madden et al. 2023a) found that teachers across the state had mixed levels of understanding when it came to marine science, and a high interest in learning more about the topic, much like findings with teachers in earlier studies (Markos et al. 2017).

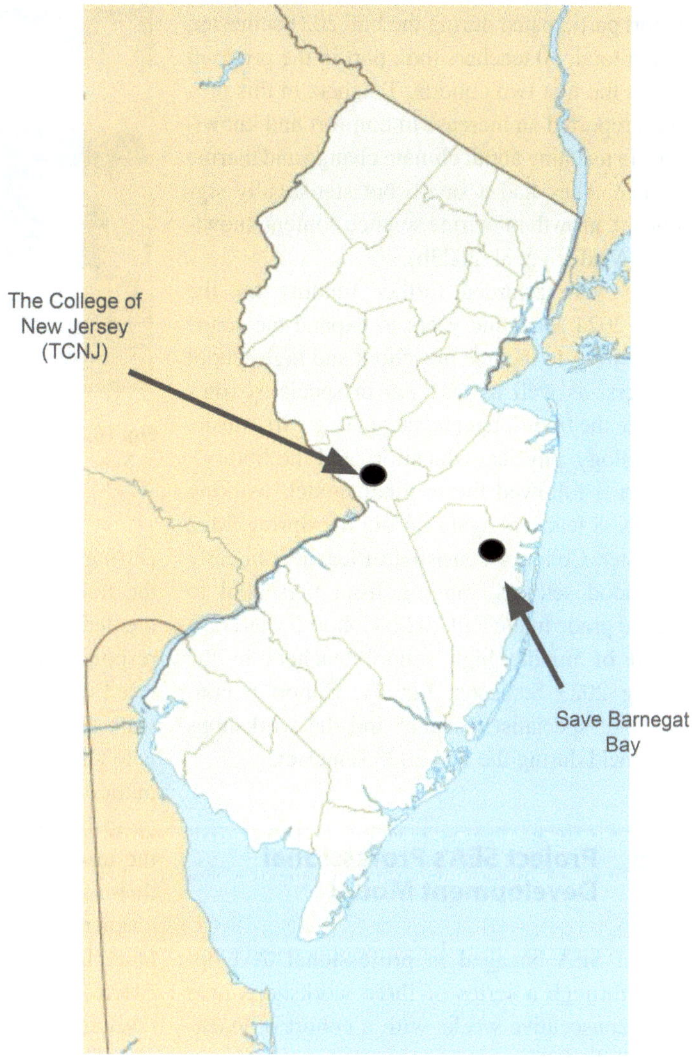

Fig. 18.1 New Jersey map. Adapted from Alexrk, *New Jersey map* [Map], Wikimedia Commons, CC BY 3.0. https://commons.wikimedia.org/wiki/File:New_Jersey_locator_map.svg

We also dedicated time during the early part of the pandemic lockdowns to revising an older curriculum developed by the Save Barnegat Bay Team, *This Way to Barnegat Bay*. Though the original lessons in the curriculum provided connections to place, they needed to integrate critical content related to climate change and its effects in coastal New Jersey. The original lessons were also not aligned with the Next Generation Science Standards (NGSS), the learning standards guiding science instruction in New Jersey (NRC 2012). Our team revised *This Way to Barnegat Bay* curriculum to ensure the lessons are up-to-date and easily accessible to teachers across the state and beyond (see resources section). During this same time frame, we also developed plans for a series of three, three-hour workshops to support teachers. We asked teachers to complete a survey before and after their participation. The survey included items describing their comfort levels teaching about climate change and marine science and their use of the Next Generation Science Standards along with a selection of items from the Survey of Ocean Literacy.

When pandemic-related restrictions were lifted, we recruited our first cohort of K-8 teachers. The first cohort's workshops took place during the Spring 2021 semester. A second larger cohort participated during the Fall 2021 semester, and in total, 30 teachers took part in the program across the first two cohorts. Teachers in this program reported an increase in comfort and knowledge in teaching about climate change and marine science. They had a small, but statistically significant, growth in marine science content knowledge (Madden et al. 2023b).

We were awarded further funding for the 2022–2024 academic years to expand the scope of our work to include preschool and high school teachers, as well as teachers in specialist roles outside the traditional classroom (e.g., art, music, technology, physical education, and the library). Cohort 3 followed the original model, working with K–8 teachers again during the Spring 2022 semester. Cohort 4 consisted of teachers in early childhood settings, ranging from preschool to second grade in the Fall 2022. Cohort 5 was composed of middle high school teachers in the Spring 2023 Semester. Finally, Cohort 6 consisted of specialist teachers and the workshops were held during the Fall 2023 semester.

18.5 Project SEA's Professional Development Model

Project SEA engaged in professional development through a series of three workshops over three consecutive weeks with a cohort of teachers. While there were variations between cohorts, we generally followed a similar plan during each round of implementation. Each workshop included time for community building and shared meals but held a different focus in terms of marine science content and pedagogy. Teacher participants completed pre- and post- workshop surveys and were also asked to teach one sample lesson and reflect on it as part of their participation in the research component of the project.

18.5.1 Workshop 1

The first workshop was held at Save Barnegat Bay's Ecocenter. The session started with a naturalist walk around the property led by Ms. Taylor.

Fig. 18.2 Project SEA teachers engaging in a discussion at Save Barnegat Bay's Ecocenter *(Photo by L. Madden, 2023)*

During the walk, she shared information about the flora, fauna, and geology of the site, commented on effects of climate change, and responded to participant questions. Teachers pictured in Fig. 18.2 are participating in this nature walk activity. An essential part of this workshop is to share tools and ideas on how to incorporate outdoor education into the classroom. We focus on instilling confidence and observation skills in the teachers so they can repeat the activities in their schoolyard with their students. On the walk, teachers select one item of interest, which they label. It is then photographed under the Scanning Electron Microscope on TCNJ's campus. The teachers were introduced to the SEM as an example of the tools and instruments used by marine scientists, allowing them to connect to the practices of science. Next, teachers engage in a demonstration lesson from the revised *This Way to Barnegat Bay* curriculum. Finally, the whole group engages in a discussion on the vision of the Next Generation Science Standards and strategies for integrating the content presented at the workshop in classroom settings.

18.5.2 Workshop 2

The second workshop was also held at Save Barnegat Bay's Ecocenter and included further examples of demonstration lessons and an interactive lecture on climate change in coastal New

Jersey. As participants stand and walk along the tidal beachfront of the Ecocenter, team members discuss key observations and data from up-to-date and recently published climate science. This session focuses on sharing a fundamental scientific understanding of physical climate-change mechanisms at the coast, especially those that are most observable and relevant to local exposures. Discussions connected to local and recent climate impacts often included sea level rise rates along the New Jersey shoreline, storm-surge flooding, flash flooding and contaminated runoff, and increasing frequency of extreme heat events. One of the foundational lessons included using a map of the watershed and exploring how urban development affects the natural water cycle on a local basis. This lesson is locally focused, place-based, and extremely adaptable for multiple grade levels and difficulty levels. Educators are always fascinated by the opportunity to examine a detailed watershed map, readily identifying the inherent connections between natural geographic features and familiar placemarks of well-known man-made infrastructure and development. All teacher participants were given maps of the Barnegat Bay watershed and of their own school's watershed to provide a tangible tool to help connect their students to the Ocean, regardless of distance from the Ocean.

18.5.3 Workshop 3

The third workshop was held on TCNJ's campus. After a final demonstration lesson from the *This Way to Barnegat Bay* curriculum, participants visited the Scanning Electron Microscope (SEM), to further build their understanding of the practices of marine scientists. The participants observed the operation of the SEM in real time, and were provided with high magnification/high resolution digital images of the items they collected from the Ecocenter nature walk (Fig. 18.3). They were also shown images taken by marine scientists at our institution, allowing them to connect their experiences to authentic marine science research practices. This workshop also included expert presenters, who share their knowledge and work with the participants. Presenters have included professors from TCNJ, teachers, and artists. Participants had the opportunity to browse children's and teacher books on all aspects of marine science. They also included time to meet with the project team to develop strategies for integrating the workshop content into their classrooms. Most importantly, educators are encouraged to stay in touch with one another.

Fig. 18.3 SEM image of moss collected by a teacher during a nature walk *(Photo by N. Magee, 2023)*

18.6 Insights and Key Takeaways

Over the course of Project SEA, we built a community focused on marine science teaching and learning. Teachers and project team members stayed connected electronically and through impromptu meetings and conferences and professional events. Community members shifted roles and perspectives, and what follows is a summary of some of the key overarching lessons learned as a result of our work together.

18.6.1 Building a Learning Community Focused on Ocean Literacy

One of the most exciting outcomes of Project SEA was the development of a community of learners and the relationships that were forged over the course of the project. Educators often work in silos. When we break down barriers and collaborate across settings and disciplines, we open the door to creative collaborations and partnerships. For many participants, rather than simply moving on after completing the workshops, participating teachers maintained contact with one another. In a few cases, the teachers co-planned additional lessons and units. The practicing teachers were also incredibly generous with the preservice teacher participants and offered their classrooms as spaces for co-teaching and exploration. Several of the preservice teacher participants went on to join the project team as a student assistant. Gaby Rivera-Peña followed this trajectory, starting as a participant in Cohort 3, then moving on to join the project team for the remainder of the funding period. In Gaby's words:

> Preservice teachers don't get to work collaboratively with practicing teachers often–the relationship is usually one where the practicing teacher mentors, and the preservice teacher learns. Project SEA's professional development was very different in that way. I gained this network of preservice teachers and practicing teachers who also found Ocean Literacy to be an important topic students should be learning about. The workshops were places where both groups worked together collaboratively and in partnership, which switched the power dynamic and types of conversations we were able to have. I also had the opportunity to work closely with an informal educator–Graceanne. As a formal educator, you don't really think about what informal educators can bring to the table when it comes to teaching and learning. People in academia don't tend to take advantage of informal educators. Being able to learn from someone who isn't bound by curriculum and standards was very refreshing and something I am definitely not used to. Collaborating with Graceanne helped me to think about more creative ways to present content and engage students.
>
> *Throughout Project SEA, not only was I fascinated by all the new information I was learning, but I also have access to several lesson plans that I can use when I have my own classroom. When I was asked to be a student worker on the project, I knew that I had a lot more to learn. I was able to take what I learned in the workshops (which were geared towards K–5 teachers) and help make more lessons for other age groups that included preschool, middle and high school.*
>
> Becoming a part of the team after participating in Project SEA was something that allowed me to continue to broaden my knowledge of Ocean Literacy and climate change. Science is not necessarily my strong subject, but I was welcomed and embraced by the team. Science is often a subject in school that does not get the attention it deserves, which transfers into teacher preparation programs. I didn't need to be an expert in the topic to be an important part of the team and this project.

As Gaby described in the vignette above, through Project SEA, we developed ongoing relationships and partnerships between formal p–12 schools, formal higher education settings in schools of both science and education, and an informal science learning center.

18.6.2 Project SEA's Equity-Focused Approach

Research demonstrates that early childhood and elementary teachers often report feeling uncomfortable with science teaching (Banilower et al. 2018). Our collaborative learning community centered an equity-focused approach to science literacy. We structured our workshops in such a way that early childhood and elementary teachers' knowledge and experience were valued and incorporated into collective sensemaking activities. These teachers were active participants in investigations and activities that built science knowledge. Further, we purposefully worked with teachers from Title 1 schools, including those located in parts of New Jersey quite far from the shore, such as the capital city, Trenton, near TCNJ's campus. In the US, Title 1 schools are defined by a high percentage of students from lower income demographics. It was critical to our team that all children in New Jersey have access to engaging and relevant strategies for teaching about the Ocean.

During our workshops with Cohort 4, focused on early childhood educators, the practicing teacher participants each worked in high-needs schools. These included a preschool teacher working at a Head Start program, a federally funded U.S. initiative that offers free early childhood education to toddlers through age five in

Fig. 18.4 Students performing *Fly, Osprey, Fly (Photo by L. Madden, 2023)*

preparation for formal schooling. Two additional educators were from a Title I lower elementary school in Trenton, where 60% of the students are Multilingual Learners (MLs). They joined the cohort together. The school serves students from kindergarten (approximately age 5) through third grade (approximately age 8). Both teachers eagerly embraced the opportunity to learn about climate change and its impact on marine ecosystems. The teachers eagerly embraced the opportunity to learn about climate change and its impact on marine ecosystems. These two teachers, a second-grade teacher and an English as a Second Language (ESL) teacher, enthusiastically took what they learned through our workshops and brought it to their students. For example, as part of the workshops, we modeled a lesson about osprey from the revised *This Way to Barnegat Bay* curriculum. The lesson covered types of nests, and where, how, and why migrating shorebirds build nests. Participants created nests out of natural materials found outdoors that osprey might also use to build nests. To conclude the lesson, we played a game where the teachers "flew" to bring materials to a nest-building area. The teachers adapted and taught the osprey lesson with their second-grade students and used it as a foundation to explore broader topics related to marine ecosystems. They integrated science with language arts to help students learn key vocabulary and concepts related to Ocean Literacy. The second graders also learned about the negative and positive impacts humans have on the environment through discussions of osprey habitat loss, and environmental organizations in the Barnegat Bay area that build osprey platforms.

The lessons at this workshop inspired the teachers so much that they decided to create a play about ospreys, which the children performed for their classmates and families. They titled the play "*Fly, Osprey, Fly*", and created a bilingual script in both Spanish and English in partnership with their students, pictured in Fig. 18.4. The teachers collaborated with the schools' art teacher, who made costumes and props with the students. For several weeks, the second graders practiced lines before their performance. This play engaged students and developed their Ocean Literacy in a way that was exciting and developmentally appropriate for young learners.

Furthermore, the second-grade teacher supervised two student teachers (who were not Project SEA participants) in her classroom. She mentored the student teachers as they developed and taught a two-week instructional unit on osprey and marine ecology. These experiences exemplify how Project SEA's impact is extended beyond the initial workshops; being part of a larger Ocean Literacy learning community can foster the sharing of knowledge and mentorship.

18.6.3 Connecting p-12 Educators with Scientific Enterprise

One central component of our workshops was engaging teacher participants in authentic science practices. By using the tools, strategies, and language of Ocean scientists, these teachers were able to actively *do* science rather than simply learn *about* it actively.

During the first visit to Save Barnegat Bay's Ecocenter, participant teachers each selected small items of interest on a nature walk to view using TCNJ's state-of-the-art Scanning Electron Microscope (SEM) during the third workshop. In addition to advanced microscopy tools used in interdisciplinary science, participants repeatedly engaged in scientific discussions and interactive explorations of up-to-date and recently emerging climate science. For example, teachers were shown how to use an electronic meter to precisely measure ambient carbon dioxide concentrations and connect those measurements to the global Keeling Curve showing the inexorable rise of carbon dioxide in the global atmosphere.

Workshops also included other just-in-time scientific discussions included local storm surge impacts of Hurricane Sandy, personal stories of losses from recent high impact flash flooding in New Jersey, and the impacts of extreme heat to school operations, student athletic events, and student learning. By engaging directly with scientists, peer-reviewed science outcomes, and cutting-edge technology used by scientists, these teachers truly participated in the scientific enterprise. New and dramatic climate impacts along the Ocean shore become recognizable often, and by observing these changes through both personal and scientific lenses, participants are able to deeply appreciate the connections across the local and global threats of climate change. This first-hand experience with scientific practice was intended to demystify Ocean science and climate change science, forging connections with scientists and science in a manner that could be shared with p–12 students and help them connect to this important enterprise as well.

Additionally, the naturalist walks during the first workshop with each cohort helped p–12 educators to see the "messiness" and humanity in the process of doing science. Though Ms. Taylor has deep knowledge and expertise on the coastal New Jersey ecosystem, there were still times that she was stumped. These moments sparked further discussions about the way scientists build knowledge and connections in the community. When the answers were not clear, we discussed how to discover answers alongside classroom students. Problem-solving and observation skills are imperative in these complicated discussions as the challenges we face due to climate change are evolving.

This exposure to the process of science and a community of learners expanded into further collaborations. For example, co-PI Dr. Magee was able to set up a weather station at one of the schools as part of a future project, providing children with further opportunities to do science. She noted, "We learned so much from the TCNJ workshops. We loved having Prof Magee here installing the weather station and providing support was phenomenal! Our students were excited about all the new ideas we brought into the classroom. The students enjoyed the weather station tremendously and were eager to check the reading each morning. They loved being on Google Meet and being the "weather person" for the entire school. We also created graphs and added vocabulary words to our weather chart each morning."

18.6.4 Curating Resources for Teachers to Adapt

The curriculum we developed, the revised version of *This Way to Barnegat Bay*, is a free resource that teachers can use anywhere. The lesson plans can serve as a starting point for teachers who want to adapt them for their own classrooms and schools. While this does not address the disparity in references to marine organisms and ecosystems compared to terrestrial examples in textbooks, it provides a starting point for helping teachers represent marine systems in their classrooms. We also developed a list of books categorized by grade level, scientific accuracy, and its relevance to the current day. Some books with less detailed information could be used to introduce a topic while more detailed books could be used to dig deeper into the topic that is being taught. Another unique aspect to the book accumulation of is that we included notes which refer to the sociocultural representation that the books may entail. In science education, it is possible to intersect other disciplines, so we dedicated time

Fig. 18.5 Kindergarteners marking up a map of their watershed and creating their map *(Photo by L. Madden, 2022)*

to finding books that were culturally appropriate and representative while still connecting each book to our science lessons. Additionally, the list includes Project SEA lessons that align with the books which allows teachers to pick books that are supplementary to the lesson.

Teacher participants also took the framework for learning about Ocean Literacy and model lesson plans to design their own based on the workshops. One Cohort 2 participant, a kindergarten teacher in a Title 1 school in coastal New Jersey used the watershed mapping activity shared by Ms. Taylor to create a similar one with her kindergarteners. In Fig. 18.5 above, kindergarteners used symbols to mark up the map of the area surrounding their school then created their own versions using paint and paper.

Exposure to Project SEA content and pedagogy, and access to the *This Way to Barnegat Bay* curriculum empowered teachers to build their own Ocean Literacy and their students' Ocean Literacy. The continued connection between practicing teachers, teachers in training, and informal educators has had a positive ripple effect on our community.

18.6.5 Educating Informal Educators of Formal Educator Needs

While this project was designed to empower practicing and preservice teachers with Ocean Literacy knowledge and resources, it also worked as professional development for the informal education program at Save Barnegat Bay. Project SEA fostered a strong and ongoing relationship where formal educators could express their needs for their science classrooms and beyond. These needs can often be met by informal education programs that have access to different resources, outdoor spaces, and alternative teaching methods. Project SEA opened opportunities in science education, enterprise, and global networking for a grassroots organization that might have yet to have made those connections.

As a result of Project SEA, grassroots environmental education is more empowered as it has allowed Save Barnegat Bay to better serve the community beyond the project. Save Barnegat Bay has since developed more targeted professional development events for formal classroom teachers. Additionally, the informal educators interning at Save Barnegat Bay were able to deliver presentations to formal educators. They extended the relationship between formal and informal learning spaces to build Ocean Literacy broadly.

18.6.6 Expanding the Conversation

Our collaborative work through Project SEA allowed us to showcase partnerships and collaboration between disciplines, institutions, and individuals. It also contributed to the global conversation on climate change education- and how critical our Ocean is to understanding this global threat. Sharing findings from our research and experiences of individual teachers allowed us to connect to and participate in the emerging field of climate change education. During the course of Project SEA, New Jersey became the first state in the United States to include climate change in its learning standards across disciplines. Our resources and tools provided starting points for teachers to begin integrating this important topic. We were able to share our work publicly through professional conferences and journal articles, but also more broadly through media features. Our efforts allowed us to join a global conversation and share tools with educators and the community at large. The relationships built between formal p–12 educators, higher education faculty, and informal learning centers also became jumping off points for further collaborations between these different silos of education.

18.7 Conclusion

We must think creatively about what it means to succeed in developing Ocean Literacy. Most individuals experience traditional school-based learning in p–12 settings. These settings can serve as springboards for Ocean Literacy when paired with purposeful curricula.

Project SEA has helped develop strong relationships between an academic institution, a grassroots nonprofit, and teachers in the classroom. Additionally, teachers have been empowered to bring their students' learning beyond the physical classroom. Partnerships like these build a foundation upon which the Ocean Literacy movement can be sustained. As illustrated in the stories we shared throughout this chapter, Project SEA- and similar programs- provide a powerful starting point for teachers to lead the way in building an Ocean-literate populace ready to tackle the challenges of our future.

18.7.1 Summary

Key concepts explored in this chapter include:

- Project SEA's place-based professional development model empowered New Jersey teachers across grade levels and disciplines to integrate marine science and Ocean Literacy into NGSS-aligned instruction, enhancing both content knowledge and instructional confidence.
- Authentic partnerships between formal and informal educators fostered a dynamic, supportive learning community, connecting classroom teachers with scientists, preservice teachers, nonprofit educators, and informal education institutions.
- Workshops immersed participants in real-world scientific research and field experiences, including interactions with live marine organisms, laboratory investigations, and use of advanced tools like scanning electron microscopes.
- An equity-centered approach ensured participation from educators in Title I schools and under-resourced districts, with intentional support for multilingual learners, early childhood educators, and inclusive practices.
- Teachers translated workshop experiences into student-centered projects, such as bilingual plays, student-led investigations, and curriculum modifications that connected local waterways and marine science to classroom learning.
- Long-term impact was achieved through sustained engagement, with teachers sharing Project SEA resources and approaches with colleagues, mentoring others, and incorporating Ocean Literacy into school-wide and district-wide planning.
- Project SEA aligned with state and national education priorities, supporting implementation of the New Jersey Climate Change Education standards and contributing to the goals of the UN Ocean Decade by fostering an Ocean-literate society.

18.7.2 Recommended Resources

The following resources provide further insights and support on this topic:

- For more information on Project SEA's work: https://sites.google.com/tcnj.edu/projectsea/
- Project Sea Curriculum: https://sites.google.com/tcnj.edu/projectsea/curriculum-resources-for-teachers

Acknowledgements The author team would like to acknowledge the incredible teachers who participated in our Project SEA workshops over the 2020–2024 school years. We have learned so much from our time working together. This publication is the result of work, research sponsored by the New Jersey Sea Grant Consortium (NJSGC) with funds from the National Oceanic and Atmospheric Administration (NOAA) Office of Sea Grant, U.S. Department of Commerce, under NOAA grant number NA24OARX417C0154-T1-01 and the NJSGC. The statements, findings, conclusions, and recommendations are those of the author(s) and do not necessarily reflect the views of the NJSGC or the U.S. Department of Commerce. NJSG-25-1030.

References

Ballantyne R (2004) Young students' conceptions of the marine environment and their role in the development of aquaria exhibits. GeoJournal 60(2):159–163

Banilower ER, Smith PS, Malzahn KA, Plumley CL, Gordon EM, Hayes ML (2018) Report of the 2018 NSSME+. Horizon Research, Inc

Fletcher S, Potts JS, Heeps C, Pike K (2009) Public awareness of marine environmental issues in the UK. Mar Policy 33(2):370–375

Ke L, Sadler TD, Zangori L, Friedrichsen PJ (2021) Developing and using multiple models to promote scientific literacy in the context of socio-scientific issues. Sci & Educ 30(3):589–607

Knapp CE (2005) The "I–thou" relationship, place-based education, and Aldo Leopold. J Exp Educ 27(3):277–285

Kumar V, Choudhary SK, Singh R (2024) Environmental socio-scientific issues as contexts in developing scientific literacy in science education: A systematic literature review. Soc Sci Humanit Open 9:100765

Madden L, Ammentorp L, Magee N, Taylor G (2023a) Marine science, climate change, and the next generation science standards: Understanding the knowledge and perspectives of K-8 New Jersey teachers. Curr J Mar Educ 38(1):3–15. https://doi.org/10.5334/cjme.74

Madden L, Ammentorp L, Magee N, Taylor G (2023b) Making waves & building knowledge: Developing K-8 teachers' capacity for teaching marine science & climate change. Sci Educ 29(2):1–8

Markos A, Boubonari T, Mogias A, Kevrekidis T (2017) Measuring Ocean Literacy in pre-service teachers: Psychometric properties of the Greek version of the Survey of Ocean Literacy and Experience (SOLE). Environ Educ Res 23(2):231–251

National Academies of Sciences Engineering and Medicine (2016) In: Snow CE, Dibner KA (eds) Science literacy. The National Academies Press, Washington, DC. https://doi.org/10.17226/23595

National Oceanic and Atmospheric Administration (NOAA) (2024) Ocean Literacy: The essential principles and fundamental concepts of Ocean sciences for learners of all ages, Silver Spring, MD

National Research Council. (NRC) (2012) A framework for K–12 science education: Practices, crosscutting concepts, and core ideas. National Academies Press

North American Association for Environmental Education (NAAEE) (2019) K-12 environmental education: Guidelines for excellence. https://dg56ycbvljkqr.cloudfront.net/sites/default/files/eepro-post-filesk-12_ee_guidelines_for_excellence_2019_4.pdf

Payne DL, Zimmerman TD (2010) Beyond terra firma: Bringing Ocean and aquatic sciences to environmental and science teacher education. In: The inclusion of environmental education in science teacher education. Springer, Dordrecht, pp 81–94

Steel BS, Smith C, Opsommer L, Curiel S, Warner-Steel R (2005) Public Ocean Literacy in the United States. Ocean Coast Manag 48(2):97–114

Wardlaw J, Benavides-Lahnstein A, Robinson L, Lorke J, Pratt-Taweh S, Ghadiri Khanaposhtani M, Ballard H, Burton V (2022) Citizen science framing and delivery models: Impacts on young people's environmental science learning. Connected Science Learning 4(5):12318643. https://www.nsta.org/connected-science-learning/connected-science-learning-september-october-2022/citizen-science

Author Contributions and Biographical Profiles

[1]*Lauren Madden: Conceptualization; Writing—Original draft, and review; Project Leadership and Supervision; Grant Writing; Photography.

[2]Louise Ammentorp: Writing—Insights and Key Takeaways Sections, review and edits; Methodology—Data Collection and Analysis.

[3]Nathan Magee: Writing—Workshop descriptions, Use of the Scanning Electron Microscope, and Connecting p–12 educators to scientific enterprise sections.

[4]Graceanne Taylor: Writing—Workshop descriptions, Use of the Scanning Electron Microscope, and Connecting p–12 educators to scientific enterprise sections.

[5]Gabriela Rivera-Peña: Writing—Participant vignette.

[6]Alexia Schianodicola: Editing—Grammar and format checking throughout, adding references.

*Corresponding Author at maddenL@tcnj.edu

Lauren Madden, PhD, is a Professor of Elementary Science Education at The College of New Jersey, USA. Her research has been supported by grants from the New Jersey SeaGrant Consortium, National Science Foundation, and U.S. Environmental Protection Agency. She holds a PhD in Science Education, MS in Marine Science, and BA in Earth Sciences with a focus on oceanography. Her current research and teaching focuses directly on climate change education, specifically in primary school settings in New Jersey. Her expertise has been featured prominently in many media outlets including the New York Times, Washington Post, The Guardian, NPR, and the Star Ledger.

Louise Ammentorp, PhD, is a Professor in the Elementary and Early Childhood Education department at The College of New Jersey, USA. Her research emphasizes innovative, collaborative approaches to teacher preparation and teacher professional development. She is the co-author of *Let Nature Be Your Teacher: Integrating Nature-Based Learning in the Elementary Classroom*, a recent work that offers practical strategies for incorporating nature-based learning in elementary education.

Nathan Magee, PhD, is currently Professor and Chair of the Physics Department at The College of New Jersey (TCNJ). He conducts research on high-altitude cloud/climate interactions, urban environmental heat and air quality impacts, and works to make up-to-date climate science accessible to undergraduate students and for K–12 instruction. Nathan is also dedicated to helping students train to become new high school physics teachers, a field with a severe shortage of highly qualified new teachers. His research has been funded by the National Science Foundation, U.S. Department of Energy, and the Research Corporation. Nathan has a B.S. in Physics from Carleton College, and a Ph.D. in Atmospheric Science from the Pennsylvania State University.

Graceanne Taylor is the Education and Outreach Director at Save Barnegat Bay. She has been an environmental educator in the Barnegat Bay Watershed for more than a decade. Graceanne holds a bachelor's degree in marine science.

Gabriela (Gaby) Rivera-Peña is a special education teacher at Paul Robeson Elementary School in Trenton, NJ. They completed her MAT in Special Education in December 2024. They hold undergraduate degrees in Elementary Special Education and Women's, Gender, and Sexuality Studies. Her research revolves around social justice education where she uses intersectional feminism, critical race theory, disability studies, queer theory, and ecojustice to explain why discussions about race, gender, disability, climate change and other social issues should be invited in K–12 classrooms.

Alexia Schianodicola is an undergraduate student at The College of New Jersey, pursuing a degree in Special Education/Elementary Education with a concentration in Psychology. She will continue her degree next Fall of 2025 to receive her MAT in special education in 2026. Alexia is passionate in research that focuses on educational practices and content, and the cognitive development in young children in classroom environments. She is very proud of Project SEA's team and its mission to further promote and accentuate environmental science in preK–12 classrooms. She will continue this mission in her own classroom in the future.

The Global Marine Curriculum: Academic Pathways for Sustainable Marine Development

19

Ninja Mueller

Abstract

How can we meet the UN Challenge 10 to restore society's relationship with the Ocean? One important group of society is the one working in Ocean-related occupations. It is therefore important to critically assess what kind of human-Ocean relationship is conveyed in academic marine education, when young adults (age 18+) prepare for their professional careers as decision makers in Ocean development. Around 3300 Academic Marine Education Programmes (AMEPs) around the world hold the potential to restore the Ocean-relationship of thousands of people. A major problem for the current destructive and exploitative human-Ocean relationship is the instrumentalising and oppressive perspective western humanity takes on the Ocean. The concepts of Sustainability, Traditional Ecological Knowledge (TEK), and Speciesism are identified as possible solutions to shift the exploitative paradigm to a respectful and valuing human-Ocean relationship. The implementation of these concepts in current academic marine education is investigated in the agricultural sector of aquaculture and fisheries. While the concept of Sustainability is present in almost all analysed curricula, TEK is rarely found, and Speciesism is entirely absent. This chapter presents an analysis of randomly selected AMEPs in the agricultural sector from around the world, identifying emerging patterns in curricular focus, with particular attention to concepts related to TEK in the curricula.

Keywords

Human-Ocean relationship · Marine education · Ocean Literacy (OL) · Speciesism · Terrestrial bias · Traditional ecological knowledge

19.1 Introduction

The UN Ocean Decade draws much needed attention to the Ocean's needs and the urgency of a sustainable human-Ocean interaction. Challenge 10 of the UN Ocean Decade calls for restoring society's relationship with the Ocean, placing a focus beyond conveying knowledge about the Ocean, but instead considering attitudes and individual mindsets about it (Intergovernmental Oceanic Commission 2024). Ocean Literacy (OL) aims to do so by increasing society's understanding of the Ocean's impact on humanity and humanity's impact on

N. Mueller (✉)
University of Vechta, Leuphana University Lüneburg, Vechta and Lunenburg, Germany
e-mail: ninja.mueller@mail.uni-vechta.de

T. J. Kennedy (ed.), *Ocean Literacy: The Foundation for the Success of the Ocean Decade, Volume I*, Sustainable Development Goals Series, https://doi.org/10.1007/978-3-031-98145-6_19

the Ocean (Ocean Literacy Network 2013). It successfully gains attention and tries to include as many people as possible, advancing research on marine education for various groups such as children, school students and adults. The necessity to push for marine education derives from the so-called terrestrial bias, meaning that issues concerning developments on land receive higher attention in all sectors of life, such as policy, education, industry, environmental protection, among others. The terrestrial bias often present in education is also prevalent in education research; only 91 publications in the Web of Science keywording "Ocean Literacy", "Marine Education" or "Ocean Education" have been found in the last two decades (Fauville et al. 2019). This illustrates the need for further marine education research in order to establish OL in educational curricula. However, while the establishment of marine education is crucial for facing the UN Ocean Decade Challenges, a critical view on the development and execution of already existing marine education should not be forgotten.

Academic Marine Education Programmes (AMEPs), such as marine biology, aquaculture, fisheries, and marine engineering, are responsible for the education of young adults (ages 18+) around the world who are interested in learning more about the Ocean. They already have a certain relationship with the Ocean that is further shaped and defined by their academic studies. The understanding of the Ocean that is conveyed in their studies is likely to be their prevalent understanding and perspective on the Ocean throughout their professional careers. Additionally, and maybe most importantly, these young people studying Ocean-related academic programmes are the decision makers of tomorrow's Ocean-related developments. Therefore, it is worthwhile to investigate the status quo of AMEPs worldwide and understand their role in mastering the 10 Challenges of the UN Ocean Decade.

Education for Sustainable Development (ESD) is an educational approach dedicated to equipping students with the knowledge, skills, and critical mindset to support sustainable development (Holst and Brock 2020). While both the practice and research of ESD are expanding, it continues to carry a strong terrestrial bias. This highlights the need to better understand what *Education for Sustainable Marine Development (ESMD)* should offer students and how it can be implemented to help restore their relationship with the Ocean. This chapter presents an initial exploration of what ESMD should aim to achieve within academic marine education and examines whether these elements are currently reflected in AMEPs around the world.

This chapter is organized into four main sections. Section 19.2, *What shall ESMD achieve?*, explores key concepts aimed at restoring students' relationship with the Ocean and fostering more sustainable human–Ocean interactions. This includes a focus on integrating Traditional Ecological Knowledge (TEK) and incorporating marine animal perspectives to move beyond the dominant anthropocentric lens in marine education. Section 19.3, *Materials and Methods*, outlines the methodology used to assess how ESMD is currently reflected in AMEPs. A newly developed global database of AMEPs provides the basis for this analysis, with a random sample examined to identify whether key concepts for sustainable marine development are included in academic curricula. Section 19.4, *Results*, presents findings on the distribution and accessibility of AMEPs across different disciplines worldwide, with special attention given to programs in the agricultural sector, particularly fisheries and aquaculture, which pose unique challenges in incorporating TEK and marine animal perspectives. The final section draws conclusions from the study and includes a summary list of the chapter's main ideas for reference.

19.2 What Shall ESMD Achieve?

The scope of this chapter does not allow to discuss all the aspects of what would be needed for a holistic ESMD. With the focus on the UN Challenge 10 to restore society's relationship with the Ocean, only the necessary measures in marine education to achieve this change in human-Ocean relationship for a sustainable development are investigated.

The large-scale exploitation and destruction of the marine environment and its devastating impacts on human and non-human animals on land and in the Ocean gave reason to the UN to calling out an Ocean Decade. Gottschlich and Bellina (2017) argue that oppression and hierarchical thinking are the main cause for large-scale exploitation of natural and human resources. Furthermore, Oakley points out that "the human-centred belief that people are superior to nature is a central part of the problem of environmental degradation" (Oakley 2019, S. 24). Therefore, the underlying convictions that cause the exploitative relationship of human society with the Ocean should be addressed in marine education.

When focussing on the human-Ocean relationship in education with special attention to internalised oppressive behaviours, the strongly western-influenced perspective of marine education and OL needs to be considered (MacNeil et al. 2021). Originating from the United States of America (USA) in the beginning of this century, OL was increasingly acknowledged and developed by western researchers and educators, with associations in Europe, Canada, Australia and the Pacific (the latter being founded through collaboration of researchers from Australia and the USA and subsequently including Pacific islands) (IPMEN 2023; McKinley et al. 2023). Investigating how the concept and wording of OL was conceived by people from different cultural backgrounds within Canada, it was found to be perceived "narrow and limiting" (MacNeil et al. 2021, S. 234), illustrating the need to critically assess around whom OL and marine education are centred.

These human-Ocean relationship-defining aspects lead to two concepts that should be addressed for a more holistic ESMD; Traditional Ecological Knowledge, and Speciesism.

19.2.1 Traditional Ecological Knowledge

While the traditional ecological Ocean knowledge of Indigenous peoples is slowly, but increasingly, accepted in Ocean governance and the management of coastal areas, and is incorporated in existing Ocean management programmes, most academic marine education does not yet draw on that knowledge to develop the curriculum (MacNeil et al. 2021; Whitehouse et al. 2014). This might be a manifestation of the greater issue of Indigenous science and knowledge not being acknowledged in western culture as legitimate science and knowledge. It is therefore rarely incorporated in academic education (Whitehouse et al. 2014) which aims to be 'scientific' as understood in a western way, where objectivity and the written word are highly valued and legitimise knowledge (Gerhard and Deborah 2019). However, as Indigenous ways of knowing evolved through the human experience with the natural world and a learning *with* rather than *about* nature, the exploration of Indigenous ways of knowing and science is worthwhile when restoring the human relationship with nature. Traditional Ecological Knowledge (TEK), as Indigenous ways of knowing are often referred to, seeks to understand the connections and relationships between humans and the natural environment (Huff 2022), therefore providing valuable approaches and perspectives to rethink and restore the human-Ocean relationship conveyed in education. For an ESMD, the incorporation of TEK in academic marine education could therefore be an enrichment in meeting the UN Challenge 10.

19.2.2 Speciesism

Drawing from the understanding that oppression and exploitation function in the same way, no matter if they concern a marginalised group of human or non-human animals, the concept of Speciesism and the role of marine animals needs to be considered for a holistic ESMD. Speciesism is the belief that the instrumentalisation of non-human animals for human use is morally justified. While multiple branches in ESD aim to include non-human animals in the same moral consideration as humans in order to avoid exploitative patterns (e.g. critical animal studies, ecopedagogy, animal liberation pedagogy, or posthumanism), they often fail to bring attention to the marine environment and its inhabitants. Therefore, marine species are disadvantaged on an intersectional level of Speciesism and terrestrialism (Norton 2003; Wadiwel 2016). Ocean Literacy aims to advance education about the marine environment, however, it mainly does so through an anthropocentric lens; The Ocean's worth is measured by its use for human health and wellbeing, and marine animals are commonly included in the term "marine resources", exposing the utility value assigned to Ocean inhabitants in OL (Santoro et al. 2017). The UN Sustainable Development Goals (SDGs) also display a highly anthropocentric perspective, justifying SDG14 "life below water" as follows: "In order for Oceans, seas and *marine resources to successfully contribute to human well-being*, ecosystem integrity, with properly functioning biogeochemical and physical processes, is required." (United Nations 2023). In order to change society's relationship with the Ocean, however, the anthropocentric lens of instrumentalising the Ocean and its inhabitants, and defining its value based on its utility for humanity needs to change too. Therefore, the concept of Speciesism as an underlying paradigm representing the justification of the instrumentalisation and utilisation of the Ocean and marine animals is considered as an important indicator for a holistic ESMD.

19.3 Materials and Methods

19.3.1 Mapping of AMEPs Around the World

In a first step, Ocean-related study programmes from around the world on bachelor and master level are listed in a database. They are derived from two study programme platforms: marinetraining.eu and studyportals.com.

Marinetraining.eu is a website listing not only academic marine study programmes, but also short courses and projects concerning marine education, however, a filtering specifically for bachelor and master programmes is possible. The list of bachelor and master programmes on marinetraining.eu comprises more than 3300 study programmes, however, some programmes are double listed.

Studyportals.com is divided into masterportals.com and bachelorportals.com and is selected after a review of various study programme platforms, since it provides the filters needed to separate Ocean-related programmes from other subjects, and furthermore reaches a higher number of Ocean-related study programmes compared to other platforms. While universities from all continents insert their study programmes, a clear western bias is evident, with most study programmes listed taking place in Europe and North America. Concerning Ocean-related study programmes, zero listings are derived from South America, even though academic marine programmes in South America exist.

Both search platforms are based in western culture, therefore, the western bias prevails. An attempt to access further AMEPs through external marine educator networks beyond the western realm did not yield any further outcomes.

The database created for the research project at hand can never claim completeness, due to the vast number of universities and study programmes worldwide and the continuous new development or omission of AMEPs. Furthermore, it is important to note that the criteria for entering the database might exclude some academic marine education programmes, since only bachelor and master programmes are considered. Therefore, all study programmes not adapted to the bachelor/master system are excluded from the database. Additionally, a prerequisite to be listed in the database is an online presence. Already cancelled programmes that still have an online presence might therefore be listed, while currently running study programmes without website or a weak digital infrastructure might be excluded.

A categorisation based on the AMEP's discipline (Applied Science, Formal Science, Humanities, Natural Science, and Social Science), geographic location (Africa, Asia, Europe, North America, Oceania, and South America), and educational level (Bachelor, Master) is carried out.

19.3.2 Curricula Analysis of AMEPs

The analysis on the presence of selected concepts in AMEP curricula is based on Holst and Brock (2020) using specific keywords to identify the concepts in question in the information provided about the AMEPs. The presence of a comprehensive overview of the AMEP contents or access to its curriculum or syllabus are an important prerequisite to further analyse it. However, information provided varies strongly throughout universities, therefore, only a random selection of AMEPs providing enough information can be analysed.

Each curriculum, syllabus or comprehensive website description is searched using a variety of keywords to identify the presence or absence of the concepts of Sustainability, Traditional Ecological Knowledge, and Speciesism. If any of the keywords is found, further reading and identification of the concept in question is performed. The keywords were identified as viable indicators to find the respective concepts throughout the work process. These concepts are understood by the author as follows:

Sustainability

Keywords: *Sustainability, Conservation, Responsibility, Care* (Note: the keywords are listed as nouns; however all alterations of these words were taken into account when searching the documents e.g. sustainability / sustainable / sustainably).

The keywords are chosen to cover different approaches to sustainability. In connection to marine education, *Conservation* usually has an ecological focus and belongs to the cognitive domain of learning. *Responsibility* and *Care* draw stronger on the moral domain which plays an important role in today's discourse around sustainability. In both cases, it is made sure that keywords counted for the analysis are indeed connected to the concept of sustainability and not mentioned in a different context.

Traditional Ecological Knowledge

Keywords: *Traditional, Indigenous, Equality.* In order to find anything related to TEK the simple keywords *Traditional* and *Indigenous* are used without further specification. When collecting the data mentions of these words not referring to anything around Indigenous or local communities in the widest sense are omitted. It is important to mention that *Equality* does not directly refer to TEK and is not suitable to pinpoint the presence of TEK-related concepts. However it is a supportive keyword as it indicates a certain awareness for concepts of Equality understood in a wider sense as opposed to oppression or exclusion.

Speciesism

Keywords: *Speciesism, Animal, Resource, Ethics*. Due to a lack of words being used throughout the curricula that indicate teaching contents about Speciesism only supportive keywords are used. As described earlier the use of the word *Resource* as description for Ocean-inhabitants is utilised in combination with the word *Animal* in order to understand how marine animals are referred to in the curricula. Furthermore the use of ethics as a basic first step towards a moral consideration of marine animals is included.

19.3.3 Keyword Placement and Curriculum Focus

Three different levels are taken into consideration in the curricula: the overall programme summary or website description; the titles of singular courses within these programmes; and the description of these courses. It is assumed that these levels show different stages of importance and acknowledgement in the AMEP. If for example a course within an academic study programme is titled "sustainable aquaculture", the concept of sustainability is officially embedded in the curriculum and anchors sustainability as a substantial key-topic of that specific course. The occurrence of the word "sustainability" in a course description instead of a title might give insight to sustainability being a cross-sectional theme that is incorporated in multiple courses with different foci. Furthermore, if sustainability is mentioned frequently in the overall programme description, it might be a special focus of the whole respective programme. If it is frequently present in the programme description but does not find mentioning in the course titles or descriptions, it might instead be used as a buzzword to attract prospective students, rather than being indeed incorporated in the programme. Hence, these three levels of mentioning the different concepts and/or keywords are taken into consideration in the analysis.

It is important to mention that the absence of a keyword in the provided information is not proof of the actual absence of the respective concept in the real study programme practice. Curricula not available in English language are translated using google document translate or google website translate, therefore, specific wordings might be lost in translation. However, no pattern of undetected concepts related to a specific language is found. The amount of AMEPs varies strongly throughout continents, therefore, a similar number of AMEPs for each continent is analysed for better comparability. The analysis focuses on the contents and learning objectives of the curricula; however, it does not consider didactic approaches, as information on these is rarely provided.

19.4 Results

19.4.1 Distribution of AMEPs Around the World

A total of 3366 AMEPs on bachelor and master level are recorded. The distribution of AMEPs around the globe shows a strong imbalance of access among continents and countries. While Asia and Europe each provide about a third of AMEPs (30% and 35% respectively), Africa, Oceania, and South America altogether comprise only 12% of all AMEPs worldwide. North America provides 23% of AMEPs, with the majority being offered in the USA and only a fifth of AMEPs in North America being offered in other countries.

The distribution of AMEPs being concentrated in the Global North and parts of Asia reveals an issue of access to academic marine education. Some countries in the Global South have only one or no AMEP to offer, denying students the opportunity to professionalise in an Ocean-related career, if they are not able to move to a different country for studying. While it is an important issue that needs to be addressed, geographic accessibility to marine education is not further discussed in this chapter, as the focus lies on analysing existing AMEPs and prevalent concepts to change the students' relationship with the Ocean.

The majority of AMEPs are offered in applied sciences and natural sciences, making up 97% of all AMEPs worldwide, while programmes offered in the social sciences, humanities and formal science comprise to 3%, meaning almost 100 programmes worldwide. Within the applied and natural sciences, a strong focus lies on 5 main disciplines making up over three quarters of all AMEPs offered globally: Engineering and Technology (22%); Earth Science (19%); Agriculture (16%); Biology (14%); and Shipping and Transportation (9%). However, the discipline foci are differently distributed on the different continents, showing for example a strong focus on Engineering and Technology in Europe, a strong focus on Earth Science in Oceania, or a strong focus on Agriculture in Asia (Fig. 19.1).

A specific focus of this book chapter is placed on the AMEPs of the agricultural sector. On the one hand, the scope of the chapter does not allow an in-depth analysis of all disciplines, and on the other hand, the agricultural sector is particularly interesting in the context of human-Ocean interaction and relationship; Fishing and aquaculture have been an important part of human culture for centuries, providing a long and rich history of traditions and practices. Indigenous communities developed various methods for fishing and up until today, fish is a major food source for many coastal communities around the globe. At the same time, fishing became a major threat to the Ocean due to overfishing, bycatch, the destruction of marine habitats through destructive fishing practices etc. Furthermore, industrial fishing and local small-scale fisheries are faced with challenges to co-exist, introducing the topic of power-relationships and oppression of marginalised local communities. As the previously introduced concepts of Sustainability, Traditional Ecological Knowledge, and the incorporation of moral consideration of marine animals all come

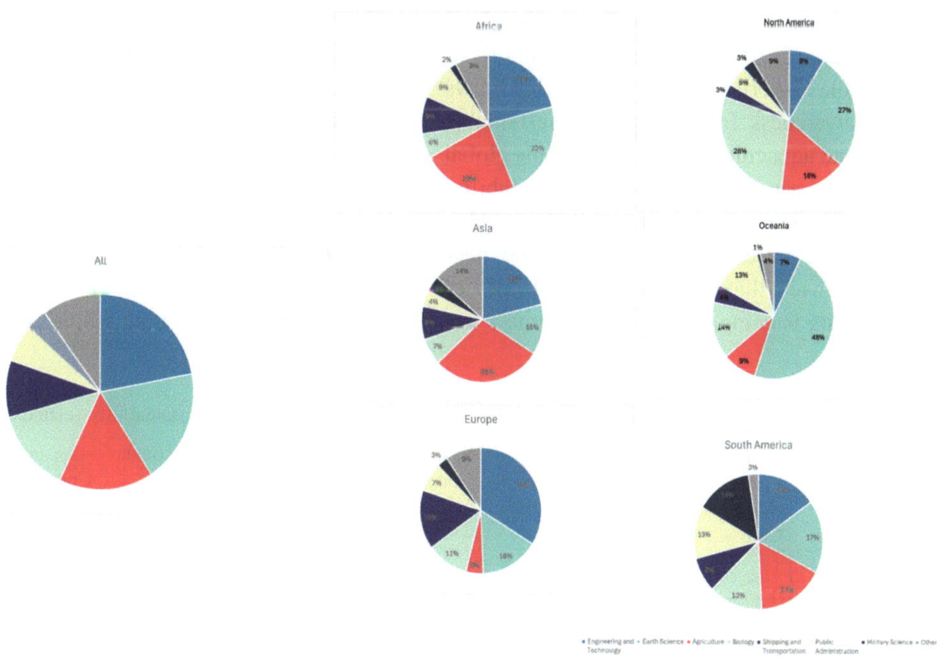

Fig. 19.1 Distribution of the 7 most prevalent disciplines in academic marine education across continents

together in the agricultural sector, it provides an interesting case for investigating AMEPs in aquacultural and fishery studies.

In the agricultural sector, a total of 542 AMEPs is found with more than half of them being offered in Asia. The syllabi of 65 AMEPs have been analysed for the presence or absence of the concepts of Sustainability, TEK, and Speciesism. No noteworthy differences have been found between bachelor and master programmes; therefore, programmes of both levels are treated within the same category in the following analysis.

19.4.2 Sustainability-related Concepts in Agricultural AMEPs

The need for a sustainable development in fisheries and aquaculture is emphasised in most of the syllabi and curricula of analysed AMEPs. Only 14 out of 65 AMEPs do not mention *Sustainability* specifically in their documents, 51 AMEPs use the word somewhere on their website or in their curricula, 5 out of those have it in the title of the programme, e.g. "Sustainable Aquaculture" or "Sustainable Fisheries Management". This shows that the awareness for the need of sustainable fisheries and aquaculture has arrived in the global agricultural marine curricula. While Sustainability needs to be incorporated in academic marine education, the presence of the word does not necessarily prove that a holistic concept of Sustainability is being taught, as it became a frequently used buzzword throughout the last years. Therefore, further keywords are searched for in order to identify if concepts are present that should be encompassed by the term Sustainability.

Compared to the word *Sustainability* being present in almost all AMEPs, the words *Conservation*, *Responsibility*, and *Care* are found less frequently, with *Conservation* being mentioned in 45 AMEPs, *Responsibility* in 29 AMEPs, and *Care* in 10 AMEPs (Table 19.1). This shows a decrease in conveyed concepts the further they deviate from subject content to concepts drawing on attitudes or mindsets related to Sustainability. Five AMEPs do not mention any of these concepts: one bachelor and 2 master programmes in Asia, 1 master programme in Europe, and 1 master programme in South America.

The levels where the different keywords are mentioned in the analysed AMEPs vary as well (Table 19.2). While *Sustainability* is mentioned in the most AMEPs overall compared to the other keywords, it does not appear as frequently in as many AMEPs' course titles or course descriptions as *Conservation* does. Only 5 AMEPs overlap with having *Sustainability* and *Conservation* both in their course titles. The keyword *Responsibility* occurs only in the course titles of 2 AMEPs, one of them also having *Sustainability* and *Conservation* in their course titles. The keyword *Care* does not occur at all in the course titles of any AMEP, summing up to 25 AMEPs that officially incorporate the concepts of Sustainability, Conservation, or Responsibility in their curricula by dedicating whole courses to it.

Table 19.1 Number of AMEPs mentioning the different keyword combinations of Sustainability-related concepts

	Sustainability Conservation Responsibility Care	Sust. Cons. Resp. –	Sust. – Resp. Care	Sust. Cons. – –	Sust. – Resp. –	– Cons. – Care	– – Resp. Care	Sust. – – -	– Cons. – -	– – Resp. –	– – – –
Africa	1	6	0	1	1	0	0	1	0	0	0
Asia	0	4	0	4	1	0	1	1	0	2	3
Europe	2	1	1	3	1	0	0	2	1	0	1
North America	2	2	0	3	0	2	0	2	2	1	0
Oceania	0	2	0	5	0	0	0	1	0	0	0
South America	0	1	0	3	0	0	0	0	0	0	1
All	**5**	**16**	**1**	**19**	**3**	**2**	**1**	**7**	**3**	**3**	**5**

Table 19.2 Number of AMEPs mentioning the keywords on different levels: Programme description/Course title/Course description

	Total no. of AMEPs analysed	Sustainabilty		Conservation		Responsibility		Care	
Africa	10	10	6/2/5	8	4/3/5	8	4/0/4	1	0/0/1
Asia	16	10	6/1/6	8	2/2/6	8	6/1/3	1	1/0/0
Europe	12	10	7/3/8	7	2/0/6	5	0/0/5	3	0/0/3
North America	14	9	4/5/6	11	6/6/8	5	2/1/3	5	3/0/3
Oceania	8	8	4/1/5	7	3/3/7	2	0/0/1	0	0/0/0
South America	5	4	2/1/3	4	1/2/3	1	0/0/1	0	0/0/0
All	65	51	29/13/33	45	18/16/35	29	12/2/17	10	4/0/7

The presence of Sustainability-related concepts in the curricula slightly varies between continents. Africa and Oceania are comparably strong in incorporating at least one of the monitored concepts, with 100% of their AMEPs mentioning one or usually multiple of the keywords in their curricula. North America also has at least one of the monitored concepts incorporated in each of the AMEPs analysed, however, around a third of these programmes only incorporate one of the concepts investigated. Europe and South America incorporate Sustainability-related concepts in almost all of the analysed AMEPs, however, both offer one AMEP each not mentioning any of these concepts. Finally, Asia offers by far the highest amount of agricultural AMEPs, so the number of randomly selected AMEPs analysed for this study is most likely not representative for the variety of programmes offered throughout Asia. Nevertheless, 3 out of the 16 analysed programmes do not incorporate any of the Sustainability-related concepts in their curricula.

The concept of Sustainability is mainly understood in an ecological dimension throughout most AMEPs. Social or economic sustainability is rarely directly addressed in the context of Sustainability, except for a few AMEPs showing a remarkable offer in courses addressing Sustainability from different angles. These AMEPs are, however, not bound to a specific continent or an academic education level but are probably rather results of a beneficial combination of institutes already engaged in Sustainability research and development. The overall lack of perspectives on Sustainability concepts beyond the ecological dimension in most AMEPs reveals the need to incorporate a more holistic understanding of Sustainability in agricultural marine education.

19.4.3 TEK-related Concepts in Agricultural AMEPs

While Sustainability-related concepts are present in almost all AMEPs, TEK is rarely incorporated in the analysed curricula (Table 19.3). Only 17 AMEPs mention at least one of the keywords, 48 AMEPs do not incorporate TEK into their official curricula. It is important to bear in mind that the keyword *Equality* is a supportive keyword to help finding TEK-related concepts but does not stand in itself for the incorporation of Indigenous ways of knowing. Therefore, more accurately, only 13 AMEPs incorporate TEK. All keywords are examined for their relatedness to the concept in question and only taken into consideration if the relation is indeed given. Even though all mentions noted in the table do relate to TEK, the majority refers to learning *about* Indigenous issues, culture, rights, policies etc., rather than learning *from* TEK. Only 4 AMEPs specifically mention TEK as a source of knowledge to draw from in their curricula. While it is an important first step to acknowledge Indigenous interests and teach about them in agricultural marine education, it still retains a westernised perspective on "the other", instead of an inclusive and equal approach to learn with different ways of knowing beyond the western paradigm.

Table 19.3 Number of AMEPs mentioning the different keyword combinations of TEK-related concepts

	Traditional Indigenous Equality	Traditional Indigenous –	– Indigenous Equality	Traditional – Equality	Traditional – –	– Indigenous –	– – Equality	– – –
Africa	0	2	0	0	0	0	1	7
Asia	0	0	0	0	0	0	1	15
Europe	0	1	1	0	0	0	1	9
North America	0	2	1	1	1	0	0	9
Oceania	0	1	0	0	1	1	0	5
South America	1	0	0	0	0	0	1	3
All	**1**	**6**	**2**	**1**	**2**	**1**	**4**	**48**

The keywords are almost entirely found in the course descriptions. Neither *Traditional* nor *Indigenous* are mentioned in any of the overall programme descriptions. In one Asian and one South American bachelor, *Equality* is mentioned in the programme descriptions. However, it both relates to Equality, independently from specifically accounting for TEK, which is also not mentioned anywhere in the respective curricula. The concept of TEK mainly finds mentioning in the course descriptions as one of multiple topics to be treated. Only in North America, 3 AMEPs dedicate own courses to the concept of TEK.

Here, the individuality of culture and location becomes evident. From the 3 AMEPs offering own courses on TEK, 2 are located in Canada and 1 in Alaska, USA, both being regions comparably advanced in restoring their Indigenous cultures and knowledge. Furthermore, the incorporation of Indigenous ways of knowing or the effort to include TEK into education shows on more levels than could be captured through this research. For example, universities in New Zealand, Oceania, provide a large number of course descriptions in English and Māori, or aim to incorporate their Indigenous culture on university level rather than in singular study programmes. Another example are some universities in South Africa, putting a focus on decolonising their curriculum, which is not detectable in the single programmes or courses, but still takes place in the offering university. However, as described in Section 19.3, *Materials and Methods*, the focus of this research project lies on detecting how the investigated concepts are truly anchored in the curricula of the analysed AMEPs, instead of understanding the overall university's agenda.

Overall, all of the continents have a high rate of AMEPs not incorporating TEK in their curricula. Mainly the AMEPs analysed from Asia and South America lack the Indigenous perspective in their curricula, except for one AMEP in South America mentioning all keywords. However, the programmes of Europe and Africa also offer only a small amount of AMEPs that consider TEK in some way. Especially with the rare occurrence of TEK in the analysed AMEPs, it is important to mention that the difference between mandatory and elective courses is not included in the analysis, as this information is not always provided, and structures differ strongly between universities. Therefore, the data only gives insight into what students can study if they choose to have a special focus on TEK, however, it does not prove that all students graduating from the respective AMEP indeed acquired TEK throughout their studies.

19.4.4 Speciesism-related Concepts in Agricultural AMEPs

The keyword *Speciesism* was not found in any of the documents provided, giving no indication on an awareness of Speciesism in any of the analysed AMEPs. However, other keywords relating to an awareness (or unawareness) of marine animals as living beings that are worth a moral consideration are found. While the previous concepts are mainly retrieved by searching keywords relating to the respective concepts, other ways of identifying the position of Speciesism in the curricula are introduced: the absence of certain words; and the presence of words possibly standing in contradiction with teaching about Speciesism.

An interesting combination is the mentioning of the words *Resource* or *Animal*. Both words are the most frequently used words among all keywords registered with *Resource* being counted 561 times, resulting in 2.85 mentions per 1000 words in the curricula and syllabi of the analysed AMEPs, and *Animal* being counted 359 times, leading to 1.82 mentions per 1000 words (Table 19.4). For comparison, the next frequent keyword is *Sustainability* with 269 counts and a frequency of 1.37 mentions per 1000 words.

Most AMEPs mention *Resource* in reference to marine animals, in total 84.56% of all AMEPs analysed. The mention of *Animal* in 66.78% of AMEPs is notably below, mainly through a strong discrepancy in Africa and South America, where *Animal* is mentioned in only 30% and 40% of AMEPs respectively. In AMEPs of the other continents, the discrepancy is less noteworthy with *Animal* usually being mentioned in less AMEPs than *Resource*. Only Asia shows a higher amount of AMEPs using the word *Animal* instead of *Resource*, when referring to marine animals.

It is meaningful that an educative sector working with animals at its centre produces academic programmes that do not mention the word *Animal* a single time in their curricula, in this random sample selection a total of 18 AMEPs (Table 19.5). While this does not provide any information about the quality of the study programmes, it gives an indication for an instrumentalising perspective on marine animals as inanimate objects, respectively resource for human use and consumption. In many cases, also in AMEPs mentioning the word *Animal*, the word *Resource* is used frequently to describe marine animals. Often, the mention of marine animals as either *Resource* or *Animal* is consistent within courses but varies between courses within one AMEP,

Table 19.4 Comparison of the keywords resource and animal

	Resource			Animal		
	Present in % of AMEPs	No. AMEPs mention in course title	Occurrence per 1000 words	Present in % of AMEPs	No. AMEPs mention in course title	Occurrence per 1000 words
Africa	100.00	3	2.41	30.00	0	0.28
Asia	75.00	6	4.17	93.75	6	5.58
Europe	91.67	4	2.69	83.33	6	1.67
North America	85.71	7	4.65	78.57	4	1.55
Oceania	75.00	2	0.97	75.00	4	2.81
South America	80.00	3	2.11	40.00	1	0.29
All	84.56	25	2.85	66.78	21	1.82

Table 19.5 Number of AMEPs mentioning the different keyword combinations of Speciesism-related concepts

	Resource Animal Ethics	Resource Animal –	Resource – Ethics	– Animal Ethics	Resource – –	– Animal –	– – –
Africa	1	2	2	0	5	0	0
Asia	4	7	0	0	1	4	0
Europe	4	5	0	1	2	0	0
North America	6	3	1	0	2	2	0
Oceania	2	3	0	0	1	1	1
South America	2	0	1	0	1	0	1
All	**19**	**20**	**4**	**1**	**12**	**7**	**2**

leading to a mixed use of words when referring to marine animals. This shows the importance of the course structure and focus for conveying specific perspectives and understandings on marine inhabitants. It is also important to mention that 8 AMEPs only use the word *Animal* and never the word *Resource*, while not using any of the other words defined to identify possible perspectives close to the animal's interests. Therefore, only the use of one wording or the other does not necessarily indicate an overall more critical or animal perspective-including understanding throughout the programme. However, it still does influence the initial view on the animal as a living being rather than a harvestable object.

It is furthermore concerning that a sector centred around the slaughter and consumption of living beings does not have a strong focus on animal ethics; 41 AMEPs do not mention ethics in their curricula, only 6 AMEPs provide one course each on ethics. These courses, however, are not focusing on animal ethics, but on research ethics in general. This reveals a lack of ethical discourse in agricultural academic marine education that would be supportive for students to critically question the instrumentalising and anthropocentric perspective on the Ocean and its inhabitants.

19.5 Conclusion

To meet the UN Challenge 10 to restore society's relationship with the Ocean, a special focus of this book chapter has been placed on the human-Ocean-relationship conveyed in academic marine education as a major driver for Ocean-related future developments. Sustainability, Traditional Ecological Knowledge, and Speciesism were identified as promising concepts to include in academic marine education in order to ensure an Education for Sustainable Marine Development evoking a respectful relationship with the Ocean.

The analysis of randomly selected AMEPs around the globe gives valuable insight into the presence and absence of agricultural AMEPs concerning the introduced concepts. The concept of Sustainability is mainly present in an ecological dimension throughout the analysed AMEPs and it becomes evident that its importance is understood around the world with few exceptions. Nevertheless, Sustainability is often not taught holistically, and especially with focus on the human-Ocean relationship, the inclusion of less cognitive approaches that require emotional investment, critical questioning of mindsets, or a change in attitudes is often not given in agricultural AMEPs.

While Sustainability arrived in most AMEPs, TEK is rarely incorporated in their curricula with only a fraction aiming to learn *from* TEK instead of *about* it. However, in order to change the relationship from an Ocean-instrumentalising perspective to an understanding of the Ocean's inherent value, a change of perspectives from a strongly westernised learning culture towards an experience-with-nature-based view could be beneficial to meet UN Challenge 10. The incorporation of TEK in academic marine education requires a sensitive handling and an awareness of the local circumstances of the university offering the AMEP.

The concept of Speciesism is not incorporated at all in the analysed AMEPs. Furthermore, the understanding of Ocean inhabitants as animals is less present than the understanding of them as resources. This is interesting in that marine animals are at the centre of agricultural AMEPs and are the component the whole agricultural marine sector is based upon. Considering marine animals as morally worthy individuals would help shifting the human-Ocean relationship from commodifying the Ocean's inhabitants towards acknowledging their intrinsic value. A stronger focus on including animal ethics in agricultural AMEPs might be a possibility to help restoring this relationship.

Overall, agricultural AMEPs show a first effort to develop towards a more Sustainability-focused education and prove that they are able to incorporate concepts that might have been contradictory to conventional commercial fishing in the first place but nevertheless found their way in the curricula. However, the vast majority of agricultural AMEPs is far from a holistic ESMD. The incorporation of further concepts enabling the students to critically question the paradigms of the sector they commence a professional career in, can therefore be a possible endeavour. Next steps resulting from this research could be to investigate if students studying one of the few special AMEPs incorporating TEK and Speciesism-related concepts on a comparably high level indeed internalise a different relationship with the Ocean. Furthermore, measures to better incorporate TEK and an understanding of the concept of Speciesism need to be understood and tested. Finally, this book chapter only gives insight into the third largest discipline of academic marine education. A similar initial analysis needs to be undertaken for the other disciplines, in order to ensure that academic marine education equips its students to master the UN Challenges and jointly work towards a healthy, balanced and valued Ocean.

19.5.1 Summary

Key concepts explored in this chapter include:

- The current westernised human-Ocean relationship is instrumentalising, leading to a destructive exploitation of the marine environment and its inhabitants.
- The human-Ocean relationship needs to change by incorporating other perspectives than an instrumentalising one.
- Academic marine education is a strong lever to meet UN Challenge 10 and restore society's relationship with the Ocean, as young adults (age 18+) manifest their relationship with and understanding of the Ocean for their professional career as decision makers in Ocean-related developments.
- Traditional Ecological Knowledge is beneficial to help students critically question existing paradigms on their individual relationship with the Ocean.
- An understanding of Speciesism and the need for moral consideration of marine animals is necessary to develop a respectful and not commodifying relationship with the Ocean and its inhabitants.

19.5.2 Recommended Resources

The following resources provide further insights and support on this topic:

- Intergovernmental Oceanic Commission. (2024). *Ocean Decade Vision 2030 White Papers: Challenge 10: Restoring society's relationship with the Ocean—UNESCO Digital Library* (No. 51.10; Ocean Decade series, S. 38). https://unesdoc.unesco.org/ark:/48223/pf0000390126
- MacNeil, S., Hoover, C., Ostertag, J., Yumagulova, L., & Glithero, L. (Diz). (2021). Coming to Terms with Ocean Literacy. *Canadian Journal of Environmental Education (CJEE)*, *24*(1), Article 1.
- Norton, B. G. (2003). Marine Environmental Ethics. Where we might start. In *Values at Sea: Ethics for the Marine Environment.* University of Georgia Press.
- Oakley, J. (2019). What Can an Animal Liberation Perspective Contribute to Environmental Education? In T. Lloro-Bidart & V. S. Banschbach (Hrsg.), *Animals in Environmental Education* (S. 19–34). Springer International Publishing. https://doi.org/10.1007/978-3-319-98479-7_2

Acknowledgements I'd like to thank marinetraining.eu and Cláudia Delgado for the great collaboration and support, as well as Prof. Dr. Marco Rieckmann, and the European Marine Science Educators Association for the fruitful discussions around this research project.

References

Fauville G, Payne DL, Marrero ME, Lantz-Andersson A, Crouch F (eds) (2019) Exemplary practices in marine science education. Springer International Publishing. https://doi.org/10.1007/978-3-319-90778-9

Gerhard PS, Deborah HW (2019) Limitations of the western scientific worldview for the study of metaphysically inclusive peoples. Open J Philos 09(03):295–317. https://doi.org/10.4236/ojpp.2019.93020

Gottschlich D, Bellina L (2017) Environmental justice and care: critical emancipatory contributions to sustainability discourse. Agric Hum Values 34(4):941–953. https://doi.org/10.1007/s10460-016-9761-9

Holst J, Brock A (2020) *Bildung für nachhaltige Entwicklung (BNE) in der Schule: Strukturelle Verankerung in Schulgesetzen, Lehrplänen und der Lehrerbildung*

Huff L (2022, February 28) Through a White Colonial Lens: A look into the US education system. School of Marine and Environmental Affairs. https://smea.uw.edu/currents/through-a-white-colonial-lens-a-look-into-the-us-education-system/

Intergovernmental Oceanic Commission (2024) Ocean Decade Vision 2030 White Papers: Challenge 10: Restoring society's relationship with the Ocean—UNESCO Digital Library (No. 51.10; Ocean Decade series, S. 38). https://unesdoc.unesco.org/ark:/48223/pf0000390126

IPMEN I. P. M. E. N (2023) *About IPMEN.* International Pacific Marine Educator Network. https://ipmen.net/ipmen-2023

MacNeil S, Hoover C, Ostertag J, Yumagulova L, Glithero L (2021) Coming to terms with Ocean Literacy. Can J Environ Educ (CJEE) 24(1):1

McKinley E, Burdon D, Shellock RJ (2023) The evolution of Ocean Literacy: A new framework for the United Nations Ocean Decade and beyond. Mar Pollut Bull 186:114467. https://doi.org/10.1016/j.marpolbul.2022.114467

Norton BG (2003) Marine environmental ethics. Where we might start. In: Values at sea: ethics for the marine environment. University of Georgia Press

Oakley J (2019) What can an animal liberation perspective contribute to environmental education? In: Lloro-Bidart T, Banschbach VS (eds) Animals in environmental education. Springer International Publishing, pp 19–34. https://doi.org/10.1007/978-3-319-98479-7_2

Ocean Literacy Network (2013) Ocean Literacy: The essential principles of Ocean sciences for learners of all ages (No. 2)

Santoro F, Santin S, Scowcroft G, Fauville G, Tuddenham P (2017) Ocean Literacy for all: A toolkit. UNESCO

United Nations (2023) Oceans and seas. Department of Economic and Social Affairs. https://sdgs.un.org/topics/Oceans-and-seas

Wadiwel DJ (2016) Fish and pain: The politics of doubt. Animal Sentience 1(3):31. https://doi.org/10.51291/2377-7478.1054

Whitehouse H, Watkin Lui F, Sellwood J, Barrett MJ, Chigeza P (2014) Sea Country: Navigating Indigenous and colonial ontologies in Australian environmental education. Environ Educ Res 20(1):56–69. https://doi.org/10.1080/13504622.2013.852655

Author Contribution and Biographical Profile

[1]*Ninja Mueller: Conceptualization; Writing—Original Draft, Review, and Editing.

*Corresponding Author at ninja.mueller@mail.uni-vechta.de

Ninja Mueller is a PhD student at the University of Vechta, Germany, investigating decolonising the marine curriculum and applying a critical Ocean pedagogy at the intersection of Ocean Literacy and critical animal studies. Ninja holds a master's degree in marine biological resources, and a master's degree in education for sustainable development. She teaches sustainability management at Leuphana University, Germany, and leads the development of a school community focusing on education for sustainable development. Furthermore, she is an active member of the Asian Marine Educators Association (AMEA) Board and member of the Critical Animal Studies group of Gothenburg University (GU-CAS).

Scientists for Ocean Literacy: Empowering Scientists as Ocean Advocates

20

Dina Eparkhina, Elisa Banchi, Ángel Borja, Donata Canu, Daria Ezgeta-Balić, Elena Giusta, Andrew King, Panayota Koulouri, Marina Locritani, Marta Manca Zeichen, Angela Pomaro, Francesca Petrera, Maria C. Uyarra, and Louise Valestrand

Abstract

The UN Ocean Decade *Scientists for Ocean Literacy* project, led by EuroGOOS, brings together marine research institutions to promote Ocean Literacy through cross and multidisciplinary collaborations and international partnerships. The project aims to integrate Ocean Literacy into the core activities of research organisations and public agencies, strengthening the use of Ocean knowledge and information in policymaking, research, and innovation. In response to the UN Ocean Decade challenges, the *Scientists for Ocean Literacy* project promotes innovative outreach and engagement strategies, fostering dialogue between marine scientists and diverse societal actors. Activities include events, resources, surveys, trainings, and showcases. Building

D. Eparkhina (✉)
European Global Ocean Observing System (EuroGOOS), Brussels, Belgium
e-mail: dina.eparkhina@eurogoos.eu

E. Banchi · D. Canu · F. Petrera
National Institute of Oceanography and Applied Geophysics (OGS), Sgonico-Trieste, Italy
e-mail: ebanchi@ogs.it; dcanu@ogs.it; fpetrera@ogs.it

Á. Borja · M. C. Uyarra
AZTI, Marine Research Division, Basque Research and Technology Alliance (BRTA), Gipuzkoa, Spain
e-mail: aborja@azti.es; mcuyarra@azti.es

D. Ezgeta-Balić
Institute of Oceanography and Fisheries, Split, Croatia
e-mail: ezgeta@izor.hr

E. Giusta · M. M. Zeichen
Italian National Institute for Environmental Protection and Research (ISPRA), Rome, Italy
e-mail: elena.giusta@isprambiente.it; marta.mancazeichen@isprambiente.it

A. King · L. Valestrand
Norwegian Institute for Water Research (NIVA), Oslo, Norway
e-mail: andrew.king@niva.no; louise.valestrand@niva.no

P. Koulouri
Hellenic Centre for Marine Research (HCMR), Gournes, Greece
e-mail: yol72@hcmr.gr

M. Locritani
Istituto Nazionale di Geofisica e Vulcanologia (INGV), Rome, Italy
e-mail: marina.locritani@ingv.it

A. Pomaro
National Research Council of Italy (CNR), Institute of Marine Sciences (ISMAR), Venice, Italy
e-mail: angela.pomaro@cnr.it

T. J. Kennedy (ed.), *Ocean Literacy: The Foundation for the Success of the Ocean Decade, Volume I*, Sustainable Development Goals Series, https://doi.org/10.1007/978-3-031-98145-6_20

on the efforts of the EuroGOOS Ocean Literacy Working Group, the project aims to bridge the gap between science and society, deepening public understanding of the Ocean's importance while promoting marine science as a career path. This chapter presents case studies from the project's partners across Europe, illustrating best practices in integrating Ocean Literacy into science, policy, and society.

Keywords

Ocean Literacy (OL) · Ocean science · Oceanography · Multi-disciplinary partnerships · Societal engagement · Science communication

20.1 Introduction: Linking Science and Society for Sustainable Ocean Stewardship

This chapter explores the role of scientists as active contributors to Ocean Literacy, which is essential for informed decision-making on environmental impacts, public policy, and maritime activities. The international Ocean Literacy framework encourages public engagement in science, fostering a well-informed, responsible relationship between society and the Ocean. The framework, built on seven core principles, highlights the Ocean's role in Earth's systems, climate, biodiversity, and the profound human-Ocean connection. Ocean Literacy also incites citizens to contribute to marine data collection and analysis through citizen science, providing solutions and supporting fit-for-purpose Ocean observations and services. The success of the UN Decade of Ocean Science for Sustainable Development 2021–2030 (UN Ocean Decade) and Agenda 2030 relies on effective Ocean observation, technological innovation, and data services, which demand strong public and policy engagement. The Essential Principles of Ocean Sciences are defined as follows:

1. Earth has one big Ocean with many features.
2. The Ocean and life in the Ocean shape the features of Earth.
3. The Ocean is a major influence on weather and climate.
4. The Ocean makes Earth habitable.
5. The Ocean supports a great diversity of life and ecosystems.
6. The Ocean and humans are inextricably interconnected.
7. The Ocean is largely unexplored.

Ocean Literacy encompasses a profound comprehension of the reciprocal relationship between the Ocean's influence on individuals and their impact on the Ocean. An Ocean-literate person grasps the fundamental principles and essential concepts governing the Ocean's functioning, communicates effectively about Ocean-related matters, and makes informed, responsible decisions concerning Ocean resources (NOAA 2024). Although the term "Ocean Literacy" may not directly translate into many languages, variations such as "sea literacy," "sea culture," "Ocean culture," or "science popularisation" convey similar concepts, reflecting its widespread recognition in contemporary Ocean science and policy circles.

Aligned with the principles of the "new societal contract of science" (Gibbons 1999), which advocates for participative science engaging both the public and policymakers through co-design, consultations, and accessible communication of results, Ocean scientists play a pivotal role. They serve as key conduits for translating scientific research into actionable insights for shaping public policies, transcending the boundaries of academic research.

European Global Ocean Observing System (EuroGOOS) highlights Ocean Literacy as a vital link between Oceanography, policy, and society, enabling public support and the sustained and fit-for-purpose operation of Oceanographic systems (Eparkhina et al. 2021; Eparkhina and Lehtonen 2017). Ocean Literacy also provides Oceanographic, environmental, and meteorological agencies with opportunities to engage diverse audiences across generations, professions, and socio-economic groups. This fosters intellectual, emotional, and creative connections with different sectors of society, helping scientific institutions resonate more deeply with the public.

Initiated and developed by EuroGOOS, the UN Ocean Decade *Scientists for Ocean Literacy* project seeks to foster synergies among Ocean Literacy initiatives undertaken by European scientific communities at national and institutional levels. By uniting Oceanographers, hydrographers, meteorologists, communicators, and policy experts, the project aims to amplify the scientific and societal impacts of these endeavours. Its overarching objective is to advocate for Ocean Literacy at institutional and funding levels, enhancing societal engagement through multidisciplinary and cross-sectoral collaborations.

The project places particular emphasis on engaging professionals from diverse sectors, including communications, social sciences, and multidisciplinary Ocean research spanning social, environmental, policy, and economic areas. By promoting the integration of Ocean Literacy into the core activities of research institutions and publicly funded agencies, the project aims to stimulate public engagement in Ocean science and help deepen collective awareness of the Ocean's fundamental role in shaping our world.

The EuroGOOS's *Scientists for Ocean Literacy* activities address pressing challenges identified by the UN Ocean Decade, such as pollution, climate breakdown and mitigation, biodiversity loss, and the need for fit-for-purpose Ocean observing systems. Of particular significance is the Ocean Decade Challenge 10, which embodies the multifaceted ambition of Ocean Literacy to restore humanity's relationship with the Ocean, fostering inclusive dialogue and co-design with all stakeholders in society (Fig. 20.1).

Fig. 20.1 *Left*: EuroGOOS policy brief "Ocean Literacy in European Oceanographic Agencies – EuroGOOS recommendations for the UN Decade of Ocean Science for Sustainable Development 2021–2030"; *Right*: EuroGOOS book "The Ocean is My Home"

Developed as a scientifically rigorous initiative with an emphasis on outreach and communication, the *Scientists for Ocean Literacy* project conducts surveys and capacity-building activities to document and enhance the evolution of Ocean Literacy as a focal area of research institute activities. Building upon the groundwork laid by the EuroGOOS Ocean Literacy Working Group, which comprises over 20 organisations, the project enhances Ocean Literacy initiatives developed within the scientific community to drive positive changes in policies, practices, and dialogue with society. This chapter describes the activities of the project partners spanning the promotion of Ocean knowledge, co-design between Oceanography and other creative and scientific disciplines, the use of Ocean technologies, innovative outreach and engagement techniques, and more.

To advance Ocean Literacy and align marine scientific community efforts with broader societal needs, particularly within the UN Ocean Decade, EuroGOOS offers the following recommendations:

1. Foster innovative Ocean observation strategies: Prioritizing innovative Ocean observation strategies is essential to effectively address the complexity of the marine environment. Integrating Ocean knowledge with science-to-policy initiatives ensures that observations are both relevant and actionable. Ocean Literacy plays a key role in this process by engaging a diverse range of stakeholders and enhancing observational strategies through broader insights and citizen science contributions.
2. Increase public engagement: Oceanographic, environmental, and meteorological agencies should actively engage the public, policymakers, and the economy through Ocean Literacy initiatives. These efforts bridge the gap between marine research and societal goals, enhancing understanding and support for sustainable Ocean practices.
3. Focus on diversity and inclusivity: Ocean Literacy underscores the importance of diversity and inclusivity as the Ocean connects us all. Activities should be accessible to all audiences, including underserved groups such as those affected by disabilities or in hospitals or prisons. Partnering with education, social services, and other sectors, as well as adopting a co-design approach with different communities, will help ensure Ocean Literacy reaches a broad and diverse audience.
4. Leverage digital tools: Digital resources and online learning are essential for expanding the reach of Ocean Literacy. These tools should be used alongside interactive, hands-on activities to maximize engagement and effectiveness.
5. Foster partnerships: Collaborations with professionals from fields such as education, social sciences, arts, and others will make Ocean Literacy initiatives more versatile and impactful. Such partnerships are key to sharing Ocean knowledge, best practices, and innovative engagement tools across disciplines and countries.
6. Strengthen national and regional collaborations: Stronger national synergies are needed to engage both coastal and inland communities. Collaborating across regions enhances the consistency and reach of Ocean Literacy, driving broader societal impact and supporting more robust and aligned marine knowledge systems.
7. Expand funding and resources: Ocean Literacy activities often remain underfunded, relying on the dedication of individual researchers. Securing adequate financial and institutional support is essential. Strengthening Ocean Literacy programmes will also attract more professionals to marine research institutions. In addition, this investment will improve marine knowledge and information services and unlock new avenues for marine data collection.
8. Enhance impact monitoring: Better assessments of Ocean Literacy activities, including their spillover effects (where engaging one group influences others), will improve programme effectiveness and help track their broader societal impact. Furthermore, monitoring these impacts will enable researchers to refine their methods, ensuring Ocean Literacy activities contribute to actionable solutions.

These recommendations help align Oceanographic research with societal needs, promoting a sustainable and well-informed relationship between humans and the Ocean. Through Ocean Literacy, research institutions can not only promote sustainable behaviour, but also access richer sources of marine data, improve science-based solutions, and design fit-for-purpose Ocean observing and services that directly benefit society. United by the vision of impactful Ocean Literacy activities in research institutions, the initiatives presented in this chapter showcase European efforts addressing these recommendations and collaborating through the *Scientists for Ocean Literacy* project.

20.2 MareDireFare Festival: Fusing Art, Science, and Public Engagement to Promote Ocean Literacy

The National Institute of Oceanography and Applied Geophysics (OGS) in Italy is committed to scientific communication and public engagement on marine issues. Among its many initiatives, the *MareDireFare* Festival (*SeaSayDo*; https://www.maredirefare.it/) has emerged as a success story dedicated to promoting Ocean Literacy across all ages and using diverse communication methods (Fig. 20.2).

The festival was launched in 2021 at the beginning of the UN Ocean Decade, focusing on "The World of Microscopic Marine Life". In its inaugural year, 18 young artists participated in a major exhibition, which included 40 artworks. Over 50 public events were held, thanks to collaborations with schools, museums, libraries, environmental non-governmental organisations (NGOs), and artists, attracting a broad and engaged audience. Building on this success, the festival returned in 2022 with the theme "The Seven Ocean Literacy Principles". This edition featured a national call for seven artists, each tasked with creating seven illustrations—one for each of the principles. A comic book telling the story of an octopus guiding a pupil to explore the Ocean was published in 2024.

The 2023 festival, themed "The Sea Starts in the City", highlighted the fusion of creative expression and scientific research. This edition included two major exhibitions: one featuring photographs of tidal environments and another along an urban eco-path marked by artistic installations and street art. The 2024 festival, themed

Fig. 20.2 Pictures of the artworks of the *MareDireFare* Festival. *Top left*: Festival logo; *Top centre*: Details of artwork from the 2022 edition "The Seven Ocean Literacy Principles"; *Right*: Sculpture depicting the movie character Alien, inspired by the marine organism *Phronima sedentaria* from the 2021 edition "The World of Microscopic Marine Life"; *Bottom left*: Sculpture of a *Mnemiopsis leidyi*; *Bottom centre*: Manhole illustration from the 2023 edition "The Sea Starts in the City": *Bottom right*: Artwork from the 2023 edition. Credit: OGS

"Chronicles from the Deep", delved into the mysteries of the seabeds from the Gulf of Trieste to the deep Ocean. The programme included a series of free events for adults and families, exploring the sea through science, art, history, and culture.

Throughout its editions, since 2021, the *MareDireFare* Festival has addressed the challenge of raising marine awareness amid intense competition from other cultural events, involving 35 artists that produced around 200 pieces of art (sculptures, illustrations, photos, etc.) and organising more than 100 public events in 4 years. The key factor has been the dynamic dialogue between art and science, significantly increasing the festival's appeal, particularly to younger audiences. By leveraging diverse communication tools and engaging a wide array of local partners—including schools, museums, bookshops, libraries, NGOs, and artists—the festival has successfully elevated its visibility and impact (Fig. 20.3). The festival has greatly helped to enhance the public's understanding of marine ecosystems, biodiversity, and the broader implications of human activities on the Ocean. Through artistic and scientific lenses, it has bridged the gap between the marine world and societal pressures, such as land use, global policies, European Union regulations, and local planning. This multi-faceted approach has been instrumental in nurturing an "Ocean generation" that is both engaged and informed.

The format of the festival has allowed exhibitions to take place not only in Trieste but also in inland locations, reinforcing the idea that our connection with the sea extends far beyond coastal regions. Looking ahead, the festival will continue its Ocean Literacy activities, increasingly addressing important issues such as equality, accessibility, and inclusivity, while also leveraging Trieste's strategic location on the border with Eastern Europe.

Fig. 20.3 Pictures from the events of the *MareDireFare* Festival in Trieste. *Top left*: Exhibited artworks in a local café; *Top right*: Seminar for the general public; *Bottom left*: Science café in a historic cafeteria; *Bottom right*: Artwork on display in a local museum. Credit: OGS

The *MareDireFare* Festival, as a cornerstone of OGS's scientific dissemination efforts, has proven to be an effective educational platform. Through partnerships with the Miramare Marine Protected Area of Trieste and other scientific institutions, it reinforces the connection between science and society, driving meaningful action toward Ocean sustainability.

20.3 Empowering Future Environmentalists: Hands-on Approach to Ocean Literacy and Marine Conservation

The Hellenic Centre for Marine Research (HCMR), a governmental research organisation in Greece, is developing educational programmes for all school levels, training courses for teachers, and workshops for stakeholders. During visits to HCMR facilities, students engage in interactive and hands-on learning activities, along with exhibitions related to marine technology and infrastructures. These activities help expand their understanding of aquatic environments, ecosystem mechanisms and functions, and the stresses these systems face due to anthropogenic pressures. Each year, approximately 5000 students participate in HCMR programmes and activities promoting Ocean Literacy and inspiring the next generation of environmentalists. Scientists at HCMR, including those at the Hydrobiological Station of Rhodes and CretAquarium, work to raise environmental awareness and inspire local action, providing information and training on climate change, natural hazards, pollution, cultural heritage, and marine resources to more than 300,000 citizens and visitors annually.

Marine and freshwater scientists also communicate HCMR's research results to students, stakeholders, and the broader public by participating in various initiatives and projects. Since 2006, the Department of Ecology and Ecosystem Management at HCMR's Institute of Marine Biology, Biotechnology, and Aquaculture has been studying a coastal marine area in northeastern Crete (Elounda Bay), focusing on local marine environmental issues such as conservation, sustainable management of marine biological resources, and Lessepsian migration of marine species due to climate change. HCMR scientists have developed and implemented a non-formal educational intervention that includes short lectures, lab activities, and fieldwork. Students visit HCMR labs to get familiar with scientific equipment and observe marine species. Fieldwork includes visits to the study area, where pupils measure abiotic parameters *in situ* using scientific instruments and have the opportunity to speak with local fishers and authorities (Fig. 20.4).

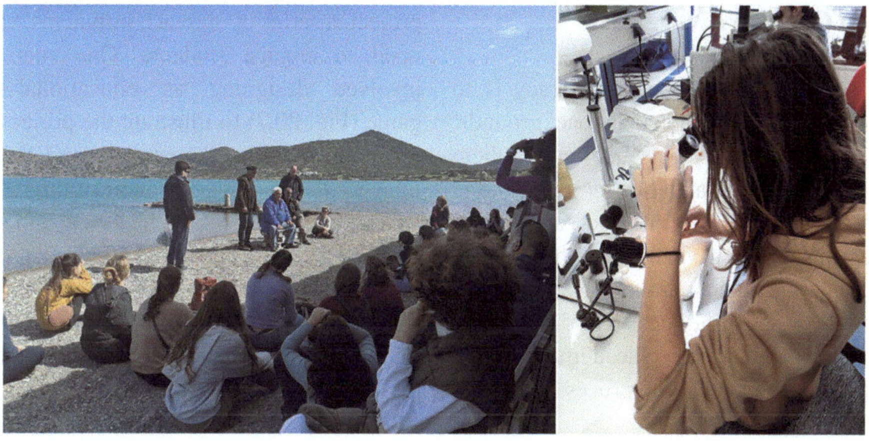

Fig. 20.4 HCMR's hands-on approach. *Left*: High school students (16–17 years old) discussing in the field with fishers in Elounda Bay (Crete, Greece). *Right*: Student observing and becoming familiar with marine macrobenthic species using a stereoscope in the HCMR lab. Credit: HCMR

Additional resources and tools, such as the MED EDUC pedagogical guide and Mediterranean Sea Literacy, are used to enhance learning. This intervention, with various adaptations (e.g., a short scientific article reviewed by high school students), has been implemented to date in three schools in the region of Crete reaching 100 students, within the framework of different European scientific projects (e.g., *GES4SEAS, ERASMUS+ BlueS_Med*), and educational programmes of the Greek Ministry of Education (e.g., *Adopt a Terrestrial/Marine Ecosystem*). It also supports schools in joining the European Blue Schools network and serves as a best-practice lesson plan for teachers (e.g., BlueMinds4Teachers). Questionnaires are given to pupils before and after the intervention to measure the impact of the Ocean Literacy activities on their knowledge, attitudes, behaviour, and emotions. Finally, this study area will be developed into an Educational Marine Area under the European project *BlueLightS*, where school students will manage a coastal maritime zone through a participatory approach, involving them in a citizen action project to protect the marine environment.

20.4 Bridging Science and Education: The Role of Ocean Literacy and Interactive Learning in Climate Awareness

One of the objectives of Ocean Literacy is to reach people of all ages and social backgrounds, simplifying scientific concepts and making them captivating and understandable for the general public. A common, universally accessible language has always been visual. For this reason, many modern and ancient authors use images as a tool to convey scientific ideas and discoveries. Since its origins, science has placed images at the centre of its communication processes—using drawings, diagrams, and later, photographs, satellite images, and maps.

The Italian National Institute of Geophysics and Volcanology (INGV), through the MACMAP, *A Multidisciplinary Analysis of Climate change indicators in the Mediterranean And Polar regions*, project, has performed a multidisciplinary analysis of climate change indicators in these regions, with an emphasis on outreach activities to engage the public with both the research process and the issues being addressed. One of the project's work packages aims to retrieve Oceanographic data from historical books, focusing on one of the earliest volumes dedicated to Oceanography: *Histoire physique de la mer*, an Oceanographic treatise on the Gulf of Lion published in 1725 by Luigi Ferdinando Marsili, which includes 40 illustrated plates (Fig. 20.5).

Historical books are a priceless heritage, containing information that is often forgotten. Data from Marsili's book, such as water density measurements collected at various geographic locations in the Gulf of Lyon between 1706 and 1707, were converted into modern units and mapped with precise coordinates for comparison with current data (Locritani and Garvani 2024). Additionally, sea level data collected at the Port of Cassis in 1707 were transformed into M2 tidal amplitude and compared with current measurements (Locritani et al. 2024) (Fig. 20.6).

Given the central role of images in science and the interdisciplinary scope of the MACMAP project, a series of outreach initiatives was developed to engage students. One such activity involved designing an educational-scientific game (Fig. 20.7) to illustrate the project's value, its multidisciplinary approach, and the role of imagery in scientific communication.

The game covers both historical and scientific topics addressed in the MACMAP project. Its strategies and objectives are designed to immerse players in the scientific research process, highlighting both the fascination of the studies and the challenges faced. In the game, players act as

Fig. 20.5 The title page of the historical volume "Histoire physique de la mer" by Luigi Ferdinando Marsili (1725) and some of the plates illustrated by him, which feature images of algae and corals. The volume contains 40 illustrated plates and 12 maps and tables. This evidence demonstrates that even in the eighteenth century, visual language was a significant aspect of scientific discourse

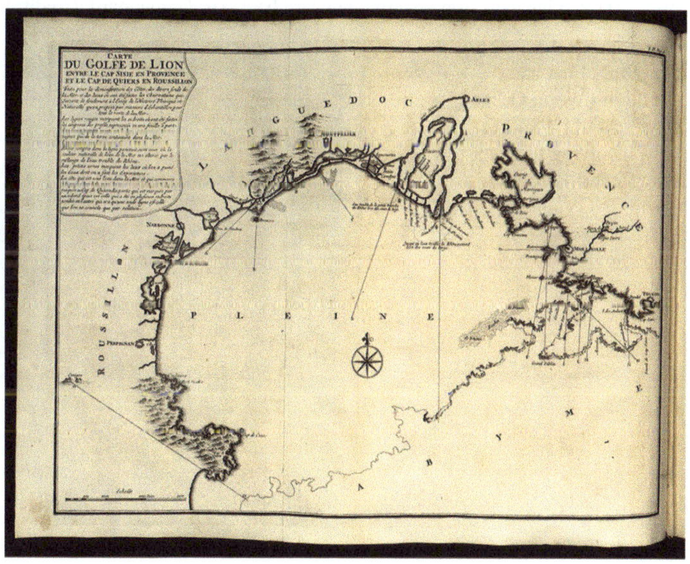

Fig. 20.6 A map of the Gulf of Lyon (Northwest Mediterranean Sea) from "Histoire physique de la mer" by Luigi Ferdinando Marsili (1725), illustrating early bathymetric profiles that delimited the continental shelf (approx., 100 m)

researchers involved in the MACMAP project, each working on a specific work package such as management, observations, modelling, or historical data analysis. The game is won if all participants achieve their work package objectives within a set deadline (timeline in Fig. 20.7c). However, if even one player fails to meet their objectives, the project fails, making the game both cooperative and competitive, simulating real scientific collaboration.

INGV has extensive experience in creating such games (Locritani et al. 2017; Locritani

Fig. 20.7 The elements of the game: (**a**) Game tokens, (**b**) Game dice, (**c**) Timeline, (**d**) Work package cards, (**e**) Game sheets, (**f**) Playing card front, (**g**) Playing card back, (**h**) Gameboard. Credit: INGV

Fig. 20.8 First game test during World Ocean Day 2024 in La Spezia, Italy, involving students from Centro Barontini. Credit: INGV

et al. 2020; Locritani et al. 2024; Locritani and Garvani 2020). However, this project is unique in involving high school students (aged 15–18) in the game's design, supervised by their science, physics, and art teachers from two schools: Parentucelli High School in Sarzana, La Spezia, Italy, and Caravillani High School in Rome, Italy, collaborating with two research centres in Italy (INGV and CNR-ISMAR). The rules, strategies, scientific and historical content, and illustrations were developed during the 2022–2023 and 2023–2024 school years. During World Ocean Day 2024 (Fig. 20.8) and Lucca Comics 2024, the game's draft version was tested to identify potential errors and improve the gameplay.

Approximately 50 students played with the game in different sessions and suggested reducing random actions (such as dice rolls) and theoretical components (such as questions) and including practical activities. These play sessions revealed that students aged 12–14 were the most engaged. The final version will be made available for download (online, printable at home on paper or cardboard via the INGV website), and printed in a larger format for use at temporary exhibitions, educational workshops, school open days, science festivals, and other events.

20.5 Summer Schools: An Effective Tool to Increase Ocean Literacy

AZTI is a science and technology centre in Spain specializing in marine and food research and innovation, with the goal of driving positive change for the future of humankind and contributing to a healthy, sustainable, and fair society. This includes a strong focus on Ocean Literacy, ensuring that outputs and developments are communicated to key stakeholders and end-users, including the general public, consumers, blue economic sectors, the scientific community, and regulators/decision-makers.

AZTI Summer Schools began in 2004 as a tool for sharing and exchanging knowledge on cutting-edge Ocean-related topics with these stakeholders and end-users. They also aim to promote engagement and professional careers in the field, support knowledge-based decision-making, and ultimately generate social change (in behaviour, research practices, regulations, etc.).

Summer schools are defined as "courses of lectures held during school and university summer vacations, taken as part of an academic course or as an independent course of study for professional or personal purposes" (Oxford Dictionary). While they are often considered academic activities, if carefully designed, they can serve as more than just tools for providing knowledge. They can also be means to network, engage, enjoy, and foster change.

Over the last two decades, AZTI has organized an annual Summer School covering a wide range of marine-related topics, reflecting the diversity of challenges and innovations in the field. Themes have ranged from climate change and invasive species to genomic tools and the restoration of marine systems. In recent years, the focus has shifted to topics such as Ocean health and human health, multiple pressures at sea, and ecosystem services. More current themes emphasize science communication, Ocean optimism, and the role of monitoring in sustainable decision-making. These courses have consistently combined scientific content with cultural events. However, over the past decade, the content has evolved from a primarily theoretical approach to a more participative and interactive format, enhancing their effectiveness in fostering dialogue and understanding.

Each Summer School edition gathers 50–60 participants from 18–20 different countries worldwide and has reached more than 1000 people (primarily from Europe, but also from America, Africa, and Asia) since the initiative began. Furthermore, it is a collaborative effort involving European projects, with 8–12 trainers from 2–5 different institutions participating each year (over 70 entities in total), facilitating a high level of knowledge exchange, engagement, and dissemination worldwide. Yet, this is not the only success of the Summer School initiative.

In 2024, AZTI organized its 20th Summer School edition under the theme *Communicating Marine Science for an Informed Society and Decision-Making*, aiming to support scientists in enhancing Ocean Literacy across society in various roles (e.g., consumers, professionals, decision-makers). Also in 2024, the EU4Ocean Coalition awarded this Summer School initiative the MakeEUBlue Award for the best Ocean Literacy initiative organized by a research organisation.

AZTI's Summer Schools continue with the 2025 edition focusing on the topic *Artificial Intelligence and Citizen Science in Marine Research* (Fig. 20.9).

Summer schools are just one of many Ocean Literacy initiatives in which AZTI is involved. The organisation is developing innovative ways to increase Ocean Literacy, such as co-

Fig. 20.9 AZTI Summer School 2023. Credit: AZTI

Fig. 20.10 European Rubik's Championship where the *Ocean Cube* has been used as a tool to increase Ocean Literacy. Credit: AZTI

developing marine board games, collaborating with Science Crunchers to create marine educational modules, and introducing the *Ocean Cube*, developed as part of the European Union GES4SEAS project. More than 400 *Ocean Cubes* have been distributed globally, serving as a tool to explain six key topics in marine management: (i) Healthy Ocean for healthy societies; (ii) Pressures in the marine environment: 1 + 1 is not always 2; (iii) Good environmental status: what it is and how it can affect your life; (iv) Tipping points: why it is best not to reach them; (v) Ecosystem-based management: managing the whole and not the hole; (vi) Ocean optimism: hope for the Ocean and societies. The *Ocean Cube* has been widely used to enhance Ocean Literacy among various audiences and sectors of society (Fig. 20.10).

20.6 Engaging Schools and Citizens in Their Relationship with the Mediterranean Sea

The Italian National Institute for Environmental Protection and Research (ISPRA) focuses its Ocean Literacy activities on enhancing scientific knowledge of the Mediterranean Sea. ISPRA employs a variety of communication tools such as exhibitions, videos, booklets, awareness events, and interactive dialogues with pupils, students, citizens, and stakeholders. These initiatives help convey scientific experiences to promote better understanding and protection of the Mediterranean Sea, fostering greater public responsibility for its valuable resources and ecosystem services.

Each year, several educational programmes are carried out, involving primary (6–11 years) and secondary (12–14 years) school pupils, as well as general awareness campaigns aimed at the broader public. ISPRA's researchers focus on key topics and threats to Mediterranean ecosystems and biodiversity, such as marine litter, coastal erosion, land-sea pollution, and climate change, along with their impacts on marine organisms and coastal ecosystems. These activities help explain the importance of Mediterranean flora and fauna and their ecosystem services, which protect marine and coastal environments while supporting a sustainable blue economy.

To raise awareness of the importance of *Posidonia Oceanica,* a protected species crucial to the Mediterranean ecosystem that acts as a natural defence against storm surges and coastal erosion when deposited on beaches, ISPRA released a cartoon titled *Banquette to the Rescue* (Fig. 20.11). The cartoon explains the role of *Posidonia* meadows and banquettes in coastal protection (Pisapia et al. 2019). This resource was shared with 250 pupils and made available on ISPRA's website for further viewing.

Fig. 20.11 ISPRA comic book "Banquette Alla Riscossa!" addressing the importance of Posidonia Oceanica for beach protection. Credit: ISPRA

To illustrate the impact of marine litter, ISPRA published widely circulated comics. *Carrie –a Brave Turtle in a Changing World* tells the story of Carrie navigating the plastic-polluted sea (Fig. 20.12) (Matiddi et al. 2021), while *Twenty Thousand Nets under the Sea* (Fig. 20.13) addresses the issue of marine litter, focusing attention on its impacts on marine organisms and ecosystems (Pisapia et al. 2023). It also depicts positive actions achieved through collaboration between researchers and fishers aimed at mitigating the problem.

The environmental education project *Mare Nostrum: Let's Discover the Mediterranean Sea Together* engaged 250 primary school pupils (6–11 years) and focused on three main themes: research at sea, life at sea, and marine environment protection. Pupils learnt about the organisms living in the Mediterranean Sea, the Ocean's importance for our lives, and various aspects of Mediterranean marine research. The project showcased the use of research vessels and under-

Fig. 20.12 ISPRA's comics. *Left*: "Carrie–A Brave Turtle in a Changing World," which tells the story of a sea turtle's adventures in a polluted sea. *Right*: Exhibition of the comic story Carrie presented during ISPRA's annual event "Science Together". Credit: Marco Pisapia—ISPRA

Fig. 20.13 ISPRA's 3D comic book "Twenty Thousand Nets under the Sea". *Left*: The book cover; *Right*: Exhibition of the comic book presented during ISPRA annual event "Science Together". Credit: Marco Pisapia–ISPRA

water scientific equipment, like Remotely Operated Vehicles (ROVs), and included a practical activity where pupils built their own cardboard ROVs (Fig. 20.14). A book was developed to complement this activity, titled *Let's Go Down to the Bottom of the Sea: Travelling with an Underwater Robot* (Giusti et al. 2021), narrating the adventures of Pollux, an underwater robot, as it explores the Mediterranean seabed. This book, aimed primarily at children but accessible to all, communicates marine research in an engaging and straightforward way.

Through these initiatives, ISPRA successfully bridges scientific research with public awareness, engaging citizens and pupils (around 2150 in 2023 and 2024) in protecting the Mediterranean

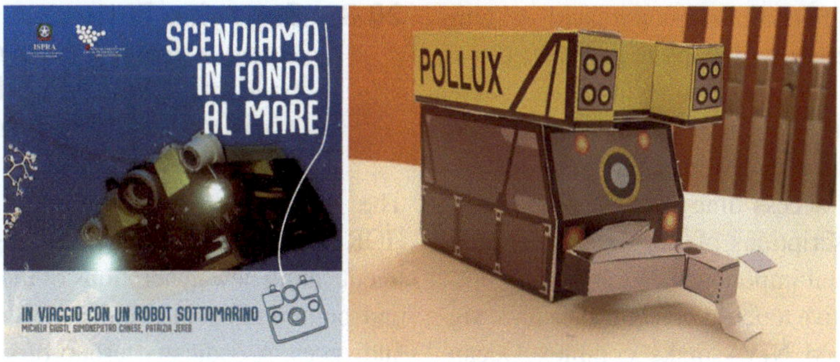

Fig. 20.14 *Left*: ISPRA book "Let's Go Down to the Bottom of the Sea: Travelling with an Underwater Robot". *Right*: Cardboard game of the Remotely Operated Vehicle (ROV) "Pollux" build by pupils. Credit: ISPRA

Sea and fostering a deeper understanding of Mediterranean marine ecosystems and their sustainable management.

20.7 Interactive Touchscreen Consoles: A Technology-Based Ocean Literacy Tool for Engaging Audiences

While Ocean Literacy activities have traditionally involved in-person interactions with school-aged children and the general public, the Norwegian Institute for Water Research (NIVA) has broadened its audience with interactive touchscreen consoles, reaching passengers on ships and visitors to science museums and national parks. For direct engagement, NIVA has partnered with local initiatives, involving children in hands-on activities focused on human impacts on the Ocean. Through collaborations with cruise companies, NIVA has also been providing Ocean Literacy experiences in the form of interactive demonstrations and presentations on Color Line, Hurtigruten, and Viking cruise ships.

The interactive touchscreen consoles have been a key Ocean Literacy tool for NIVA over the past 7 years. Initially developed through the European ResponsSEAble project (2015–2019; Grant no. 652643), the consoles feature a graphical user interface designed for touchscreens ranging from approximately 20 to over 50 inches. The Ocean Literacy content has been continuously updated and expanded, with translations into six languages: English, German, Norwegian, French, Spanish, Polish, and Russian. This development has been supported by several European Union projects, including INTAROS (2016–2022; Grant no. 727890) and JERICO-S3 (2020–2024; Grant no. 871153), as well as the Research Council of Norway's SATVIEW project (2021–2022; Grant no. 322714).

The consoles offer two main modules of Ocean Literacy content. The first, *Key Ocean Literacy Stories*, focuses on human impacts and Ocean-related issues, including pollution, overfishing, eutrophication, renewable Ocean technologies, invasive species, and coastal tourism. This content is presented through videos (live-action and animated), photos, data plots, text, and a simple video game. Although designed with school-aged children in mind, it also appeals to adults. The second module, *Ocean Data Products and Near-Real-Time Observations*, includes an interactive global viewer of Ocean surface temperature, salinity, currents, and ice cover using data from the European Copernicus programme and the US National Oceanic and Atmospheric Administration (NOAA). It also features an interactive viewer for satellite remote sensing products in European seas and near real-time data from NIVA's FerryBoxes—automated systems installed on ferries and ships that collect Ocean data, such as temperature, salinity, and water quality, while the vessels are traveling their regular routes. Users can view measurement locations on a map and access time-variable plots, along with short descriptions of key Ocean variables and their societal importance.

The interactive touchscreen consoles have significantly helped NIVA make complex Ocean-related topics more accessible and engaging for a broad audience. Installed in high-traffic locations, these consoles reach thousands of users each month, promoting awareness of critical issues like pollution, overfishing, and climate change. Viewing statistics, assuming two viewers per interaction, averaged 4387 views per month on a Color Line ship and 3820 views per month on a Hurtigruten cruise ship over a three-month period. During this same period, content was accessed in German (44%), Norwegian (25%), English (19%), and French (12%), with Spanish, Polish, and Russian translations not yet implemented. By combining interactive tools, real-time data, videos, and educational games, the consoles effectively engage both children and adults, deepening their understanding of human impacts on the Ocean. Continuous feedback from user interviews and surveys further refines the content, ensuring it resonates with diverse audiences and strengthens their connection to Ocean knowledge and conservation (Fig. 20.15).

20.8 Engaging Future Generations: A Commitment to Marine Science Education and Ocean Conservation

The Institute of Oceanography and Fisheries (IOR) in Croatia has a long-standing commitment to educating new generations of marine biologists and popularizing marine science among diverse audiences. A dedicated group of scientists at the institute actively engages in science communication, targeting various demographics through multiple channels. This approach enables the broad dissemination of marine knowledge, with scientists tailoring their methods to focus on youth, the general public, or media.

Recently, IOR has implemented several projects specifically aimed at science popularisation. Two notable initiatives are the *Blue-connect* and *Blue-connect 2.0* projects, in which IOR is a partner, and the *More budućnosti* (*Future Sea*) project, which is exclusively managed by IOR. Over the past two and a half years, IOR scientists have organized a wide range of activities, including lectures and workshops in schools and kindergartens, on-site institute tours, exhibitions, culinary workshops, and more. Between 2022 and 2024, these two projects alone facilitated over 100 activities, engaging approximately 2500 children, young people, and their teachers across Croatia. These activities are vital, as formal education often lacks comprehensive coverage of

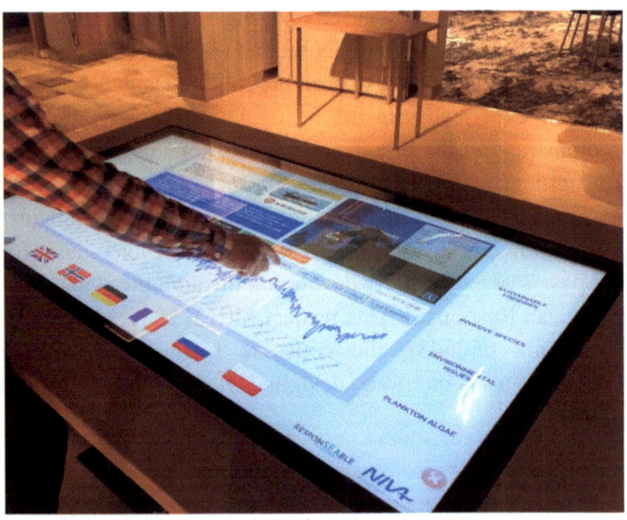

Fig. 20.15 Touchscreen console on MS *Roald Amundsen* displaying near real-time data from FerryBox ship of opportunity observations. Credit: Kai Sørensen

key Ocean Literacy principles and concepts (Ezgeta-Balić and Balić 2024).

Under the motto "I Explore," as part of the *Blue-connect* project, IOR scientists participated in the European Researchers' Night in 2022, 2023, and 2024, bringing Ocean-related topics closer to the public in the city centre of Split (Fig. 20.16). In addition to the main events, IOR conducted numerous workshops, such as *Marine Diaries—Bivalve Shells and Their Secrets*, *Life Under the Surface, Marine Litter Art*, and *Life in a Drop of Water.* A culinary workshop, *Small Pelagic Fish in Gastronomy,* was organized in collaboration with a professional chef, where students prepared healthy dishes and learned about the benefits of seafood. Moreover, in collaboration with an art school, an exhibition titled *Earth without Ocean* was held, where students created sculptures from waste materials, presenting a dystopian vision of a sea-less Earth to raise awareness about the Ocean's importance. Various creative approaches were used to engage children and youth in Ocean-related topics, with most activities focused on Split, Croatia's second-largest city, and its surroundings.

In contrast, the *Future Sea* project emerged from the recognition that children in rural areas and smaller towns have fewer opportunities to engage with marine science compared to their urban counterparts. To address this disparity, the project aimed to enhance Ocean Literacy among primary school students in underserved areas along the Adriatic coast and inland Croatia. These workshops covered various aspects of Ocean conservation, including the importance of a healthy Ocean, environmental changes, and necessary conservation actions (Fig. 20.17).

An interesting aspect of this project was the series of workshops conducted in small, remote schools with only five to ten students. Although the reach of these workshops was relatively limited, their impact was significant, fostering inclusivity and engagement in these isolated areas. Students from inland schools, many of whom had never seen the sea, expressed amazement as they discovered the Ocean's profound impact on their lives.

In total, more than 45 workshops were organized at 25 primary schools across Croatia in 1 year. The project culminated in a creative exhibition featuring over 140 student works during European Researchers' Night in Split in 2023 and the publication of *The Ocean in Our Hands: A Guide to Ocean Literacy* in Croatian (Fig. 20.18). The *Future Sea* project has gained both local and national recognition as a platform

Fig. 20.16 Presentation of the Institute of Oceanography and Fisheries at European Researchers' Night in Split under the theme *I Explore…*. The event featured interactive presentations, games, and educational materials developed by IOR scientists. Part of the Reconnecting Science with the Blue Society–*Blue-connect 2.0* project, funded by the European Union's HORIZON-MSCA-2023-CITIZENS-01 programme. Credits: IOR and CROPIX

Fig. 20.17 Workshops in schools held as part of the project *More budućnosti* (*Future Sea*), funded by the Adris Foundation. During the workshops, pupils explored topics such as plankton, shellfish, and diving through virtual reality after a brief presentation. Students created Ocean-inspired artwork, which was exhibited during European Researchers' Night 2023 in Split. Credit: IOR

Fig. 20.18 Ocean Literacy resources developed through the "Blue-connect" and "Future Sea" projects for Croatian teachers and pupils. *Left*: A guide outlining key Ocean Literacy principles and concepts for teachers, including ideas for hands-on experiments. *Middle*: An educational classroom poster explaining the main Ocean Literacy principles. *Right*: An illustrated picture book created with contributions from children at the Rusulica kindergarten. All resources are freely available in Croatian at www.futuresea.eu

for spreading marine knowledge among students and teachers, facilitating information exchange.

Through these diverse initiatives, the Institute continues to play an important role in educating the public about marine science and fostering a deeper understanding of Ocean conservation across various demographics in Croatia.

20.9 Accessibility as a Key Enabler Towards More Effective Scientific Dissemination

The Ocean is a fundamental resource for all living beings. For humans, it serves as a social, economic, political, and linguistic connector, as well as an aesthetic and physical one, covering more than 70% of our planet. Today, global challenges highlight the complexity of the Ocean in terms of environmental variables, stakeholders, governance, and economic assets. Complex systems are often unpredictable, with rapid changes occurring faster than our ability to adapt or control them. In this context, scientists, policymakers, and the general public must collaborate to assess the impact of human activities on the Ocean and achieve sustainable resource management.

Additionally, the overwhelming influx of data and news adds further complexity to a chaotic global scenario where science is increasingly called upon to support policy decisions and effective solutions. Dissemination and scientific communication are crucial enablers in this process and statutory missions of many research institutions, such as the National Research Council of Italy (CNR). CNR promotes, enhances, and applies research results while contributing to the transfer of knowledge to society.

Effective science dissemination, communication, and education, particularly when developed with a long-term strategy, play a crucial role in transferring research-derived knowledge to the public, especially younger generations. By providing access to information and knowledge, these tools empower individuals to participate in the global debate and reconnect with the environment.

The growing interest in environmental issues, climate change, and their evident effects on daily life has placed increased pressure on scientists to disseminate their findings promptly and respond to inquiries from the media, citizens, and institutions. Furthermore, the complexity of these issues necessitates collaboration with professionals from other sectors to combine expertise, perspectives, and skills, ensuring initiatives are impactful.

Developing accessible products and collaborative, co-designed initiatives is essential to effectively disseminate knowledge and ensure inclusiveness across diverse audiences without losing meaning or content. A notable example is *Abecedarium: The Ocean in Sign Language*, a long-term participatory project launched in 2020 by Ocean Space and CNR-Institute of Marine Sciences, in collaboration with ENS-Ente Nazionale Sordi ONLUS (The Italian National Agency for the Deaf) and CNR-Institute of Cognitive Sciences and Technologies.

The project aims to create an eco-glossary in Italian Sign Language (LIS) on environmental and marine-related issues through a participatory process involving the deaf community. The initiative reflects the urgency of making scientific communication and Ocean Literacy accessible to all, adopting a multidisciplinary and multisensory approach that enables engagement beyond national boundaries.

The project has benefited from contributions across various disciplines, offering diverse perspectives and encouraging critical reflection. By fostering dialogue between the deaf community and Ocean scientists to develop a shared and recognized LIS vocabulary for Ocean-related terms, *Abecedarium* leveraged digital media during the pandemic to connect participants from different regions and backgrounds, incorporating their Ocean experiences and knowledge. Participants engaged in 15 training sessions led by scientists, artists, and architects, exploring issues related to Ocean preservation and focusing on the linguistic refinement of existing and known signs for marine concepts.

Abecedarium contributes to the development of educational resources that prioritize accessibility and inclusion. Originating within the deaf community and employing an artistic and aesthetic approach, it addresses the needs of a broader audience while offering valuable insights to scientists. Art and culture possess transformative power, inspiring individuals to reconnect

Fig. 20.19 *Abecedarium*. Credit: Ocean Space and CNR-Institute of Marine Sciences

with nature and deepen their commitment to its preservation. The project also explores the interplay between nature and language, uncovering the contradictions that emerge from their interaction (Fig. 20.19).

20.10 Conclusion

The key recommendations from the introduction are vividly reflected in the diverse European initiatives showcased throughout the chapter. These efforts demonstrate the vital role that scientific institutions play in promoting a deeper understanding of the Ocean's significance in shaping our environment and sustainable future.

Ocean Literacy is a critical tool for translating scientific research into societal impact, especially when embedded within research institutions. As global challenges like climate breakdown, pollution, and resource management demand urgent solutions, scientific institutions must bridge the gap between Ocean science and society. Through Ocean Literacy initiatives, they transform complex data into accessible knowledge, fostering informed decision-making among the public, policymakers, and stakeholders. This shift from purely academic research to socially engaged science underscores the importance of integrating Ocean Literacy into the core missions of scientific organisations.

The case studies presented in this chapter illustrate how European research institutions are leading by example. AZTI's Summer Schools provide a platform for exchanging Ocean knowledge among scientists, policymakers, and professionals, encouraging informed societal actions. HCMR in Greece combines scientific research with community-based marine conservation, directly involving schools, local authorities, and fishers in Ocean Literacy activities. The MACMAP project in Italy empowers high school students to co-design educational games, fostering experiential learning. Meanwhile, ISPRA demonstrates how creative outreach tools such as comics and ROV-building workshops, effectively communicate marine science to diverse audiences. NIVA in Norway leverages technology through interactive touchscreen consoles, offering real-time Ocean data to audiences in high-traffic locations and enhancing public engagement with marine science.

The importance of sustained funding and resources is another key takeaway. Scaling and sustaining such initiatives require adequate financial and institutional support. Many Ocean Literacy activities currently depend on short-term project funding, posing significant challenges to their long-term continuity and impact. Expanding and diversifying resources for Ocean Literacy will be critical to broadening societal reach and ensuring lasting impact, enabling initiatives to thrive beyond the limitations of temporary funding cycles.

Ocean Literacy activities serve to increase the visibility and relevance of marine science in everyday life. By collaborating with sectors like art and education, as exemplified by the *MareDireFare* Festival in Italy, scientific institutions extend the cultural footprint of Ocean Literacy, making marine science more accessible and inspiring to broad audiences. These interdisciplinary collaborations transcend traditional academic boundaries, promoting creativity and inclusivity.

The chapter's initiatives further emphasize the need for inclusivity in Ocean Literacy efforts. Projects like *Abecedarium* in Italy, which creates a marine-related sign language eco-glossary for the deaf community, and *Future Sea* in Croatia, which focuses on underserved rural communities, ensure that Ocean Literacy reaches a broader audience. Such initiatives foster inclusive dialogues and equitable access to marine education, making sure that all groups, regardless of physical, social, or geographic barriers, can engage with Ocean science.

These case studies underscore the transformative role scientific institutions play in advancing Ocean Literacy and fostering sustainable Ocean stewardship. By embedding Ocean Literacy into their core activities, institutions contribute to the UN Ocean Decade's challenges and the 17 Sustainable Development Goals of Agenda 2030. They bring critical Ocean issues to the attention of policymakers and empower society to take informed, responsible actions, recognizing that the Ocean's future is intrinsically tied to our own. This alignment of scientific research with public engagement is essential for building an Ocean-conscious society, driving meaningful change, and supporting the long-term sustainability of Ocean ecosystems on both regional and global scales.

20.10.1 Summary

Key concepts explored in this chapter include:

- Bridging science and society. Ocean Literacy helps turn marine science into knowledge that drives policy and public action. It is essential for informed decision-making and sustainability.
- Art and science working together. The *MareDireFare* Festival of the National Institute of Oceanography and Applied Geophysics (OGS) blends creativity with research, using art to make Ocean topics more engaging and accessible.
- Hands-on learning inspires action. The Hellenic Centre for Marine Research (HCMR) connects the public with marine conservation efforts through interactive experiences, sparking respect and action for our Ocean.
- Learning through games and visuals. Istituto Nazionale di Geofisica e Vulcanologia (INGV) makes Ocean and climate science fun and interactive, inviting young people learn through engaging games and co-design.
- Summer schools as hubs of Ocean knowledge. AZTI's summer schools bring together experts and learners of all ages, offering immersive experiences that explore key marine issues and build capacity.
- Connecting people to the sea. The Italian National Institute for Environmental Protection and Research (ISPRA) raises awareness about the Mediterranean Sea ecosystems using comics, games, and interactive exhibits.
- Technology bringing Ocean Literacy to new spaces. The Norwegian Institute for Water Research (NIVA) uses touchscreen consoles in ferries, cruise ships, parks, and museums, providing real-time Ocean data to thousands of visitors.
- Inspiring future marine scientists. The Institute of Oceanography and Fisheries (IOR) engages students in both coastal and inland areas, making marine science exciting and accessible through hands-on activities.
- Making Ocean Literacy truly inclusive. The National Research Council of Italy's *Abecedarium* project breaks barriers by creating Ocean Literacy resources in sign language, showing how inclusion can empower everyone to engage in Ocean sustainability.
- A roadmap for stronger Ocean Literacy. EuroGOOS outlines eight key recommendations to align Oceanographic research with societal needs, driving sustainable behaviours, improved observations, and science-based solutions. Through the *Scientists for Ocean Literacy* project, European initiatives are working together to turn Ocean knowledge into action.

20.10.2 Recommended Resources

The following resources provide further insights and support on this topic:

- Blue-connect and Blue-connect 2.0 projects: https://jaistrazujem.hr/
- BlueLightS project: https://blue-lights.eu/
- BlueMinds4Teachers: https://maritime-forum.ec.europa.eu/empowering-teachers-train-trainer-network-european-blue-schools_en
- BlueS_Med project: https://www.blueschoolsmed.eu/
- EuroGOOS Scientists for Ocean Literacy project: https://eurogoos.eu/Ocean-literacy/
- EuroGOOS Ocean Literacy resources library: https://eurogoos.eu/Ocean-literacy-resources/
- GES4SEAS Ocean Cube: https://www.youtube.com/watch?v=E7qFN2AxIdY
- GES4SEAS project "Achieving Good Environmental Status for maintaining ecosystem services by assessing integrated impacts of cumulative pressures": https://www.ges4seas.eu/
- MED EDUC project: https://www.mededuc.eu/en/
- Mediterranean Sea Literacy: https://www.emseanet.eu/fins-water-library
- More budućnosti (Future Sea) project: https://futuresea.eu/
- ResponSEAble project: https://responseable.acteon-environment.eu/

Acknowledgements AZTI's contribution was supported by the European Union's Project GES4SEAS—Achieving Good Environmental Status for maintaining ecosystem SErvices, by ASsessing integrated impacts of cumulative pressures (grant agreement No. 101059877).

References

Eparkhina D (2022) Scientists for Ocean Literacy: An Ocean Decade project to empower scientists as Ocean champions. Mar Technol Soc J 56:114. https://doi.org/10.4031/MTSJ.56.3.22

Eparkhina D, Lehtonen K (2017) The Ocean is my home: For children, parents and our Ocean. (ed: Eparkhina D, Nolan G). EuroGOOS, Brussels

Eparkhina D, Pomaro A, Koulouri P, Banchi E, Canu D, Uyarra M, Burke N (2021) Ocean Literacy in European oceanographic agencies: EuroGOOS recommendations for the UN decade of Ocean science for sustainable development 2021–2030. EuroGOOS Policy Brief, Brussels, Belgium. https://doi.org/10.25607/OBP-1076

Ezgeta-Balić D, Balić N (2024) Navigating through Ocean Literacy gaps: An analysis of elementary school textbooks in Croatian education. Mediterr Mar Sci 25(1):1–13. https://doi.org/10.12681/mms.35378

Gibbons M (1999) Science's new social contract with society. Nature 402(Suppl 6761):C81–C84. https://doi.org/10.1038/35011576

Giusti M, Canese S, Jereb P (2021) Scendiamo in fondo al mare. ISPRA. ISBN 978-88-448-0985-0

Locritani M, Garvani S (eds.) (2020). MAREOPOLI. Edizioni Caracol. ISBN 978-88-32240-19-1

Locritani M, Garvani S (2024) Water density data extrapolated by water weight measurements reported in Histoire Physique de la Mer (1725) by L.F. Marsili. Istituto Nazionale di Geofisica e Vulcanologia (INGV). https://doi.org/10.13127/histoiremarsili/waterdensity

Locritani M, Garvani S, Di Laura F, Merlino S, Talamoni R (2017) Giocando verso uno sviluppo sostenibile: Il contributo della sede INGV di Portovenere nella realizzazione di giochi didattico-scientifici. Miscellanea INGV 39:22. https://doi.org/10.13127/misc/39

Locritani M, Merlino S, Garvani S, & Di Laura F (2020) Fun Educational and Artistic Teaching Tools for Science Outreach. Geosci Commun 3(2):179–190.

Locritani M, Garvani S, Manzella G (2024) Sea level data reported in Histoire Physique de la Mer (1725) by L.F. Marsili and transformed to calculate the M2 tidal amplitude. Istituto Nazionale di Geofisica e Vulcanologia (INGV). https://doi.org/10.13127/histoiremarsili/sealevel

Matiddi M, Pisapia M, Poponessi S (2021) Carrie: A brave turtle in a changing world. (illus: Pisapia M, Poponessi S). ISPRA

National Oceanic and Atmospheric Administration (NOAA) (2024) Ocean Literacy: The essential principles and fundamental concepts of Ocean sciences for learners of all ages. Silver Spring, MD

Pisapia M, Rotini A, Poponessi S (2019) Banquette alla riscossa. ISPRA

Pisapia M, Angiolillo M, Izzi A, Genta D, Poponessi S (2023) Twenty thousand nets under the sea. (illus: Pisapia M, Poponessi S). ISPRA

Author Contributions and Biographical Profiles

[1]*Dina Eparkhina: Co-conceptualization, Writing—Original Draft, Review, and Editing, Methodology—Data Collection and Analysis, Project Administration, Supervision.
[2]Elisa Banchi: Writing—Original Draft, Review, and Editing.
[3]Ángel Borja: Writing—Original Draft, Review, and Editing.
[4]Donata Canu: Co-conceptualization, Writing—Original Draft, Review, and Editing.
[5]Daria Ezgeta-Balić: Writing—Original Draft, Review, and Editing.
[6]Elena Giusta: Writing—Original Draft, Review, and Editing.
[7]Andrew King: Writing—Original Draft, Review, and Editing, Methodology—Data Collection and Analysis.
[8]Panayota Koulouri: Writing Methodology—Original Draft, Review, and Editing.
[9]Marina Locritani: Writing—Original Draft, Review, and Editing.
[10]Marta Manca Zeichen: Co-conceptualization, Writing—Original Draft, Review, and Editing.
[11]Angela Pomaro: Writing—Original Draft, Review, and Editing.
[12]Francesca Petrera: Writing—Original Draft, Review, and Editing, Methodology—Data Collection and Analysis.
[13]Maria C. Uyarra: Writing—Original Draft, Review, and Editing.
[14]Louise Valestrand: Writing—Original Draft, Review, and Editing, Methodology—Data Collection and Analysis.
*Corresponding Author at dina.eparkhina@eurogoos.eu

Dina Eparkhina is a science-policy mediator promoting the uptake of Ocean science in policymaking, connecting stakeholders, and enhancing expert networks and best practices. She is the instigator of the EuroGOOS Ocean Literacy Network and co-chair of the resulting Working Group. Dina coordinates activities as an Ocean Decade Implementing Partner and leads the *Scientists for Ocean Literacy* project. She authored The Ocean Is My Home (2017), translated into six languages and used internationally in schools and Ocean Literacy programmes in marine institutions, and is the lead author of the EuroGOOS policy brief on Ocean Literacy in Oceanographic agencies.

Elisa Banchi is a biologist with a PhD in genetics. She works as a researcher at the National Institute of Oceanography and Applied Geophysics (OGS) and studies marine biological communities using molecular approaches. She authored more than 30 scientific publications in ISI peer-reviewed journals. She is involved in various scientific dissemination and public engagement activities and is a member of the EuroGOOS Ocean Literacy Working Group and of the OGS Oceanography division communication team.

Ángel Borja holds a PhD in Biology (1984, Basque Country University, Spain) and a Doctorate in Sciences (Honoris Causa, 2015, Hull University, UK). He is the Principal Investigator at AZTI. His research focuses on making marine ecology relevant for policymakers and managers, studying the impacts of human activities on marine ecosystems, and monitoring recovery efforts. Ángel also works on the protection of the Ocean, the effects of climate change, and European directives like the Water Framework and Marine Strategy Framework Directives. He has authored over 450 papers and is a Chief Editor at Frontiers in Ocean Sustainability.

Donata Canu is a senior scientist at OGS, National Institute of Oceanography and Applied Geophysics, Italy, where she has been working since 2002. She holds a PhD in Environmental Sciences from the University of Venice, Ca′ Foscari. As a marine ecologist and modeller, she focuses on the interconnection between humans and marine ecosystems to understand the impacts of human activities and climate change on marine ecosystems to support conservation and restoration efforts and coastal resilience. She has authored more than 50 ISI scientific publications and is actively engaged in dissemination and outreach activities. Donata is a member of the EuroGOOS Ocean Literacy Working Group.

Elena Giusta is a senior technologist specializing in public communication and management, with extensive experience in international cooperation and European initiatives. She has contributed to the Water Framework Directive's implementation and worked on many European Union projects, including FP7 AWARE, PSI-Connect, Water JPI, H2020 PIANO, MARINA, and others. Elena was also involved in the implementation of the Tsunami Directive and the IOC-UNESCO's early warning system. She is a member of the EuroGOOS Ocean Literacy Working Group.

Daria Ezgeta-Balić is a Senior Research Associate at the Institute of Oceanography and Fisheries, holding a PhD in Applied Marine Science. In addition to her scientific work, she is deeply committed to science communication and actively engages with children and young people to raise awareness of the Ocean's vital role in sustaining life on Earth. She received the National Annual Award for Science Popularization in 2022 from the Croatian Ministry of Science and Education. Daria is a member of the EMB Communication Panel and the EuroGOOS Ocean Literacy Working Group.

Andrew King is a Senior Research Scientist and Department Head of the Oceanography section at the Norwegian Institute for Water Research. He also holds an Adjunct Professor appointment at University of Illinois Chicago. The primary focus of King's research is characterizing the Oceanic distribution and supply of nutrient elements, especially bioactive trace elements, and elucidating their interaction with primary producers. His interests also include the effects of Ocean acidification on marine biogeochemical processes and incorporating science on ships of opportunity. He is a member of the EuroGOOS Ocean Literacy Working Group.

Panayota Koulouri a marine biologist and Senior Researcher at IMBBC, HCMR in Crete. She specializes in marine biodiversity, benthic ecology, coastal management, artificial reefs, and Ocean Literacy. Yolanda co-chairs the EuroGOOS Ocean Literacy Working Group, represents HCMR in the EU4Ocean Coalition, and is CEO of the spin-off company "Artificial Reef Innovative Applications" (ARIA).

Marina Locritani is an environmental scientist specializing in marine science, with a PhD in the field. She is a researcher at the Istituto Nazionale di Geofisica e Vulcanologia (INGV), where she focuses on studying marine litter using a citizen science approach. She has authored approximately 20 scientific publications in ISI peer-reviewed journals and is involved in various scientific dissemination and public engagement activities. Marina is a member of the EuroGOOS Ocean Literacy Working Group and the INGV Ambiente Working Group for outreach events and communication.

Marta Manca Zeichen is a Senior Researcher. She is a marine biologist with a deep experience in biological, biogeochemical and satellite Oceanography. She has been involved in several national and EU projects and European Directives related to monitoring and assessment of marine status, fisheries, Marine Protected Areas, Ocean acidification, Maritime Spatial Planning and tsunami detection. She is a member of the EuroGOOS Ocean Literacy Working Group.

Angela Pomaro holds a degree in civil engineering from the University of Padua. She is a researcher in physical Oceanography at the National Research Council of Italy Institute of Marine Sciences (CNR-ISMAR), focusing on wave climate studies and the effects of meteo-Oceanographic conditions on circulation patterns. She coordinates a project on the impact of underwater noise on marine ecosystems and manages the "Acqua Alta Oceanographic Tower" in the North Adriatic Sea. Angela also leads CNR-ISMAR's marine observational system group and co-chairs the EuroGOOS Ocean Literacy Working Group.

Francesca Petrera, biotechnologist by training, after a PhD in Biomolecular Sciences, approached the world of scientific communication thanks to the master's in science communication at the International School for Advanced Studies (SISSA) in Trieste. She is now the coordinator of the press office at the National Institute of Oceanography and Applied Geophysics—OGS. She is a member of the Steering Group of the European Marine Board Communication Panel.

Maria C. Uyarra holds a PhD in Biology and an MSc in Applied Ecology and Conservation from the University of East Anglia, UK. A Senior Researcher at AZTI (Spain) since 2011, she has over 20 years of interdisciplinary research experience, co-authoring 40+ scientific papers. Her research focuses on Ocean Literacy, biodiversity conservation, marine recreation, and marine policy. Maria is a member of EuroGOOS and REEDUCAMAR's Ocean Literacy Working Groups, advises the PROBLEU marine education project, and leads the Basque Country network for blue schools. She also collaborates with master courses and STEAM initiatives.

Louise Valestrand is a head engineer at the Norwegian Institute of Water Research (MS, Univ. of Oslo in education of science, specializing in marine biology). Her expertise lies in marine phytoplankton taxonomy, including the identification of toxic algae. She is actively engaged in national monitoring programmes and has been working with citizen science and Ocean Literacy projects, including media consoles showcasing key narratives in Ocean Literacy. She is a member of EuroGOOS Ocean Literacy Working Group.

21 Co-Designing Transdisciplinary Science for the Ocean We Want: Actions for Scientists to Enrich Ocean Literacy

Lei Li, Minyan Zhao, Min Wei, Shuangfang Hu, Baki Billah, and Ziyi Lin

Abstract

Marine pollution, particularly from nano- and microplastics (NMP), poses urgent and complex environmental challenges and health concerns. Addressing this issue necessitates a holistic, transdisciplinary approach that involves not only experts from various scientific fields but also the broader community. While interdisciplinary collaboration among scientists is critical for understanding the impacts of NMP on marine ecosystems and human health, transdisciplinary efforts extend beyond the scientific community to engage policymakers, educators, and the public. Public understanding and engagement—key components of Ocean Literacy (OL)—are essential for translating scientific knowledge into action. The EPICD framework (Engagement with the public and social media to counter misinformation; Partnerships with educators, media, and community organizations; Interdisciplinary collaboration to enrich Ocean Literacy; Communication training for scientists; Digital platforms to disseminate open-access resources) provides a structured approach to bridge the gap between scientific discoveries and public awareness. By integrating these efforts, EPICD fosters a positive feedback loop, where enhanced public knowledge drives further research, informs policy decisions, and promotes more sustainable behaviors, ultimately contributing to the health of our Ocean.

Keywords

Actions for scientists · Ocean Literacy (OL) · Marine plastic pollution · Knowledge dissemination · Co-design research · Ocean partnerships · EPICD framework

L. Li (✉)
Shenzhen Institutes of Advanced Technology, Chinese Academy of Sciences; Shenzhen University of Advanced Technology; Shenzhen-Hong Kong Institute of Brain Science, Shenzhen, China
e-mail: saralilei@siat.ac.cn

M. Zhao
Institute of Tibetan Plateau Research, Chinese Academy of Sciences, Beijing, China
e-mail: minyan.zhao@anso.org.cn

M. Wei
Shenzhen Longhua Innovative Experimental School, Shenzhen, China

S. Hu
High School of Shenzhen University of Advanced Technology, Shenzhen, China

B. Billah
Jahangirnagar University, Savar, Bangladesh

Z. Lin
University of Chinese Academy of Sciences, Beijing, China
e-mail: zy.lin@siat.ac.cn

T. J. Kennedy (ed.), *Ocean Literacy: The Foundation for the Success of the Ocean Decade, Volume I*, Sustainable Development Goals Series, https://doi.org/10.1007/978-3-031-98145-6_21

21.1 Introduction

Marine pollution, particularly from nano- and microplastics (NMP), has emerged as one of the most pressing environmental challenges of our time, infiltrating everything from deep-sea sediments to our dinner plates. The complexity of NMP pollution, affecting marine ecosystems, food chains, and human health, demands a multifaceted approach to tackle it effectively. Interdisciplinary collaboration, where scientists from diverse fields work together, is essential to explore the intricate dynamics between pollution, marine life, and human health. However, the problem is not only scientific—effective solutions require the engagement of broader communities, including policymakers, educators, and the public. This is where a transdisciplinary approach becomes critical. Transdisciplinary efforts bridge the gap between scientific research and real-world action by integrating insights from scientific experts and other stakeholders, ensuring that solutions are relevant, actionable, and inclusive.

Ocean Literacy plays a central role in this process by educating and empowering the public to understand the importance of Ocean ecosystems and the impact of their actions on marine environments. Scientists must not only conduct research but also actively share their findings with the public. For this purpose, we proposed the EPICD framework to provide a practical structure for achieving this goal, promoting interdisciplinary collaboration among scientists while fostering communication with broader communities to enhance Ocean Literacy and promote informed action.

21.1.1 Marine Pollution, Particularly NMP Pollution, Is a Deeply Complex and Enduring Challenge

The Ocean faces a range of interconnected threats that jeopardize its health and sustainability. Climate change, overfishing, habitat destruction, and pollution are among the most significant challenges, each compounding the impact of the others. Rising sea temperatures and acidification disrupt marine life and coral reefs, while overfishing depletes key species, upsetting ecological balance. Habitat destruction, including coastal development and seabed mining, further degrades critical ecosystems. Pollution, in its many forms, adds another layer of stress, impacting water quality and marine organisms. Together, these multifaceted threats pose significant risks to biodiversity, food security, and the overall health of the Ocean.

Among the various challenges to Ocean health, marine pollution stands out due to its pervasive nature and immediate impacts on ecosystems. Pollution encompasses a broad range of contaminants, including plastics, chemicals, oil spills, and agricultural runoff, all of which degrade water quality and harm marine life (Allen et al. 2022). Plastic pollution, in particular, is a growing concern as it accumulates in marine environments, affecting even the most remote areas of the Ocean (Law and Thompson 2014). Pollutants can disrupt food webs, introduce toxic substances into the ecosystem, and pose direct threats to marine species. Addressing marine pollution is crucial for restoring Ocean health and protecting biodiversity.

Nano- and microplastics, defined as plastic particles smaller than 5 millimeters and nanoplastics measuring even smaller, have become pervasive pollutants in marine ecosystems (Thompson et al. 2004). These particles originate from various sources, including the degradation of larger plastic debris, the shedding of synthetic fibers from textiles, and the release of microbeads from personal care products (Zhang et al. 2022). These pollutants have become pervasive in marine environments, detected in surface waters, deep-sea sediments, and even Arctic ice (Lim 2021). Due to their small size, nano- and microplastics are easily ingested by marine organisms, from plankton to fish, and can be transported throughout the food web. Their widespread presence and persistence in the environment make them a pressing issue in marine pollution.

Nano- and microplastics pose significant threats to marine ecosystems and biodiversity. These particles can be ingested by a wide range of marine organisms, from the smallest zooplankton to large marine mammals, often leading to physical harm, reduced feeding efficiency, and impaired growth and reproduction. Microplastics are also proposed to be vectors for toxic chemicals, such as persistent organic pollutants, which can accumulate on their surfaces and be transferred to organisms upon ingestion. This can disrupt the health of individual species and alter the structure and function of entire marine food webs, ultimately threatening the biodiversity and stability of Ocean ecosystems.

The pervasive nature of nano- and microplastics in the Ocean has implications for human health, primarily through the consumption of contaminated seafood. As microplastics are ingested by marine organisms, they can move up the food chain, potentially ending up on our plates (Cox et al. 2019). Ingested microplastics may carry harmful chemicals, such as plastic additives, heavy metals, and persistent organic pollutants, which pose health risks when accumulated in the human body (Sun et al. 2023). Although research on the long-term effects is still ongoing, the presence of microplastics in seafood raises concerns about their potential impact on human health, including inflammation, endocrine disruption, reproductive diseases, cardiovascular diseases, emotional disorders and other health issues (Lin et al. 2024; Marfella et al. 2024).

21.2 The Multifaceted Nature of NMP

Interdisciplinary research is crucial to understand the multifaceted nature of NMP pollution, while a transdisciplinary approach is needed to address it effectively, involving collaboration not only across scientific disciplines but also with the broader community.

Tackling the complex challenge of marine nano- and microplastic pollution demands mission-driven, co-designed scientific research that aligns scientific efforts with societal needs. This approach requires a thorough understanding of the sources, chemical properties, ecological impacts, health effects, associated molecular mechanisms, and potential solutions to address marine plastic pollution, with a particular focus on nano- and microplastics.

Interdisciplinary collaboration is essential in tackling marine pollution, bringing together diverse fields such as biology, chemistry, environmental science, and social science. Biologists study the impacts of plastics on marine life, accessing their impact on health using lab animal models or clinical data, while chemists analyze the breakdown, persistence, and toxicity of plastic particles. Environmental scientists assess the pathways and distribution of pollutants, identifying key areas of concern. Social scientists explore the human dimensions, including public awareness, behavior change, and policy implications. By integrating these diverse perspectives, interdisciplinary research could provide a comprehensive understanding of the multifaceted nature of plastic pollution, paving the way for more holistic and effective management strategies.

By setting clear, mission-driven goals, co-designed research ensures that scientific outcomes are not only academically rigorous but also practical and applicable to environmental management and policy. This collaborative approach helps bridge the gap between science and action, accelerating the development of innovative solutions that can effectively mitigate plastic pollution and promote Ocean health.

21.2.1 Breaking the Negative Cycle of Marine Microplastic Pollution

Marine microplastic pollution represents a pressing global challenge, with detrimental impacts on ecosystems and potential implica-

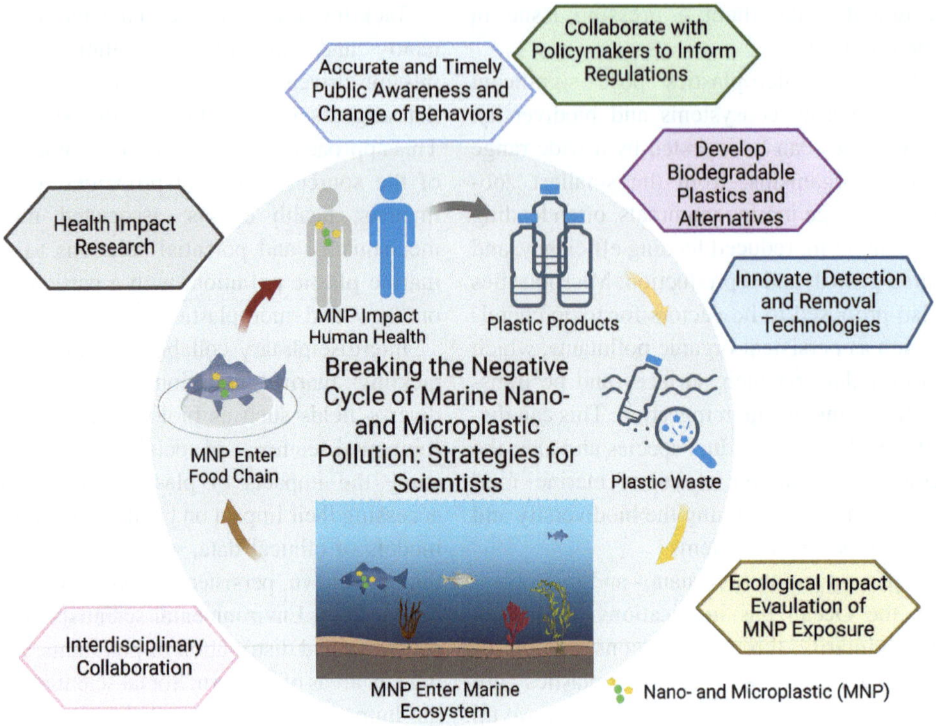

Fig. 21.1 Cycle of marine nano- and microplastic pollution (created with Biorender.com)

tions for human health. This negative cycle begins with plastic waste in the environment, which breaks down into micro- and nanoplastics that enter marine ecosystems. These particles accumulate in aquatic organisms, bioaccumulate through the food chain, and ultimately pose risks to human health. Microplastics can act as vectors for harmful chemicals, disrupt endocrine functions, and induce oxidative stress and inflammation upon ingestion. Addressing this issue requires a comprehensive, science-based approach that targets pollution sources, assesses ecological and health impacts, and develops effective mitigation strategies. Figure 21.1 shows the central cycle depicting the movement of marine nano- and microplastics from plastic waste in the environment, entering marine ecosystems, accumulating in aquatic organisms, and eventually impacting human health through the food chain. Key strategies for scientists are represented in hexagons surrounding the central cycle, including Health Impact Research, Interdisciplinary Collaboration, Accurate and Timely Public Awareness and Behavior Change, Collaborate with Policymakers to Inform Regulations, Develop Biodegradable Plastics and Alternatives, Innovate Detection and Removal Technologies, Ecological Impact Evaluation of MNP Exposure.

21.2.2 Innovative Strategies for Scientists

Innovative scientific strategies are essential for addressing the pervasive issue of marine microplastic pollution. By integrating research across disciplines and leveraging technological advancements, scientists can develop holistic solutions to mitigate the environmental and health impacts of microplastics. Solving this problem demands a toolbox of strategies. Key approaches include interdisciplinary collaboration, ecological impact assessment, health impact research, detection and removal technologies, and development of biodegradable alternatives.

Interdisciplinary Collaboration

Innovative scientific strategies are essential for addressing the pervasive issue of marine microplastic pollution. Rather than treating it as a narrow technical problem, scientists increasingly recognize it as a complex systems challenge that links the Oceans, ecosystems, and human health. By integrating research across disciplines and leveraging advances in sensing, data analytics, and materials science, scientific programs can move from merely documenting microplastic contamination to testing concrete solutions and informing policy and practice. Solving this problem therefore demands a coordinated toolbox of strategies that mutually reinforce one another. Key elements of this toolbox include co-designed interdisciplinary collaboration, ecological impact assessment, health impact research, innovative detection and removal technologies, and the development and implementation of biodegradable alternatives, which are elaborated in the sections that follow.

Ecological Impact Assessment

Comprehensive evaluations of the ecological impacts of microplastic exposure are crucial for understanding the long-term consequences on marine ecosystems (Zhang et al. 2024). Research must go beyond identifying the presence of microplastics and instead focus on how these particles interact with aquatic organisms, influence food webs, and disrupt ecosystem functions. This includes assessing the physical damage to organisms, such as ingestion and entanglement, and the biochemical impacts, such as oxidative stress and immune responses. These insights are critical for conservation strategies aimed at protecting vulnerable species and habitats, as well as predicting future ecosystem responses to ongoing plastic pollution.

Health Impact Research

Research into the health impacts of microplastics is vital, particularly their effects on human health through the food chain (Habumugisha et al. 2024). Microplastics can carry toxic chemicals, including persistent organic pollutants (POPs), heavy metals, and endocrine-disrupting compounds (Sun et al. 2023; Zhang et al. 2024). Studies are needed to explore more about microplastics: how they are absorbed, distributed, metabolized, and excreted in the body. Research should also investigate the cellular and molecular mechanisms underlying microplastic-induced health effects, such as inflammation, oxidative stress, and disruption of cellular processes. Findings from these studies can inform public health guidelines and contribute to global health policies that mitigate exposure risks.

Detection and Removal Technologies

Developing advanced technologies for detecting and removing microplastics from marine environments is critical (Chukwuka et al. 2024; Junaid et al. 2024). Innovations such as nanotechnology, enhanced filtration systems, and bio-based materials for adsorption and degradation offer promising solutions. Remote sensing technologies, including satellite imagery and drones, can be utilized to monitor microplastic hotspots, while AI-driven analytics can improve the accuracy and efficiency of detection. In-situ removal technologies, such as floating booms and skimmers, can help clean up microplastics from waterways before they reach the open Ocean, reducing the overall ecological burden.

Development of Biodegradable Alternatives

Advancing research into biodegradable and sustainable plastic alternatives offers a proactive approach to reducing microplastic pollution at its source. Scientists are exploring polymers derived from natural materials like starch, cellulose, and algae, which degrade more easily in marine environments without releasing harmful microplastics (Deng et al. 2022; Smidova et al. 2024). By developing and promoting these alternatives, the reliance on conventional plastics can be reduced, thereby minimizing future environmental contamination.

21.2.3 Transdisciplinary Approaches Foster a Deeper Level of Collaboration

Transdisciplinary research goes beyond traditional interdisciplinary approaches by actively involving non-academic stakeholders, from identifying problems to putting solutions into action. According to the theory of planned behavior, it is crucial to enhance public behavioral intentions and guide responsible behavior in the marine environment (Panwanitdumrong and Chen 2021; Zhao et al. 2022). If there is only marine science knowledge available, it cannot arouse public interest in learning, let alone enhance public participation in marine conservation. Purposeful and interesting marine environment interpretation through transdisciplinary collaboration is an important means to help the public participate in marine knowledge learning and generate conservation behaviors.

This approach will foster a deeper level of collaboration, integrating knowledge and efforts from scientists, policymakers, industry, and communities. By combining technical expertise with practical insights from those directly impacted by marine pollution, transdisciplinary collaboration ensures that solutions are more targeted, innovative, and actionable. This dynamic exchange of knowledge enhances the ability to address the complex challenges of plastic pollution and fosters sustainable interactions between humans and the Ocean.

It will increase public awareness and behavior change. Educating the public about the dangers of microplastics and promoting behavior change are crucial areas where scientists can play a supportive yet impactful role. While public outreach is often led by communication specialists and environmental organizations, scientists provide the essential data and insights that underpin these efforts. By translating complex research findings into accessible, clear messages, scientists can help shape educational campaigns that effectively convey the links between plastic use, pollution, and health impacts. Their involvement ensures that public awareness initiatives are grounded in scientific evidence, making them more credible and compelling. Scientists can also collaborate with educators, journalists, and community leaders to develop resources that promote responsible plastic use, proper waste disposal, and engagement in local conservation activities, amplifying the reach and impact of these efforts.

It will influence policy engagement. Scientists play a vital yet often behind-the-scenes role in shaping policies to combat microplastic pollution. While policymakers and advocacy groups drive the regulatory process, scientific input is essential to ensure that policies are evidence-based and effective. Scientists provide the foundational research that informs regulations on plastic production, waste management, and pollution control. Their role includes conducting risk assessments, modeling environmental and health impacts, and providing expert testimony that can guide decision-making. By collaborating with policymakers, scientists help translate research findings into actionable policy recommendations, offering data-driven insights that enhance the effectiveness of regulations. Although supportive, their involvement is critical in ensuring that policy measures are not only scientifically sound but also adaptable to emerging research and technological advancements.

21.3 Communicating Science for Ocean Literacy

Scientific findings must be actively shared with the public to enhance Ocean Literacy, ensuring that the significance of Ocean health is widely understood and acted upon, which is a challenging task. Effective communication between scientists from diverse fields is essential for advancing our understanding of microplastic pollution and developing effective solutions. Traditional methods of knowledge dissemination, such as academic conferences, journals, and expert reviews, have long been the cornerstone of scientific communication. Seminars inspired by historical traditions, like those of Pavlov, provide critical platforms for researchers to share findings, debate methodologies, and collaborate across disciplines. These traditional avenues facilitate deep academic engagement and ensure the rigor and credibility of scientific knowledge. However, they often operate within the confines of academic circles, limiting the broader impact of scientific discoveries.

Despite their value, conventional dissemination methods face significant challenges, including time delays, accessibility issues, and language barriers. The publication process in high-profile journals can be lengthy, and conference participation often requires travel, which can be cost-prohibitive and environmentally unsustainable. Moreover, language barriers can exclude valuable contributions from non-English-speaking researchers, while geographic limitations can restrict access to cutting-edge discussions. These factors create a gap between knowledge creation and its application in policy and public awareness, hindering the timely translation of research into actionable solutions.

To bridge these gaps, scientists must embrace modern platforms and innovative approaches to knowledge dissemination. Digital platforms, such as open-access journals, preprint repositories, and virtual conferences, offer more immediate and inclusive alternatives to traditional publishing. These platforms allow for rapid sharing of research findings, fostering real-time collaboration and feedback among scientists from diverse fields. Virtual forums and webinars enable global participation, breaking down geographic barriers and promoting cross-disciplinary exchange that is crucial for tackling complex issues like microplastic pollution.

Social media and other digital communication tools also play a transformative role in disseminating scientific knowledge beyond the academic community. By leveraging platforms like Twitter, LinkedIn, WeChat and ResearchGate, scientists can share research summaries, engage in public discussions, and collaborate with colleagues worldwide. Engaging with non-scientific audiences through simplified communication methods, such as infographics, videos, and storytelling, can further enhance public understanding and inspire behavioral change. These formats make complex scientific concepts more accessible and relatable, helping to build public support for evidence-based policies and environmental initiatives.

Scientists can also collaborate with media outlets and content creators to amplify their research findings, ensuring that key messages reach policymakers, educators, and the public. By working with journalists to produce clear, accurate, and engaging content, researchers can help shape the public narrative around microplastic pollution, emphasizing the urgency of the issue and the need for collective action.

These enhanced dissemination strategies are not just about spreading information but also about fostering a dynamic dialogue between science and society. By making research more accessible and actionable, scientists can play a pivotal role in driving societal action, influencing policy, and ultimately, breaking the negative cycle of marine microplastic pollution. The integration of traditional and modern communication methods can bridge the gap between knowledge creation and its real-world application, ensuring that scientific advancements translate into meaningful environmental and health outcomes.

21.4 Three Critical Dimensions of Connecting Science to Ocean Literacy

Addressing the complex issue of marine nano- and microplastic pollution requires a multifaceted approach that integrates transdisciplinary research, enhances Ocean Literacy, and fosters collaboration across various levels. The conceptual framework depicted in Fig. 21.2 illustrates the interconnection of these three critical dimensions, represented by the X, Y, and Z axes, which together form a 3D model of Transdisciplinary Science and Ocean Literacy, aiming to visually capture the dynamic interplay between transdisciplinary research, Ocean Literacy, and multi-level collaboration, highlighting the systematic approach required to address the growing challenge of marine nano- and microplastic pollution. The 3D model consists of three axes, each representing a key dimension in the fight against Ocean pollution: X-axis (depicted in pink)

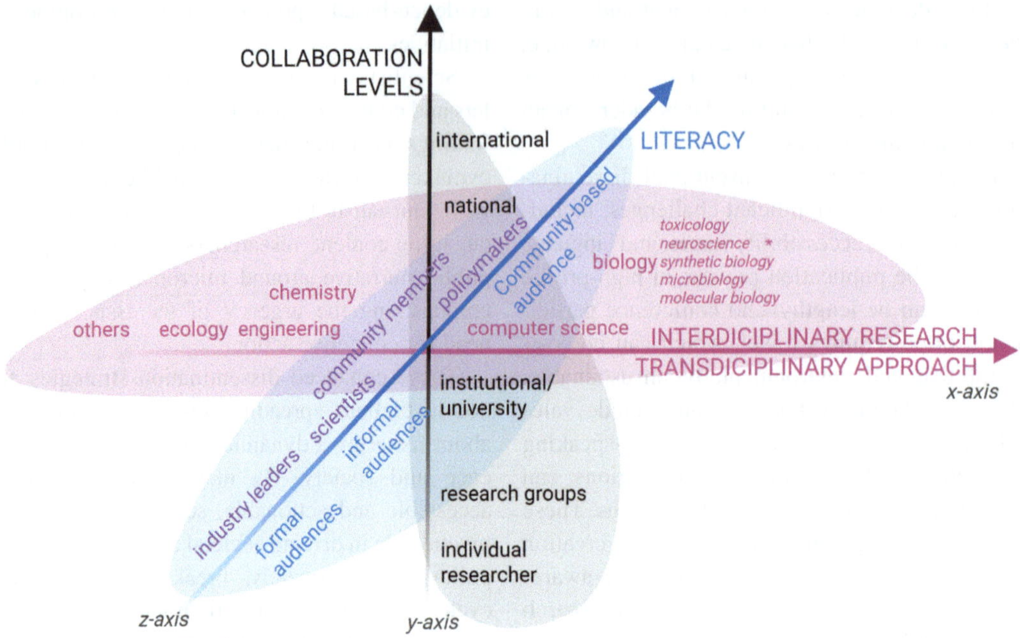

Fig. 21.2 Three critical dimensions of connecting scientific research to Ocean Literacy

emphasizes the blending of multiple scientific disciplines and involvement of transdisciplinary approaches to tackle the complex issue of nano- and microplastic pollution. Y-axis: (depicted in blue) represents different audiences of Ocean Literacy. Z-axis: (depicted in grey) shows the hierarchical levels of collaboration needed, ranging from individual researchers to international effort.

21.4.1 Co-designed Interdisciplinary Research

Interdisciplinary research emphasizes the integration of multiple scientific perspectives, including chemistry, ecology, engineering, computer science, and biology, to tackle the intricate problem of nano- and microplastic pollution. This dimension reflects the need to go beyond traditional disciplinary boundaries to develop comprehensive solutions that address the full scope of plastic pollution. For example, chemists and material scientists work on the development of biodegradable plastics (Deng et al. 2022; Smidova et al. 2024), while ecologists study the impacts of microplastics on marine organisms and ecosystems (Zhang et al. 2024). Engineers create innovative detection and removal technologies (Chukwuka et al. 2024; Junaid et al. 2024), and computer scientists develop models to predict the spread of microplastics in Ocean currents (Cottom et al. 2024). Artificial intelligence serves as a cross-cutting enabler, integrating data and insights from chemistry, ecology, engineering, biology, and Ocean science to advance high-resolution monitoring, automated microplastic identification, and predictive modeling. Beyond these capabilities, AI can also facilitate early-warning systems for pollution hotspots, optimize cleanup strategies, enhance the design of biodegradable materials, and strengthen biological and ecological assessments by identifying biomarkers of exposure, modeling organismal responses, and supporting decision-making frameworks that align scientific solutions with societal and policy needs. This blending of expertise fosters a holistic understanding of marine plastic pollution, driving the development of integrated solutions that are scientifically robust and socially relevant.

Co-design is a collaborative approach to research that actively involves stakeholders, including scientists, policymakers, industry leaders, and community members, in the research design process. This inclusive approach is particularly valuable in transdisciplinary research, where the blending of multiple scientific disciplines is necessary to tackle complex problems like nano- and microplastic pollution. Co-design ensures that research questions, methodologies, and outcomes are aligned with the needs and priorities of all stakeholders, leading to more relevant and impactful results. By integrating diverse perspectives from the outset, co-design helps to identify gaps in existing knowledge, explore innovative solutions, and enhance the applicability of research findings to real-world challenges. The benefits of co-design include increased stakeholder buy-in, improved data quality through diverse input, and a greater likelihood that research outcomes will inform policy, industry practices, and public behavior.

A notable example of co-design in action is the collaborative project on marine microplastic pollution led by a team of young researchers. Recognizing the urgent need to understand the impact of microplastics on human health, five diverse research teams from disciplines including marine toxicology, neurobiology, reproductive endocrinology, and clinical medicine united for a shared purpose. The collaboration brought together researchers from institutions such as Jahangirnagar University and Bangladesh University of Engineering and Technology in Bangladesh, Kyushu University in Japan, the Shenzhen Institutes of Advanced Technology at the Chinese Academy of Sciences, Hong Kong Polytechnic University, and Guilin Medical College in China.

The teams collectively developed a research proposal titled "The Impact of Microplastics on Brain Health and Reproductive Health and the Damage of Barrier Function (2021–2024)." Led by Prof. Li Lei from the Shenzhen Institutes of Advanced Technology of the Chinese Academy of Sciences, the project received guidance from UNESCO-IOC/WESTPAC, which aims at leading global efforts to understand the sources, fate and effects of plastics and microplastics in the marine environment and funding from the Alliance of National and International Science Organizations for the Belt and Road Regions (ANSO), a member of the Ocean Decade NGOs. Significant progress has been achieved, including the publication of peer-reviewed articles that document the distribution and characteristics of microplastics and highlight their detrimental effects on pulmonary, reproductive, and brain health. Notably, one of the articles from this same interdisciplinary collaborative team was cited in the 2022 WHO report on nano- and microplastic exposure and its health implications (Lu et al. 2021; WHO 2022).

21.4.2 Building Partnerships in Ocean Literacy

Ocean Literacy focuses on enhancing public understanding and engagement with marine pollution issues, making it a crucial component in the fight against nano- and microplastic pollution. Educational and outreach efforts target diverse audiences, ranging from elementary and high school students to university communities and informal learners. Educational programs, public seminars, and digital platforms are used to disseminate knowledge, encouraging behavioral change and supporting the adoption of sustainable practices. By fostering Ocean Literacy, these efforts help bridge the gap between scientific research and societal action, ensuring that the public is not only aware of the issues but also actively involved in addressing them.

Promoting Ocean Literacy through collaborative efforts is akin to the functioning of a neural circuit—just as neurons work together in networks to process information, collaborative partnerships in Ocean Literacy bring together diverse groups to create a collective impact. Ocean Literacy initiatives require the concerted effort of educators, scientists, policymakers, and community leaders working in tandem to foster a deeper understanding of marine pollution issues. These collaborations help to build a more informed society capable of enacting positive change.

At the local level, schools, universities, and community organizations can collaborate to deliver educational programs that inspire the audiences and raise awareness about Ocean health. Regional partnerships can involve governments and non-profits working together to implement policies that promote Ocean-friendly practices and community engagement. At the global level, international collaborations allow for the exchange of best practices, the development of unified strategies, and the amplification of outreach efforts.

By working across these levels, partnerships in Ocean Literacy create a ripple effect—empowering communities, influencing policy, and driving societal action. Such a multi-tiered approach ensures that Ocean Literacy initiatives are comprehensive, inclusive, and capable of making a lasting impact.

21.4.3 Funding as Hub for Collaborative Science

In Fig. 21.2, the Z-axis represents the hierarchical levels of collaboration needed to tackle marine pollution, ranging from individual researchers to large-scale international efforts. Collaboration across disciplines, sectors, and borders is vital for developing and implementing effective solutions. The interconnected nature of scientific research, public engagement, and collaborative partnerships demonstrates that breaking the negative cycle of marine plastic pollution requires coordinated action at every level. From local community initiatives to global alliances, fostering collaboration ensures that diverse perspectives are integrated, resources are pooled, and efforts are synergized to maximize impact.

Funding is a critical enabler of transdisciplinary research, collaboration, and Ocean Literacy initiatives. Financial support not only sustains research projects but also facilitates the integration of diverse expertise, the dissemination of findings, and the implementation of solutions. For transdisciplinary projects that span multiple scientific domains and involve complex stakeholder interactions, adequate funding is essential to cover the costs of research activities, communication, training, and outreach.

One notable example of funding support for collaborative science is the ANSO, which plays an active role in supporting the co-design and co-creation of science knowledge. As a member of the Ocean Decade NGOs, ANSO promotes international cooperation among its 78 members from 52 countries, encouraging partnerships that address global challenges, including marine pollution (Kennedy et al. 2024, p. 31). Funding bodies like ANSO and the International Partnership Program of the Chinese Academy of Sciences are crucial for bridging the gaps in transdisciplinary science, ensuring that collaborative efforts are well-resourced and capable of making a tangible impact on marine stewardship and public engagement.

21.5 EPICD in Action: Creating a Cycle of Knowledge and Engagement

The EPICD framework serves to connect scientific research with public engagement, creating a positive feedback loop that fosters both knowledge dissemination and informed action, thereby advancing marine stewardship efforts.

Research on nano- and microplastics has become one of the most rapidly expanding and highly scrutinized fields in environmental and Ocean sciences, reflecting growing scientific and societal concern (Thompson et al., 2024). The abundance of research also offers opportunities to enhance Ocean Literacy. By curating and effectively communicating scientific insights, stakeholders can empower the public to make informed decisions and take action towards Ocean conservation. Leveraging digital platforms, social media, and collaborative partnerships can further help translate scientific findings into meaningful, actionable knowledge that fosters a more Ocean-literate society.

Another significant challenge is the complexity of scientific language, which can create a communication gap between researchers and the public. Technical jargon and detailed data can be intimidating and difficult to understand, limiting the accessibility of scientific information. Geographic and socio-economic disparities also play a role; remote and underfunded communities often lack access to quality educational resources, digital tools, or opportunities for direct engagement with scientists. Moreover, the lack of interdisciplinary collaboration can lead to fragmented and incomplete messaging, where the broader impacts of Ocean issues are not fully conveyed. Misinformation and competing narratives, especially on social media, further complicate the public's understanding of Ocean health. Additionally, time constraints and a lack of incentives for scientists to engage in outreach activities can limit their involvement in Ocean Literacy initiatives, reducing the potential for impactful communication.

Addressing these barriers of knowledge creation and dissemination requires a strategic approach that prioritizes accessibility, inclusivity, and collaboration with actionable steps, which we proposed to be "EPICD" (pronounced as "Epic-D") for the scientists. EPICD stands for:

- **E**ngagement with the public and social media to counter misinformation.
- **P**artnerships with educators, media, and community organizations.
- **I**nterdisciplinary collaboration to enrich Ocean Literacy.
- **C**ommunication training for scientists.
- **D**igital platforms to disseminate open-access resources.

By implementing these solutions, the Ocean science community can break down barriers to effective literacy, creating a more informed and empowered society ready to tackle the challenges facing our Ocean.

21.5.1 Engagement with the Public and Social Media to Counter Misinformation

Despite growing efforts to promote Ocean Literacy, several barriers hinder the effective dissemination of Ocean science knowledge. A key challenge arises from the overwhelming increase in scientific publications on critical topics like microplastics (Thompson et al. 2024). While this surge in research highlights the growing attention to microplastic pollution, it also complicates the effective dissemination of information to the public. The sheer volume of new scientific data can be difficult to navigate, even for experts, making it challenging for educators, policymakers, and the public to stay informed about the latest findings. This information overload can hinder efforts to communicate key messages about Ocean health and the impacts of human activities, such as plastic pollution. Bridging the gap between scientific research and public understanding requires targeted strategies that simplify complex data, such as creating accessible summaries, visual aids, and engaging storytelling that resonates with diverse audiences.

As Ocean research advances at an exponential rate, it is crucial to counter misinformation and ensure the public receives accurate information. Scientists should actively engage with social media, participate in public discussions, and provide trustworthy sources of information. Don't just publish—participate. Under the framework of EPICD, we have disseminated our latest research on the impact of microplastics on brain and reproductive health through various channels, including school lectures, public talks at libraries, and online platforms. In addition to mainstream social media, platforms such as TEDx and SELFtalks provide valuable venues for scientists to communicate cutting-edge research to the public. Professional academic societies are also beginning to support such outreach by hosting events where researchers share and interpret the latest scientific findings. These initiatives represent promising models for strengthening public engagement and enhancing Ocean Literacy. Our science communication efforts have reached approximately 2800 students from Grades 3–12 (ages 8–15) across 16 schools in China, as well as an online audience of over 100,000 people of all ages.

21.5.2 Partnerships with Educators, Media, and Community Organizations

Developing partnerships with educators, media professionals, and community organizations can help disseminate scientific knowledge more effectively and reach diverse populations. The purposeful and interesting marine environment interpretation is provided through partnerships with interdisciplinary collaboration, that could raise public's awareness and enhance their environmentally responsible behavior.

Scientists play a pivotal role in fostering Ocean Literacy by connecting with educators and students to bring real-world scientific insights into the classroom. Approaches include guest lectures, school visits, and collaborative projects that directly engage students with marine science topics. Scientists can partner with teachers to develop interactive lesson plans, provide hands-on experiences such as field trips or lab demonstrations, and integrate current research findings into classroom discussions. Involving scientists in curriculum development ensures that educational content is accurate, up-to-date, and aligned with the latest scientific advancements. Another effective approach is to offer mentorship programs, where scientists guide students through projects or science fairs, helping to ignite a passion for Ocean conservation. By building these direct connections, scientists not only share knowledge but also inspire students to pursue careers in marine science, fostering a new generation of Ocean advocates.

For instance, the Shenzhen Institutes of Advanced Technology, Chinese Academy of Sciences, has pioneered a "Science+ education alliance—the first of its kind in China. By pooling resources and expertise, the alliance has established a network that connects schools and educational institutions across diverse regions. This collaborative platform supports educational development through shared resources such as specialized courses, state-of-the-art equipment in mentorship programs, and interdisciplinary knowledge exchanges. Regular principal forums, school visits, and educational activities foster a collaborative learning environment, enriching science education at the primary and secondary school levels with support from more than 230 researcher including professors, associated professor, post-doc trainees, and postgraduate students from the Shenzhen Institutes of Advanced Technology and Shenzhen University of Advanced Technology. The alliance's initiatives have reached over 30,000 students, disseminating cutting-edge scientific knowledge to schools across China. Through this partnership, the "Science+" alliance exemplifies how collaborative models can effectively promote science literacy, engage young learners, and inspire the next generation of stewards.

21.5.3 Interdisciplinary Collaboration to Enrich Ocean Literacy

Fostering interdisciplinary collaboration can enhance the depth and breadth of Ocean Literacy messages, ensuring they encompass ecological, social, and economic dimensions. Establishing dedicated funding and recognition programs for outreach efforts can incentivize scientists to participate more actively in public engagement.

The complex and interdisciplinary nature of the research project "The Impact of Microplastics on Brain Health and Reproductive Health and the Damage of Barrier Function" demonstrates the power of setting common goals and fostering collaborative environments—an approach that would have been unfeasible for individual teams working in isolation, especially for young researchers. The project also facilitated two international conferences and the training of over 250 university and postgraduate students from Bangladesh and Thailand.

21.5.4 Communication Training for Scientists

To bridge the communication gap, scientists should be trained in science communication skills, learning how to distill complex information into clear, engaging, and relatable messages, utilizing storytelling, visual aids, and simplified language can make Ocean science more approachable for non-expert audiences.

The dissemination of scientific knowledge has evolved significantly, blending traditional methods such as academic journals, books, and conferences with modern digital tools like social media, podcasts, and online platforms. Classical approaches, while authoritative and rigorous, often have limitations in accessibility and speed; peer-reviewed publications can take months to reach the public, and conference presentations are typically limited to academic audiences. In contrast, digital tools enable immediate, wide-reaching, and interactive engagement with diverse audiences. Social media platforms like WeChat, Toutiao, Bilibili, YouTube, and Twitter (X) allow scientists to share research findings quickly, break down complex topics, and engage directly with the public in real-time. Digital tools also offer opportunities for visual storytelling through infographics, videos, and animations, making scientific information more digestible and appealing. While traditional methods provide credibility and depth, modern tools enhance reach and engagement, highlighting the need for a balanced approach that leverages the strengths of both to maximize the impact of Ocean Literacy efforts.

Scientists can employ engaging and innovative strategies that make Ocean science accessible and exciting to reach a broader audience, formal, informal or community based. Storytelling is a powerful tool; sharing real-life research adventures, discoveries, and challenges can capture the audience' imaginations and make abstract scientific concepts relatable. Scientists can also utilize multimedia resources, such as videos, virtual reality experiences, and interactive apps, offering immersive experiences that traditional ways cannot provide. In addition, emerging AI tools—particularly large language models—can assist scientists in preparing interactive, audience-tailored outreach materials, helping to overcome technical barriers and reduce the time required to produce high-quality science communication content.

The advent of digital technology offers unprecedented opportunities to enhance public understanding of Ocean science. Interactive websites, mobile apps, and virtual reality (VR) simulations can transform the way people learn about the Ocean, making complex scientific concepts more accessible and engaging. For instance, VR experiences can take users on virtual dives into coral reefs or deep-sea environments, providing an immersive understanding of marine ecosystems that would otherwise be out of reach. Augmented reality apps can bring marine life into homes and classrooms, offering interactive learning experiences that captivate audiences of all ages.

Citizen science projects are another strategy to actively involve people in Ocean conservation. Programs like beach clean-ups, biodiversity monitoring, and plastic pollution tracking empower participants to see the impact of their actions firsthand, reinforcing the connection between learning and real-world environmental stewardship.

21.5.5 Digital Platforms to Disseminate Open-access Resources

Digital platforms can be leveraged to offer open-access educational resources, ensuring that all communities, regardless of location or socio-economic status, have access to Ocean science information.

Online platforms such as webinars, MOOCs (Massive Open Online Courses), and digital libraries make high-quality Ocean science education available to global audiences, breaking down geographic and economic barriers. Social media channels enable scientists to communicate directly with the public, correct misinformation, and build a community of informed Ocean advocates. By integrating these technologies into their outreach efforts, scientists can create a more inclusive, dynamic, and responsive Ocean Literacy landscape, empowering the public with the knowledge needed to make informed decisions about Ocean conservation.

21.5.6 The Positive Feedback Loop: Ocean Literacy and Ocean Science

We propose that a positive feedback loop between Ocean Literacy and Ocean science occurs when advancements in scientific research enhance public understanding of Ocean-related issues, which in turn stimulates greater public engagement and support for further scientific inquiry. This mutually reinforcing relationship drives increased awareness, funding, and action towards marine conservation, creating a cycle where Ocean Literacy fuels scientific discovery, and science informs and deepens public literacy. As the public becomes more knowledgeable, their advocacy and behavioral changes contribute to better research opportunities and conservation outcomes, amplifying the overall impact on Ocean health.

The dynamic interplay between Ocean Literacy and scientific research is characterized by mutual reinforcement, where each element strengthens the other. As scientific research uncovers new insights into marine ecosystems, pollution impacts, or human health implications, these findings can be communicated to the public through educational programs, media, and community outreach. Enhanced Ocean Literacy empowers individuals to make informed decisions, advocate for sustainable practices, and support Ocean-friendly policies. In turn, a well-informed public can provide valuable feedback, participate in citizen science initiatives, and push for increased funding and prioritization of marine research. This public support not only aids in shaping research agendas that address real-world concerns but also helps in the validation and dissemination of scientific knowledge. As awareness and understanding grow, the public's willingness to engage in conservation efforts strengthens, creating a virtuous cycle that continuously uplifts both Ocean science and literacy. This loop ultimately fosters a culture where science and society collaborate seamlessly to protect and restore marine environments.

A well-informed public plays a crucial role in driving marine conservation efforts by making sustainable choices, supporting environmental initiatives, and advocating for the protection of marine ecosystems. Awareness of the impacts of marine pollution, especially marine nano- and microplastics pollution, empowers individuals to minimize plastic use and promote plastic recycle. Informed communities are more likely to participate in conservation activities, such as beach clean-ups and habitat restoration, amplifying collective action. Public knowledge of marine issues also generates grassroots support for marine protected areas, sustainable policies, and conservation funding, helping to safeguard Ocean biodiversity for future generations.

An informed public can significantly influence policy-making and environmental stewardship by advocating for stronger environmental laws, holding policymakers accountable, and supporting science-based decision-making. When citizens understand the science behind marine issues, they are more likely to demand policies that address threats such as plastic pollution, and habitat destruction. This civic engagement pushes governments and institutions to prioritize marine conservation in their agendas. Public pressure can lead to the adoption of sustainable fisheries management, stricter regulations on pollutants, and investment in green technologies (Qiang et al. 2022). Additionally, informed citizens often serve as environmental stewards within their communities, leading local conservation efforts and educating others. This ripple effect extends beyond individual actions, fostering a societal culture that values and actively protects the Ocean. Consequently, the positive feedback loop between Ocean Literacy and Ocean science is not just beneficial for advancing knowledge; it plays a transformative role in shaping a resilient and sustainable future for our Ocean.

21.6 Conclusion

Addressing marine pollution, especially from nano- and microplastics, requires more than just scientific inquiry—it demands a transdisciplinary approach that brings together diverse scientific disciplines, policymakers, educators, and the public. While interdisciplinary collaboration allows scientists to investigate the complex dynamics of marine pollution and its relationship with human, transdisciplinary efforts ensure that this knowledge is shared with the broader community to foster public engagement and informed decision-making. EPICD is an effective framework for achieving this integration. By emphasizing Engagement, Partnerships, Interdisciplinary Collaboration, Communication Training, and Digital Platforms, EPICD facilitates a dynamic exchange of knowledge between scientists and the public—an approach that has become increasingly essential in an era of rapidly expanding scientific discoveries and accelerating information flow. This continuous cycle of scientific discovery, public awareness, and informed policymaking will contribute to more effective solutions for Ocean conservation and a healthier, more sustainable Ocean.

- The EPICD framework facilitates the connection between science and public engagement. By emphasizing Engagement, Partnerships, Interdisciplinary Collaboration, Communication Training, and Digital Platforms, the EPICD framework ensures that scientific research is effectively communicated and utilized by the public.
- EPICD offers actionable steps for scientists. The EPICD framework provides clear strategies for scientists to engage with the public, collaborate across disciplines, and improve communication, supporting the goal of a more sustainable Ocean. The EPICD framework enables a cycle where scientific discovery leads to public awareness, which in turn drives policy changes and additional research, ultimately contributing to better protection of the Ocean.

21.6.1 Summary

Key concepts explored in this chapter include:

- Marine pollution is a complex issue. Nano- and microplastic pollution represent significant threats to marine ecosystems and human health, requiring urgent and multifaceted solutions.
- Interdisciplinary collaboration among scientists is essential. Scientists from various fields must work together to understand the impact of pollution on marine environments and human health, as the problem involves complex, interconnected systems.
- A transdisciplinary approach is necessary. A transdisciplinary approach that includes not just scientists but also policymakers, educators, and the broader public is crucial for developing actionable solutions to marine pollution.
- Ocean Literacy (OL) is critical for public awareness. Enhancing public understanding of Ocean issues through OL is essential for fostering responsible behaviors and empowering individuals to take action toward conservation.

21.6.2 Recommended Resources

The following resources provide further insights and support on this topic:

- UNESCO IOC Microplastics: https://www.ioc.unesco.org/en/microplastics
- Ocean Literacy Portal—Ocean Plastic Pollution An Overview: Data and Statistics: https://ocean-literacy.unesco.org/plastic-pollution-Ocean/
- World Health Organization (WHO)—Dietary and inhalation exposure to nano- and microplastic particles and potential implications for human health: https://www.who.int/publications/i/item/9789240054608
- WHO Microplastics in drinking water: https://cdn.who.int/media/docs/default-source/wash-documents/microplastics-in-dw-information-sheet190822.pdf
- Joint Group of Experts on the Scientific Aspects of Marine Environmental Protection (GESAMP)—How to monitor plastics in the Ocean: http://www.gesamp.org/news/how-to-monitor-plastics-in-the-oceans
- GESAMP—Sources, Fate and Effects of plastics and micro-plastics in the marine environment: http://www.gesamp.org/work/groups/40

- IOC Sub-Commission for the Western Pacific (WESTPAC) Riverine Plastic https://ioc--westpac.org/decade-actions/riverine-plastics/team/

Acknowledgments This work was supported by the Collaborative Research Project (Project Number: ANSO-CR-KP-2021-12) of the Alliance of National and International Science Organizations for the Belt and Road Regions (ANSO) and further inspired a project supported by the International Partnership Program of the Chinese Academy of Sciences, "Transboundary Microplastic Transport and Its Multi-Dimensional Impacts on Gut-Brain Health: Towards Global Coordination Collaboration". We extend our gratitude to Dr. Jianping Fan, Ms. Shuwen Xu, Mr. Rujiao Song, Mr. Haozhao Liu, Ms. Ningning Ding, Dr. Kengpo Lai, and Dr. Lixin Zhu for their invaluable insights and support in coordinating interdisciplinary research collaboration and promoting science literacy toward sustainable development.

References

Allen D, Allen S, Abbasi S, Baker A, Bergmann M, Brahney J, Paul McGinnity IO, Prather KA, Prospero JM, Revell LE, Sander SG, Shim WJ, Slade J, Stein A, Tarasova O, Wright S (2022) Microplastics and nanoplastics in the marine-atmosphere environment. Nat Rev Earth Environ 3:393–405. https://doi.org/10.1038/s43017-022-00292-x

Chukwuka AV, Adegboyegun AD, Oluwale FV, Oni AA, Omogbemi ED, Adeogun AO (2024) Microplastic dynamics and risk projections in West African coastal areas: Developing a vulnerability index, adverse ecological pathways, and mitigation framework using remote-sensed oceanographic profiles. Sci Total Environ 953:175963. https://doi.org/10.1016/j.scitotenv.2024.175963

Cottom JW, Cook E, Velis CA (2024) A local-to-global emissions inventory of macroplastic pollution. Nature 633(8028):101–108. https://doi.org/10.1038/s41586-024-07758-6

Cox KD, Covernton GA, Davies HL, Dower JF, Juanes F, Dudas SE (2019) Human consumption of microplastics. Environ Sci Technol 53(12):7068–7074. https://doi.org/10.1021/acs.est.9b01517

Deng C, Seidi F, Yong Q, Jin X, Li C, Zhang X et al (2022) Antiviral/antibacterial biodegradable cellulose nonwovens as environmentally friendly and bioprotective materials with potential to minimize microplastic pollution. J Hazard Mater 424(Pt A):127391. https://doi.org/10.1016/j.jhazmat.2021.127391

Habumugisha T, Zhang Z, Uwizewe C, Yan C, Ndayishimiye JC, Rehman A, Zhang X (2024) Toxicological review of micro- and nano-plastics in aquatic environments: Risks to ecosystems, food web dynamics and human health. Ecotoxicol Environ Saf 278:116426. https://doi.org/10.1016/j.ecoenv.2024.116426

Junaid M, Liu S, Liao H, Yue Q, Wang J (2024) Environmental nanoplastics quantification by pyrolysis-gas chromatography-mass spectrometry in the Pearl River, China: First insights into spatiotemporal distributions, compositions, sources and risks. J Hazard Mater 476:135055. https://doi.org/10.1016/j.jhazmat.2024.135055

Kennedy TJ, Dufourneaud O, Castillo MP, Deane S, Durand J, Giaccone T, Jonda S, Mascarenhas T, Mtonga C, Ng K, Palazot S, Paul S, Salazar ER, Santisteban R, Shuwen X, Silva A, Solis Rivera V, van't Land H, Zhou Y (2024) Ocean Decade manual for Non-Governmental Organizations (NGOs). UNESCO-IOC, Paris. [The Ocean Decade Series, 54]. https://unesdoc.unesco.org/ark:/48223/pf0000392198

Law KL, Thompson RC (2014) Oceans. Microplastics in the seas. Science 345(6193):144–145. https://doi.org/10.1126/science.1254065

Lim X (2021) Microplastics are everywhere - but are they harmful? Nature 593(7857):22–25. https://doi.org/10.1038/d41586-021-01143-3

Lin Z, Li Z, Ji S, Lo HS, Billah B, Sharmin A et al (2024) Size-dependent deleterious effects of nano- and microplastics on sperm motility. Toxicology 506:153834. https://doi.org/10.1016/j.tox.2024.153834

Lu K, Lai KP, Stoeger T, Ji S, Lin Z, Lin X et al (2021) Detrimental effects of microplastic exposure on normal and asthmatic pulmonary physiology. J Hazard Mater 416:126069. https://doi.org/10.1016/j.jhazmat.2021.126069

Marfella R, Prattichizzo F, Sardu C, Fulgenzi G, Graciotti L, Spadoni T et al (2024) Microplastics and nanoplastics in Atheromas and cardiovascular events. N Engl J Med 390(10):900–910. https://doi.org/10.1056/NEJMoa2309822

Panwanitdumrong K, Chen CL (2021) Investigating factors influencing tourists' environmentally responsible behavior with extended theory of planned behavior for coastal tourism in Thailand. Mar Pollut Bull 169:112507

Qiang X, Guo C, Gu W, Song Y, Zhang Y, Gong X et al (2022) The complex of phycobiliproteins, fucoxanthin, and krill oil ameliorates obesity through modulation of lipid metabolism and antioxidants in obese rats. Nutrients 14(22):4815. https://doi.org/10.3390/nu14224815

Smidova K, Selonen S, van Gestel CAM, Fleissig P, Hofman J (2024) Microplastics originated from agricultural mulching films affect enchytraeid multigeneration reproduction and soil properties. J Hazard Mater 479:135592. https://doi.org/10.1016/j.jhazmat.2024.135592

Sun N, Shi H, Li X, Gao C, Liu R (2023) Combined toxicity of micro/nanoplastics loaded with environmental pollutants to organisms and cells: Role, effects, and mechanism. Environ Int 171:107711. https://doi.org/10.1016/j.envint.2022.107711

Thompson RC, Olsen Y, Mitchell RP, Davis A, Rowland SJ, John AW et al (2004) Lost at sea: Where is all the plastic? Science 304(5672):838. https://doi.org/10.1126/science.1094559

Thompson RC, Courtene-Jones W, Boucher J, Pahl S, Raubenheimer K, Koelmans AA (2024) Twenty years of microplastic pollution research—what have we learned? Science 386(6720):eadl2746

WHO (2022, August 30). Dietary and inhalation exposure to nano-and microplastic particles and potential implications for human health. World Health Organization. https://www.who.int/publications/i/item/9789240054608

Zhang M, Lin Y, Booth AM, Song X, Cui Y, Xia B et al (2022) Fate, source and mass budget of sedimentary microplastics in the Bohai Sea and the Yellow Sea. Environ Pollut 294:118640. https://doi.org/10.1016/j.envpol.2021.118640

Zhang Y, Wang C, Jia R, Long H, Zhou J, Sun G et al (2024) Transfer from ciliate to zebrafish: Unveiling mechanisms and combined effects of microplastics and heavy metals. J Hazard Mater 479:135645. https://doi.org/10.1016/j.jhazmat.2024.135645

Zhao M, Li Z, Xia B, Chen W, Tang T, Meng Z, Ding Y (2022) Enhancing residents' environmentally responsible behavioral intentions: The role of awe and place attachment in Potatso National Park communities, Tibet. Forests 13(8):1251

Authors and Affiliations

[1]*Lei Li: Conceptualization; Writing—Original Draft, Review, and Editing; Supervision.
[2]Minyan Zhao: Writing—Original Draft, Review, and Editing.
[3]Min Wei: Writing—Original Draft, Review, and Editing.
[4]Shuangfang Hu: Writing—Original Draft, Review, and Editing.
[5]Baki Billah: Methodology—Data Collection and Analysis.
[6]Ziyi Lin: Methodology—Data Collection and Analysis.
*Corresponding Author at saralilei@siat.ac.cn

Lei Li PhD, is a Professor and research group leader at the Shenzhen Institutes of Advanced Technology (SIAT), Chinese Academy of Sciences (CAS), and Shenzhen University of Advanced Technology (SUAT), China. She earned her PhD in Departments of Physiology and Anatomy from the University of Hong Kong and pursued neuroscience research as a visiting scholar at Massachusetts Institute of Technology (MIT). Her laboratory studies brain-peripheral organs communication and the effects of emotions and environmental pollutants on health, with a strong emphasis on international collaborations. Li has published over 30 peer-reviewed papers in leading journals, including Current Biology and Neuron, and actively promotes science literacy through public talks, reaching over 1.34 million people.

Minyan Zhao PhD, is the Program Manager in the Division of Science Programs at the Secretariat of the Alliance of National and International Science Organizations for the Belt and Road Regions (ANSO) and an Associate Professor at the Institute of Tibetan Plateau Research, Chinese Academy of Sciences (CAS), Beijing. She earned her PhD and later worked as a postdoctoral researcher at the Institute of Geographic Sciences and Natural Resources Research, CAS. Her research specializes in Environmental Interpretation of Protected Areas and Regional Sustainable Development.

Min Wei is the Director of Educational Innovation and Science Literacy Programs (Grades 1–9) at Shenzhen Longhua Innovative Experimental School in Shenzhen, China. She holds a Master of Education in Chemistry Education from South China Normal University. Her primary responsibilities include coordinating and planning science popularization activities, enhancing science education facilities, organizing professional development programs for teachers, and integrating advanced resources to promote students' scientific literacy. The school serves 3939 students and employs 240 teachers. Min Wei is also involved in compiling science popularization resources and educational materials for grades 1–9 (ages 6–15).

Baki Billah PhD, is a Professor in the Department of Zoology at Jahangirnagar University, Bangladesh, with over 15 years of research experience addressing diverse environmental issues. He earned his PhD in Biology from Hong Kong Baptist University. Dr. Billah utilizes genomics and physiological techniques to study the cellular responses of fish at early life stages to abiotic, biotic, and anthropogenic stressors in Bangladesh's aquatic ecosystems. He is actively involved in interdisciplinary research initiatives, particularly focusing on Ocean microplastic pollution.

Shuangfang Hu PhD, is a Biology and Science Teacher at the High School of Shenzhen University of Advanced Technology, China. She earned her PhD from South China University of Technology of China and served as a visiting scholar at Rutgers University in the United States. Currently, she oversees the science education program for approximately 2000 high school students, focusing on integrating innovative thinking into foundational education. Dr. Hu is involved in contributing to the compilation of science popularization resources and educational materials for grades 10–12 (ages 15–18).

Ziyi Lin is currently a master's student in Electronic Information – Biomedical Engineering at the University of Chinese Academy of Sciences (supervior: Lei Li). He earned a Bachelor of Science in Neuroscience and Pharmacology from the University of Toronto (2015–2019) and subsequently worked at the Shenzhen Institutes of Advanced Technology, Chinese Academy of Sciences, from 2019 to 2023. His research focuses on brain–peripheral communication, examining how factors such as emotions and microplastics influence health.

Ocean Literacy at a Global Turning Point: Reflections, Foundations, and Future Pathways

22

Teresa J. Kennedy

Volume I of the three-volume series *Ocean Literacy: The Foundation for the Success of the Ocean Decade* (*Transforming Education, Research and Engagement*) examines how education, curriculum development, and research contribute to advancing Ocean Literacy and strengthening society's connection to the Ocean. Across 21 chapters, over 90 authors from 19 countries/regions present diverse geographic, disciplinary, and institutional perspectives, highlighting innovative educational models, teacher training programs, national curricula, and community-based initiatives that expand Ocean Literacy across civil society. Collectively, these contributions demonstrate that Ocean Literacy is not only a foundational element of science education, but also a catalyst of sustainability, inclusivity, and informed action within the framework of the Ocean Decade. By amplifying underrepresented voices, engaging marginalized communities, and integrating Indigenous and local knowledge systems, the volume illustrates how Ocean Literacy supports more equitable and sustainable decision-making. Many chapters further underscore the importance of interdisciplinary collaboration across scientific and educational domains, as well as transdisciplinary approaches that engage communities, policymakers, and civil society in co-creating sustainable solutions.

A recurring theme throughout this volume is the importance of embedding Ocean Literacy into both formal and informal education to ensure equitable access for all learners, regardless of geographic location, socioeconomic background, or systemic barriers, and to deepen understanding of the Ocean's role in shaping climate, biodiversity, and human well-being. Many chapters highlight hands-on, experiential learning approaches, ranging from place-based marine education to integrating technology, data, and citizen science in real-world oceanographic research. Other chapters examine how Ocean Literacy can be integrated into national education policies and teacher training programs, equipping educators with the tools and knowledge needed to inspire systemic change and informed decision-making across current and future generations. In particular, several chapters emphasize that engaging youth as change agents and future leaders is essential to fostering long-term Ocean sustainability and ensuring intergenerational continuity in Ocean knowledge and stewardship.

While education remains at the heart of this volume, the final chapters broaden the discussion, emphasizing the role of scientists, policymakers, NGOs, civil society, and interdisciplinary collaborations in advancing Ocean Literacy beyond the traditional classroom setting. This transition reflects a growing recognition that building an Ocean-literate society requires engagement across multiple sectors, uniting education with gover-

T. J. Kennedy (✉)
University of Texas at Tyler, Tyler, TX, USA

International Council of Associations for Science Education (ICASE), Cork, Ireland
e-mail: tkennedy@uttyler.edu

T. J. Kennedy (ed.), *Ocean Literacy: The Foundation for the Success of the Ocean Decade, Volume I*, Sustainable Development Goals Series, https://doi.org/10.1007/978-3-031-98145-6_22

nance, research, and community action. Cross-sector collaboration also requires a conceptual shift in how we understand the Ocean, viewing it not as a series of fragmented basins, but as a single, dynamic system that connects and sustains all life on Earth. The historical use of "ocean" as a common noun reflects the perception of multiple, separate bodies of water. Today, however, scientific consensus affirms that the Ocean functions as one vast, interconnected entity, with ecological and physical processes that transcend geopolitical and geographic boundaries. Just as most languages capitalize "Earth" when referring to the planet, the Ocean merits recognition as a proper noun, symbolizing not only its singular identity, but also its foundational role in maintaining planetary health. Only by recognizing the Ocean as one can society truly address the challenges it faces.

This volume aligns closely with the goals of the Ocean Decade, particularly Challenge 10, which seeks to restore humanity's relationship with the Ocean by fostering more inclusive, just, and accessible pathways for engagement of civil society. The Challenge 10 White Paper underscores the importance of knowledge-sharing, strategic communication, education, and cultural connections in achieving this transformation. It ensures equitable opportunities for everyone, including historically marginalized communities, to shape a more sustainable Ocean future. The insights presented in Volume I serve as a foundation for these efforts, illustrating how Ocean Literacy can bridge the gap between scientific discovery and societal action.

As the first installment of this three-volume collection, Volume I establishes the foundation for advancing Ocean Literacy through innovative teaching practices, scholarly inquiry, and inclusive engagement. It sets the stage for exploring broader societal and global applications in Volumes II and III. Building on this foundation, Volume II focuses on fostering community, the Sustainable Blue Economy, and the Arts, highlighting public engagement, creative expression, and economic strategies that strengthen sustainable relationships with the Ocean. Volume III examines communication, technology, and global initiatives, exploring cutting-edge tools, digital platforms, and global collaborations that drive Ocean Literacy in the twenty-first century.

These volumes provide a comprehensive perspective on how Ocean Literacy can inspire meaningful action toward a more informed, resilient, and sustainable world. By deepening our understanding of the Ocean's influence on Earth's systems, and our reciprocal impact on the Ocean, we take meaningful steps toward restoring the balance needed for a healthy future for humanity and the planet.

On behalf of the more than 250 authors contributing to this three-volume project, I invite readers to continue this journey through Volume II: *Fostering Community, the Sustainable Blue Economy, and the Arts* and Volume III: *Accelerating Communication, Technology, and Global Initiatives*. In doing so, the series strengthens connections between people, science, society, and the living Ocean, advancing the vision and goals of the United Nations Decade of Ocean Science for Sustainable Development.

Volume I: Contributors (Alphabetical Order)

Foreword (2)
Leinen, Margaret S.
Ryabinin, Vladimir E.

Reflections (3)
Bassan, Niccolò
Kennedy, Declan
Soapi, Katy

Chapter Authors (91)
Açık, Şermin (Chapter 17)
Ammentorp, Louise (Chapter 18)
Banchi, Elisa (Chapter 20)
Barbeito Morandeira, Paula (Chapter 3)
Barord, Gregory J. (Chapter 6)
Bell, Hannah (Chapter 10)
Billah, Baki (Chapter 21)
Bjelvenmark, Jens (Chapter 15)
Borja, Ángel (Chapter 20)
Bridge, Nicola (Chapter 2)
Bueno, Marilia (Chapter 11)
Cackowski, Celia (Chapter 9)
Canu, Donata (Chapter 20)
Caputo, Anna (Chapter 9)
Carluccio, Paola (Chapter 3)
Carretti, Claire (Chapter 10)
Çavaş, Bülent (Chapter 17)
Christofoletti, Ronaldo Adriano (Chapter 11)
Coulton, Jack (Chapter 3)
da Silva, Alexandre Augusto (Chapter 5)
De Abreu, Amalia (Chapter 10)
Eparkhina, Dina (Chapter 20)
Ezgeta-Balić, Daria (Chapter 20)
Fauville, Géraldine (Chapter 15)
Frensley, B. Troy (Chapter 10)
Gardiner, Greta Olsen (Chapter 9)
Giusta, Elena (Chapter 20)
Glass, Bonnie (Chapter 14)
Glithero, Diz (Chapter 2)
Gonçalves, Leandra R. (Chapter 11)
Gottfried Burguett, Esteban (Chapter 3)
Gozzo Andreotti, Andrezza Justino (Chapter 11)
Greene, Thomas F. (Chapter 4)
Hart, Natalie (Chapter 2)
Hu, Shuangfang (Chapter 21)
Ignacio, Barbara Lage (Chapter 11)
Kennedy, Teresa J. (Chapters 1, 7, 22)
King, Andrew (Chapter 20)
Koç, Simge (Chapter 17)
Kolaç, Mısra (Chapter 17)
Koulouri, Panayota [Yolanda] (Chapter 20)
Kveven, Ardi (Chapter 13)
Lawrence, Lisa (Chapter 9)
Lee, Tammy D. (Chapter 14)
Lemiec, Gail (Chapter 14)
Li, Lei (Chapter 21)
Lin, Ziyi (Chapter 21)
Lippi, Adriana (Chapter 11)
Locritani, Marina (Chapter 20)
Lodes, Katie M. (Chapter 7)
Lovat, Valentina (Chapter 3)
Madden, Lauren (Chapter 18)
Magalhães, Ana Vitória Tereza (Chapter 3)
Magee, Nathan (Chapter 18)
Magnusson, Jennifer L. (Chapter 7)
Manca Zeichen, Marta (Chapter 20)
Marrero, Meghan E. (Chapter 8)
Marshall, Peter (Chapter 10)

T. J. Kennedy (ed.), *Ocean Literacy: The Foundation for the Success of the Ocean Decade, Volume I*, Sustainable Development Goals Series, https://doi.org/10.1007/978-3-031-98145-6

Martelli Mazzo, Tatiana (Chapter 11)
Matsumoto, George I. (Chapter 7)
McRuer, Jen (Chapter 2)
Mueller, Ninja (Chapter 19)
Napolitano, Kristen V. (Chapter 8)
Nuss, Sarah (Chapter 9)
Orsolini, Karina F. (Chapter 5)
Pedroso Conti, William Remo (Chapter 11)
Petrera, Francesca (Chapter 20)
Pomaro, Angela (Chapter 20)
Reis, Pedro (Chapter 12)
Rivera-Peña, Gabriela (Chapter 18)
Santoro, Francesca (Chapter 3)
Sasaki, Tsuyoshi (Chapter 16)
Sawada, Kota (Chapter 16)
Scagnolatto, Giovanna (Chapter 5)
Schianodicola, Alexia (Chapter 18)
Schio, Caroline (Chapter 12)
Searle, Josh (Chapter 13)
Skantz, Ewa (Chapter 15)
Smith, Bethany (Chapter 9)
Sole Intonti, Gaia Maria (Chapter 3)
Solli, Anne (Chapter 15)
Strang, Craig (Chapter 15)
Takahashi, Camila Keiko (Chapter 11)
Taylor, Graceanne (Chapter 18)
Tombolini, Alessia (Chapter 3)
Uyarra, Maria C. (Chapter 20)
Valestrand, Louise (Chapter 20)
Velazco, Fernando Martín (Chapter 14)
Wallis, Katherine (Chapter 10)
Wei, Min (Chapter 21)
Zhao, Minyan (Chapter 21)

Volume I: Contributors (Chapter Order)

Forwards
Ryabinin, Vladimir E.
Leinen, Margaret S.

Reflections
Kennedy, Declan
Bassan, Niccolò
Soapi, Katy

Chapter 1
Kennedy, Teresa J.

Chapter 2
Glithero, Diz
Bridge, Nicola
McRuer, Jen
Hart, Natalie

Chapter 3
Santoro, Francesca
Barbeito Morandeira, Paula
Lovat, Valentina
Magalhães, Ana Vitória Tereza
Sole Intonti, Gaia Maria
Gottfried Burguett, Esteban
Tombolini, Alessia
Carluccio, Paola
Coulton, Jack

Chapter 4
Greene, Thomas F.

Chapter 5
da Silva, Alexandre Augusto
Scagnolatto, Giovanna
Orsolini, Karina F.

Chapter 6
Barord, Gregory J.

Chapter 7
Matsumoto, George I.
Lodes, Katie M.
Kennedy, Teresa J.
Magnusson, Jennifer L.

Chapter 8
Marrero, Meghan E.
Napolitano, Kristen V.

Chapter 9
Nuss, Sarah
Caputo, Anna
Lawrence, Lisa
Gardiner, Greta Olsen
Smith, Bethany
Cackowski, Celia

Chapter 10
Frensley, B. Troy
Bell, Hannah
Carretti, Claire
Wallis, Katherine
De Abreu, Amalia
Marshall, Peter

T. J. Kennedy (ed.), *Ocean Literacy: The Foundation for the Success of the Ocean Decade, Volume I*, Sustainable Development Goals Series, https://doi.org/10.1007/978-3-031-98145-6

Chapter 11
Martelli Mazzo, Tatiana
Lippi, Adriana
Pedroso Conti, William Remo
Gozzo Andreotti, Andrezza Justino
Takahashi, Camila Keiko
Ignacio, Barbara Lage
Bueno, Marilia
Christofoletti, Ronaldo Adriano
Gonçalves, Leandra R.

Chapter 12
Schio, Caroline
Reis, Pedro

Chapter 13
Searle, Josh
Kveven, Ardi

Chapter 14
Lee, Tammy D.
Glass, Bonnie
Lemiec, Gail
Velazco, Fernando Martín

Chapter 15
Fauville, Géraldine
Strang, Craig
Skantz, Ewa
Bjelvenmark, Jens
Solli, Anne

Chapter 16
Sasaki, Tsuyoshi
Sawada, Kota

Chapter 17
Çavaş, Bülent
Açık, Şermin
Koç, Simge
Kolaç, Mısra

Chapter 18
Madden, Lauren
Ammentorp, Louise
Magee, Nathan
Taylor, Graceanne
Rivera-Peña, Gabriela
Schianodicola, Alexia

Chapter 19
Mueller, Ninja

Chapter 20
Eparkhina, Dina
Banchi, Elisa
Borja, Ángel
Canu, Donata
Ezgeta-Balić, Daria
Giusta, Elena
King, Andrew
Koulouri, Panayota [Yolanda]
Locritani, Marina
Manca Zeichen, Marta
Pomaro, Angela
Petrera, Francesca
Uyarra, Maria C.
Valestrand, Louise

Chapter 21
Li, Lei
Zhao, Minyan
Wei, Min
Hu, Shuangfang
Billah, Baki
Lin, Ziyi

Chapter 22
Kennedy, Teresa J.

Authors' Institutional Affiliations

Associação TransforMAR, Campinas, Brazil
AZTI, Marine Research Division, Basque Research and Technology Alliance (BRTA), Gipuzkoa, Spain
Blue Alliance Marine Protected Areas (BA), Marseille, France
Canadian Ocean Literacy Coalition (COLC), Halifax, Nova Scotia, Canada
Central Campus Regional Academy, Des Moines, Iowa, USA
Communications Inc., London, England, United Kingdom
Dalhousie University, Halifax, Nova Scotia, Canada
Dokuz Eylül University, Izmir, Türkiye
East Carolina University, North Carolina, USA
European Global Ocean Observing System (EuroGOOS), Brussels, Belgium
Everett Community College, Everett, Washington, USA
Experimental School of Shenzhen Educational Sciences Institute Guangming, Shenzhen, China
Federal University of São Paulo (UNIFESP), Santos, Brazil
Gullmarsgymnasiet, Lysekil, Sweden
Hellenic Centre for Marine Research, Institute of Marine Biology, Biotechnology & Aquaculture, Former American Base of Gournes, Crete, Greece
I AM WATER: Ocean Conservation, Cape Town, South Africa
Institute of Oceanography and Fisheries, Split, Croatia
Institute of Tibetan Plateau Research, Chinese Academy of Sciences, Beijing, China
Intergovernmental Oceanographic Commission (IOC) of UNESCO, Paris, France
Intergovernmental Oceanographic Commission (IOC) of UNESCO Project Office for Ocean Literacy, UNESCO Regional Bureau for Science and Culture in Europe, Venice, Italy
International Council of Associations for Science Education (ICASE), University College Cork, Cork, Ireland
Istituto Nazionale di Geofisica e Vulcanologia (INGV) Rome, Italy
Italian National Institute for Environmental Protection and Research (ISPRA), Rome, Italy
Jahangirnagar University, Savar, Dhaka, Bangladesh
Kingsborough Community College, Brooklyn, New York, USA
Leuphana University, Lueneburg, Germany
Mercy University, New York, USA
Monterey Bay Aquarium Research Institute (MBARI), Moss Landing, California, USA
National Institute of Oceanography and Applied Geophysics (OGS), Sgonico-Trieste, Italy
National Research Council of Italy (CNR), Institute of Marine Sciences (ISMAR), Venice, Italy
North Carolina Aquarium at Fort Fisher, North Carolina, USA
Norwegian Institute for Water Research (NIVA), Oslo, Norway

T. J. Kennedy (ed.), *Ocean Literacy: The Foundation for the Success of the Ocean Decade, Volume I*, Sustainable Development Goals Series, https://doi.org/10.1007/978-3-031-98145-6

Ocean Conservation Trust, Coxside, England, United Kingdom
Pacific Community Centre for Ocean Science, Suva, Fiji Islands
Paul Robeson Elementary School, Trenton, New Jersey, USA
Phillips 66, Ferndale, Washington, USA
Save Barnegat Bay, Toms River, New Jersey, USA
Scripps Institution of Oceanography, California, USA
Shenzhen Institutes of Advanced Technology, Chinese Academy of Sciences, Shenzhen, China
Shenzhen Longhua Innovative Experimental School, Shenzhen, China
Shenzhen University of Advanced Technology, Shenzhen, China
Shenzhen University of Advanced Technology High School No. 22, Shenzhen, China
St. Joseph's Academy, St. Louis, Missouri, USA
Stony Brook University, New York, USA
Stultifera Navis Institutom, Baja California Sur, Mexico
The College of New Jersey, Ewing, New Jersey, USA
The Lawrence Hall of Science, University of California, Berkeley, USA
Tokyo University of Marine Science and Technology, Tokyo, Japan
United Nations Educational, Scientific and Cultural Organization (UNESCO), Paris, France
University College Cork, Ireland
Universidade de Lisboa, Portugal
University of California–San Diego, USA
University of Chinese Academy of Sciences, Beijing, China
University of Gothenburg, Gothenburg, Sweden
University of North Carolina Wilmington, North Carolina, USA
University of Texas at Tyler, Tyler, Texas, USA
University of Vechta, Vechta, Germany
Virginia Institute of Marine Science, Gloucester Point, Virginia, USA

Volume I: Expert Reviewers (44)

****Reviewed 2 chapters**
*****Reviewed 3 chapters**

Achkar, Roger
Almeida, Mariana
Baylon, Angelica
Berger, Carolyn
Bhalla-Ladd, Maya
***Bottega, Caroline
Brewer, Elaine
Brock, Jennifer
Cook, Megan
Chilton, Linda
Custer, Fawn
David, Laura
Darrouy Bassani, Paulette
Eslopor, Jose Marie A.
Fackler, Claire
**Fassina, Caroline
Finley, Savanna
Fischer, Kristian
Fuentes-Tauber, Yajaira
Gaglioti, Martina
**Grant, Dave
Grogan, Amy
Holland, Kurt
Hil, Abigail
Lang, Frances
Mack, Jamie
Manzo, Lyndsey
**Modzelesky, Alexis
Muñoz-Muga, Pilar
Peters, Michele A.
Putri, Mutiara Rachmat
Ramalho, Débora
Realdon, Giulia
Risch, Tiffany
Ritter, Julia
Robertshaw, Jason
Russell, Katie
Sawant, Ajay
**Stevens, Russell
Stone, Debbi
Vallejo, Ben
**Verhofstad, Selene
VishnuRadhan, Renjith
Whitman, Barbara Carr

T. J. Kennedy (ed.), *Ocean Literacy: The Foundation for the Success of the Ocean Decade, Volume I*, Sustainable Development Goals Series, https://doi.org/10.1007/978-3-031-98145-6

Volume I: Acronym List

The following acronyms related to Ocean Literacy were included in this volume.

Acronym	Full Name
AAEE	Australian Association for Environmental Education
ADCP	Acoustic Doppler Current Profiler
AMEA	Asia Marine Educators Association
AMEP(s)	Academic Marine Education Programme(s)
ANSO	Alliance of National and International Science Organizations for the Belt and Road Regions [China]
ARGO / Argo	Global array of ~3,800 free-drifting robotic profiling floats supporting large-scale Ocean circulation studies
ASLO	Association for the Sciences of Limnology and Oceanography
BGC	Biogeochemistry or BioGeoChemical
BGC-Argo	Biogeochemistry-Array (profiling float array)
BIMS	Black in Marine Science
BIPOC	Black, Indigenous, and People of Color
BlueMinds4Teachers	Digital Toolkit for empowering educators in Ocean Literacy
BlueS_Med	Blue Schools in the Mediterranean Sea Basin
BNE	Bildung für nachhaltige Entwicklung (Education for Sustainable Development) [Germany]
BRIDGE-Black Sea	EU Horizon 2020 Black Sea Young Ambassadors Project

Acronym	Full Name
BSN	Blue Schools Network
BUIK	Bacteria Uchishiro Ichinose Kondo (high-temperature fermentation bacteria) [Japan]
CaNOE	Canadian Network for Ocean Education
CEOSE	Committee on Equal Opportunities in Science and Engineering
CBNERR-VA	Chesapeake Bay National Estuarine Research Reserve in Virginia
CIMAS	Cooperative Institute for Marine and Atmospheric Studies
CIRM	Interministerial Commission for Marine Resources [Brazil]
CNPq	Conselho Nacional de Desenvolvimento Científico e Tecnológico (National Council for Scientific and Technological Development) [Brazil]
CNR	Consiglio Nazionale delle Ricerche (National Research Council) [Italy]
CNRS	Centre National de la Recherche Scientifique (National Centre for Scientific Research) [France]
CSC	Course Specific Competency (outcomes)
CTD	Conductivity, Temperature, and Depth
CTE	Career and Technical Education
DAMST	Draw-A-Marine-Scientist Test
DAST	Draw-A-Scientist-Test
DBR	Design-Based Research

T. J. Kennedy (ed.), *Ocean Literacy: The Foundation for the Success of the Ocean Decade, Volume I*, Sustainable Development Goals Series, https://doi.org/10.1007/978-3-031-98145-6

Acronym	Full Name
DCO	Decade Coordination Office
DOE	Department of Education
EARTH	Education and Research: Testing Hypotheses (curriculum and workshop series conducted by the Monterey Bay Aquarium Research Institute (MBARI) [USA]
ECOP(s)	Early Career Ocean Professional(s)
EcoLab	Environmental science labs
EE21	Environmental Education for the 21st Century
EMSEA	European Marine Science Educators Association
ENSO	El Niño Southern Oscillation
EP(s)	Essential Principle(s)
EPICD	Engagement, Partnerships, Interdisciplinary Collaboration, Communication Training, Digital Platforms
ERASMUS+	European Region Action Scheme for the Mobility of University Students plus education, training, youth, and sport across Europe and beyond
ESD	Education for Sustainable Development
ESG	Environmental, Social, and Governance
ESMD	Education for Sustainable Marine Development
ESRA	Explore, Feel, Reconnect, and Act
FC(s)	Fundamental Concept(s)
FCT	Fundação para a Ciência e a Tecnologia (Foundation for Science and Technology) [Portugal]
FORMAS	Swedish Research Council for Sustainable Development
GDC	Global Diversity Competency
GEOPAths	Pathways into the Earth, Ocean, Polar and Atmospheric & Geospace Sciences
GEOPAths-EXTRA	Engaging students in the geosciences through extra-curricular experiences and training
GES	Good Environmental Status

Acronym	Full Name
GES4SEAS	Achieving Good Environmental Status for Maintaining Ecosystem Services by Assessing Integrated Impacts of Cumulative Pressures (European Union project)
GIS	Geographic Information System
GLOBE	Global Learning and Observations to Benefit the Environment (NASA, NOAA, NSF, U.S. Department of State Program)
GO-BGC	Global Ocean Biogeochemistry Array
GOBAI-O2	GO-BGC Oxygen dataset
GO-SHIP	Global Ocean Ship-based Hydrographic Investigations Program
GOOS	Global Ocean Observing System
GU-CAS	Gothenburg University Critical Animal Studies group
Gy11	Swedish Upper Secondary Curriculum
ICASE	International Council of Associations for Science Education
ICES	International Council for the Exploration of the Sea [Denmark]
ICMBio	Instituto Chico Mendes de Conservação da Biodiversidade (Chico Mendes Institute for Biodiversity Conservation) [Brazil]
ICT	Information and Communication Technology
Ifremer	Institut Français de Recherche pour l'Exploitation de la Mer (French Research Institute for Exploitation of the Sea) [France]
IMED	It's My Estuary Day (project) [USA]
IMPAC5	Fifth International Marine Protected Areas Congress
IOC	Intergovernmental Oceanographic Commission
IOC-UNESCO	Intergovernmental Oceanographic Commission of the United Nations Educational, Scientific, and Cultural Organization

Acronym	Full Name
IPCC	Intergovernmental Panel on Climate Change
IPBES	Intergovernmental Science-Policy Platform on Biodiversity and Ecosystem Services [Germany]
IPMEN	International Pacific Marine Educator Network
IRB	Institutional Review Board
JEDI	Justice, Equality, Diversity, and Inclusion
JSPS	Nihon Gakujutsu Shinkōkai (Japan Society for the Promotion of Science) [Japan]
KMO	Kaiser-Meyer-Olkin (sampling adequacy test)
K–5	Kindergarten through 5th grade (ages 5 through 11)
K–12 / p–12	Kindergarten through 12th Grade / Preschool through 12th Grade (typically ages 5 through 18)
LDEO	Lamont-Doherty Earth Observatory [USA]
LG(s)	Learning Goal(s)
LL(s)	Living Lab(s)
MACMAP	Multidisciplinary Analysis of Climate Change Indicators in the Mediterranean and Polar Regions (Italian Research Initiative) [Italy]
MCTI	Ministério da Ciência, Tecnologia e Inovações (Ministry of Science, Technology, and Innovation) [Brazil]
MEXT	Monbu-kagaku-shō (Ministry of Education, Culture, Sports, Science and Technology) [Japan]
MIA	Minorities in Aquaculture [USA]
MISS	Minorities in Shark Sciences [USA]
MMC	Monitoramento Mirim Costeiro (Junior Coastal Monitoring) [Brazil]
MNP	Marine Nano- and Microplastic
MOOC	Massive Open Online Course
MPA(s)	Marine Protected Area(s)
MRV(s)	Marine Robotic Vehicle(s)
MSFD	Marine Strategy Framework Directive
MSP	Marine Spatial Planning

Acronym	Full Name
NMEA	National Marine Educators Association [USA]
NAME	Northwest Aquatic and Marine Educators [USA]
NASA	National Aeronautics and Space Administration [USA]
NASA PACE	NASA's Plankton, Aerosol, Cloud, Ocean Ecosystem Mission
NEBS	Network of European Blue Schools [Europe]
NGO(s)	Non-Governmental Organization(s)
NGO(s)–UNESCO	NGOs in Official Partnership with UNESCO
NGSS	Next Generation Science Standards [USA]
NMEA	National Marine Educators Association [USA]
NMP	Nano- and Microplastics
NOAA	National Oceanic and Atmospheric Administration [USA]
NRC	National Research Council [USA]
NSC	National Sciences Competency
NSF	National Science Foundation [USA]
NSTA	National Science Teaching Association [USA]
NYSMEA	New York State Marine Education Association [USA]
O2	Olimpíada do Oceano (Ocean Olympics) [Brazil]
OAC	Ocean Affairs Council [Taiwan]
OCP	Ocean & Climate Platform (an alliance between NGOs and research institutions)
OCV	Ocean & Climate Village (IOC-UNESCO traveling exhibition)
OD	Ocean Decade (United Nations Decade of Ocean Science for Sustainable Development)
ODC(s)	Ocean Decade Challenge(s)
ODO(s)	Ocean Decade Outcome(s)
OECD	Organisation for Economic Co-operation and Development
OFAA	Oceans for All Alliance [South Africa]

Acronym	Full Name
OGW	Ocean Guardians Workshop [South Africa]
OL	Ocean Literacy
OLD(s)	Ocean Literacy Dialogue(s)
OLP(s)	Ocean Literacy Principle(s)
OLR	Ocean Literacy Research
OLRC	Ocean Literacy Research Community
OLWA	Ocean Literacy With All Programme (IOC)
ONG(s)	Organisation(s) Non-Gouvernementale(s) (French) [equivalent of NGO(s) in English]
ORCA	Ocean Research College Academy [USA]
OSS	Ocean and Society Survey
OTGA	Ocean Teacher Global Academy
PACE	Plankton, Aerosol, Cloud, Ocean Ecosystem (NASA mission)
PADI	Professional Association of Diving Instructors
PD	Professional Development
PISA	Programme for International Student Assessment
PL	Professional Learning
PBL	Project-Based Learning
PMEL	Pacific Marine Environmental Laboratory [USA]
POP	Persistent Organic Pollutant
Project SEA	Project Science, Education and Action [USA]
RELATO	*Red de Educación Latinoamericana para el Océano* (Latin American Education Network for the Ocean)
R/V	Research Vessel
RVs	Relational Values
ROV(s)	Remotely Operated Vehicle(s)
S.M.A.R.T.	Specific, Measurable, Achievable, Relevant, and Time-bound
SCG	Strategic Communication Group (UNESCO Ocean Decade)
SCI	Science Citation Index
SCUBA	Self-Contained Underwater Breathing Apparatus
SDG(s)	Sustainable Development Goal(s)
SDG 4	Sustainable Development Goal 4 Quality Education
SDG 5	Sustainable Development Goal 5 Gender Equity
	Sustainable Development Goal 13 Climate Change
SDG14	Sustainable Development Goal 14 Life Below Water
SDG 17	Sustainable Development Goal 17 Partnerships for the Goals
SEABEYOND	Sea Beyond Educational Programme (project by Prada Group conducted in partnership with UNESCO-IOC since 2019 to raise awareness of sustainability and Ocean preservation)
SEM	Scanning Electron Microscope
SIDS	Small Island Developing State(s)
SLO	Student Leaning Objectives
SOA	Sustainable Ocean Alliance
SOCCOM	Southern Ocean Carbon and Climate Observations and Modeling
SOLE	Survey of Ocean Literacy and Experience
SOS	ScienceOnShore Program [USA]
STEAM	Science, Technology, Engineering, Arts, and Mathematics
STEM	Science, Technology, Engineering, and Mathematics
TEK	Traditional Ecological Knowledge
UIDEF	Unidade de Investigação e Desenvolvimento em Educação e Formação [Portugal]
UN	United Nations
UNDP	United Nations Development Programme
UNEP	United Nations Environmental Programme
UNESCO	United Nations Educational, Scientific and Cultural Organization

Acronym	Full Name
UNESCO-IOC	Intergovernmental Oceanographic Commission of the United Nations Educational, Scientific, and Cultural Organization's Intergovernmental Oceanographic Commission
UNESCO-IOC/ WESTPAC	IOC Sub-Commission for the Western Pacific
UNOC3	United Nations Ocean Conference (3rd edition)
VBN	Values–Beliefs–Norms (Theory)
VR	Virtual Reality

Acronym	Full Name
VUCA	Volatility, Uncertainty, Complexity, Ambiguity
WEBLUE	"Blue Gender Equity for Ocean" effort to achieve gender equality within the blue economy (the sustainable use of Ocean resources) and Ocean science
WESTPAC	IOC Sub-Commission for the Western Pacific
WHOI	Woods Hole Oceanographic Institution
WIC	Writing Intensive Competency

2021 United Nations Decade
2030 of Ocean Science
for Sustainable Development

MIX
Papier aus verantwortungsvollen Quellen
Paper from responsible sources
FSC® C105338

If you have any concerns about our products,
you can contact us on
ProductSafety@springernature.com

In case Publisher is established outside the EU,
the EU authorized representative is:
Springer Nature Customer Service Center GmbH
Europaplatz 3, 69115 Heidelberg, Germany

Printed by Libri Plureos GmbH
in Hamburg, Germany